Neuroscience of Yoga

Akshay Anand
Editor

Neuroscience of Yoga

Theory and Practice: Part 1

Springer

Editor
Akshay Anand
Neuroscience Research Lab, Department of Neurology
Post Graduate Institute of Medical Education and Research
Chandigarh, India

CCRYN-Collaborative Centre for Mind Body Intervention, PGIMER
Chandigarh, India

Editorial Assistant
Pooja Nadholta
Neuroscience Research Lab, Department of neurology, Postgraduate Institute
of Medical Education and Research
Chandigarh, India

ISBN 978-981-97-2853-4 ISBN 978-981-97-2851-0 (eBook)
https://doi.org/10.1007/978-981-97-2851-0

This Springer imprint is published by the registered company Springer Nature Singapore Pte Ltd.
The registered company address is: 152 Beach Road, #21-01/04 Gateway East, Singapore 189721, Singapore

Foreword

THE DALAI LAMA

I firmly believe we live at a time when ancient Indian knowledge can contribute to greater human happiness and a happier, more peaceful world. India's long-standing traditions of doing no harm—'ahimsa'—motivated by compassion—'karuna'—can help us find peace of mind.

Exploring ways to achieve a calmly abiding mind—'shamatha'—and a clear insight into reality—'vipashyana'—has led to a remarkable understanding of the workings of the mind. This allows us to recognise and tackle our destructive emotions such as anger, fear, attachment and jealousy.

Then there is yoga with its physical training and breathing exercises that help us relax and achieve a suppleness of body and mind, the ideal preparation for cultivating deeper meditation.

This comprehensive book presents evidence of the positive effects of yoga in the context of neuroscience and in social situations.

I congratulate the editor Prof. Akshay Anand, as well as the researchers and contributing authors, for shining the light of modern science on the benefits the practice of yoga can bring.

Dharmshala, India
17 December 2022

His Holiness the Dalai Lama

Preface

Part I

This book covers experimental theory and practice of Yoga that has enhanced its neuroscientific understanding. This is an excellent handbook for the researchers in the field of evidence-based integrative health. It encompasses traditional and modern tools used in neuroscience. It explores the various mechanisms underlying the neuroscience of Yoga among different populations such as prisoners, pregnancy, and healthcare workers, in various dimensions of biology, for example, immune modulation, neurobiology, palliative care, computer vision, endocrinology, Ayurgenomics, cardiac health, tele-Yoga for promoting science of wellness. It provides the basic understating of Yoga and its possible effects with evidence-based mechanisms in various physiological states. It also provides information for the modern biologists, physicians, Yoga academia, and policymakers of how mind-body complexities in neuroscience, mental health, and preventive healthcare can be useful for health and disease. It serves as a guide for integrative health practitioners, patients, educationists, philosophers, graduate students, and faculty pursuing or starting research in the field of biology, complementary medicine, and other integrative health. The chapters in this book also serve as a comprehensive resource for clinical trials in Yoga. In addition, flow charts and illustrations have been provided to understand how healthy brain ageing can be achieved. It is recommended to read this book in combination with Part II.

Chandigarh, India Akshay Anand

The original version of the book has been revised. The correction to this book can be found at https://doi.org/10.1007/978-981-97-2851-0_16

Acknowledgements

We extend our heartfelt gratitude to our esteemed reviewers for their invaluable contributions to the peer-review process of this book. Their critical insights and suggestions have significantly enhanced the quality of the content. This book could not have achieved such high standards without their diligent and selfless efforts. We sincerely appreciate their dedication, critical thinking, and thoughtful feedback, which have been instrumental in shaping this work.

Special thanks to Dr. Pooja Nadholta for her Editorial Assistance in the book publication.

1. Dr. Bala Subramaniam
2. Dr. Partha Roy
3. Dr. Amol Patil
4. Dr. Gaurav Trivedi
5. Dr. Rima Dada
6. Dr. Rakesh Biswas
7. Dr. Satbir Khalsa
8. Dr. Mahendra Bind
9. Dr. Vijaya
10. Dr. Kanupriya Sharma
11. Dr. Anumeha Bhagat
12. Dr. Radhika Khosla
13. Dr. Hemant Bhargav
14. Dr. Monika Gautam
15. Dr. Neelam Dahiya

Contents

Chapter 1
Psycho-Neuro-Immune Modulation by *Yoga*

Surabhi Gautam, Taruna Arora, and Rima Dada

Contents

Abstract *Yoga* is a popular Indian origin mind-body medicine for physical development, mental growth, and spiritual elevation. It enhances physical efficacy, emotional resilience, awareness and helps us to understand at deeper realms the purpose of life and aims to achieve the union of individual and universal consciousness. Various therapeutic advantages of *Yoga* include modulation of the epigenome by upregulation of genes which promote oxidative stress, mitochondrial and nuclear genome integrity, reducing the rate of aging and the biological age, stress reduction, hypothalamus–pituitary–adrenal (HPA) axis alteration, autonomic nervous system

S. Gautam
Laboratory for Molecular Reproduction and Genetics, Department of Anatomy, All India Institute of Medical Sciences (AIIMS), New Delhi, India

Department of Orthopaedics, Emory Musculoskeletal Institute, Emory University School of Medicine, Atlanta, GA, USA

T. Arora
Division of Reproductive Biology, Maternal and Child Health, Indian Council of Medical Research, New Delhi, India

R. Dada (✉)
Laboratory for Molecular Reproduction and Genetics, Department of Anatomy, All India Institute of Medical Sciences (AIIMS), New Delhi, India
e-mail: dr.rimadada@aiims.gov.in

A. Anand (ed.), *Neuroscience of Yoga*,
https://doi.org/10.1007/978-981-97-2851-0_1

balance, increased heart rate variability, physiological, biochemical, and anatomical changes in the brain, and immunological modulation. Despite several research studies, the specific biochemical pathways underlying the benefits of *Yoga* and similar activities are still unclear. *Yoga* sets various physiological events in the body that aids in the reduction of the stress response. Stress is body's natural tendency to respond to a challenging situation. Chronic stress alters the functioning of immune system by reducing immune cells functioning and increasing active immunosuppressive pathways in the body. The HPA axis, hypothalamic–pituitary–gonadal (HPG) axis, and sympathetic–adrenal–medullary (SAM) system are all thought to have a role in the immunological response. This chapter focuses on research that demonstrates changes in gene expression by modulation of methylome and epigenome, neurotransmitter, cytokine, and hormone levels as a result of yogic therapies which maintain a homeostatic metabolic profile of a *Yoga* practitioner. Hence, we summarize the mode of action of *Yoga* on multiple pathways and its role in positively modifying the psychoneuroimmunology of the body.

Keywords *Yoga* · Inflammation · Immune-ageing · Immune-modulation · Psycho-neuro-immune axis

1.1 Introduction

Yoga is an Indian origin mind-body medicine that promotes balance in all aspects of health, including physical, mental, emotional, and spiritual well-being (Woodyard 2011). The philosophy of *Yoga* is based on the concept of personal transformation. *Yoga*'s ultimate purpose is to achieve self-realization, which may be accomplished via persistent practice (Gard et al. 2014). Its goal is to achieve a balance of physical and mental health via illness prevention, health maintenance, and mental serenity. The Ministry of AYUSH describes *Yoga* as a practical discipline which is evolutionary in nature{Citation}. According to *Patanjali*, the eight limbs of *Yoga* consist of *Yamas, Niyma, Asana, Pranayama, Pratyahara* (first five limbs the *bahiranga Yoga*), *Dharana, Dhyana*, and *Samadhi* (last three components are *antarang Yoga*). The ultimate purpose of *Yoga* is to achieve inner well-being, and there are various schools of *Yoga*, each with its unique beliefs and practices (Gautam et al. 2021a; Tolahunase et al. 2021). *Yoga* is a spiritual discipline that results in the fusion of one's own consciousness with the consciousness of the universe and thus establishes harmony between mind and body and enables an individual to thrive rather than just survive.

1.1.1 Yoga*: Role in Neuropsychiatric Disorders*

Various research studies have documented that *Yoga* promotes health, prevents the onset of disease, is a powerful adjunct to modern medicine in disease management, and has rehabilitative potential (Woodyard 2011; Bernardi et al. 2001; Pilkington et al. 2005; Ross and Thomas 2010; Khalsa et al. 2012). It modifies the physiology behind the stress response (Gautam et al. 2020; Gautam and Dada 2021). Due to stressful conditions, the body works harder to maintain homeostasis and optimize various ongoing mechanisms. Allostasis describes the physiological changes that a live body goes through in order to preserve equilibrium during times of stress (Sterling and Eyer 1988). The neuroendocrine system regulates the release of hormones and other chemical messengers, which mediates the stress response (Tsigos et al. 2000). When the sympathetic nervous system (SNS) is activated in response to threats/external stress, a cascade of neurophysiological and metabolic alterations occurs to preserve homeostasis. This response, known as allostatic overload, is maladaptive under conditions of chronic stress but is adaptive to the organism following an acute stress. This maladaptive response can result in a number of comorbid chronic complex physical and mental health issues (Herman et al. 2016). Chronic stress and allostatic overload may be the underlying etiology in a number of stress-related mental illnesses, including anxiety, stress, and depression (McEwen 2006). Numerous studies show that physical exercise and *Yoga* practice, two forms of non-pharmacological therapy, help reduce the symptoms of psychosomatic diseases by mitigating the negative consequences of chronic stress (Gautam et al. 2019, 2020, 2021b, Vijayaraghava et al. 2015; Tolahunase et al. 2017, 2018; Bisht et al. 2020). Although both physical activity and *Yoga* generate a rise in encephalin levels, too little or too much physical exercise causes oxidative stress and exhaustion, whereas *Yoga* energizes the body and normalizes free radical levels. *Yoga* optimizes the stress response by acting on hypothalamic–pituitary–adrenal (HPA) axis, by normalizing the circadian rhythm, autonomic nervous system and reduces inflammation (Gard et al. 2014; Bernardi et al. 2001; Gautam et al. 2020; Herman et al. 2016).

Yoga regulates the emotions, behaviors, cognition, and peripheral physiology through top-down and bottom-up mechanisms, as well as improves the efficiency and integration of the processes that support self-regulation (Gard et al. 2014; Goldin and Gross 2010). *Yoga*'s top-down and bottom-up techniques are both accessible and adaptable, allowing for self-regulation of the nervous system and hence a reduction in allostatic stress (Gard et al. 2014). Meditation, attention-control techniques, ethical conduct, and self-monitoring are examples of top-down *Yoga* practices, whereas prolonged *Asana* (movement) and *Pranayama* (breath) practices are examples of bottom-up *Yoga* activities. This concept is similar to modern neuroscience, which claims that our brain may change as a result of experience and training. The physiology of brain plasticity is an important finding that has paved the way for rehabilitation research and the recovery of lost functions following a neurologic injury (Mateos-Aparicio and Rodríguez-Moreno 2019). Neuroplasticity is the capacity of neural structures to restore/modulate their functions and rearrange structures through neural structural flexibility. *Yoga* is a mental exercise that changes the neurocognitive processes that drive our personalities (Tolahunase et al. 2018; Halappa et al. 2018). Both motor movements and sensory experiences are involved in *Yoga Asana* and yogic practices. As a result, it's possible that yogic practices make use of neuronal plasticity to bring about long-term changes in brain structure and function. Although genetics and development have a larger role in forming brain structures, experience-related neuronal plasticity appears to play a larger role in remodeling neural structures (Villemure et al. 2015). *Yoga* has also been proven to enhance levels of brain-derived neurotropic factor (BDNF), a neuroplasticity marker, in individuals with severe depression (Tolahunase et al. 2021). In depressed individuals, 12-week *Yoga* intervention significantly increased systemic BDNF, melatonin, and DHEA levels, which is associated with improvements in indicators of mind-body connection and cellular health (Tolahunase et al. 2017, 2018). This chapter highlights the efficacy of *Yoga* as a therapeutic intervention and a health model.

1.2 Stress and Psycho-Neuro-Immunology

The study of the relationships between behavioral, neurological, endocrine, and immunological systems is known as psychoneuroimmunology. The old belief that the mind may play a role in health and disease has been given a scientific foundation due to psychoneuroimmunology (Segerstrom and Miller 2004; Tausk et al. 2008). The brain communicates with the immune system through autonomic nervous system (ANS) and neuroendocrine system (NES). The immune system detects signals by NES and ANS via receptors on the surface of immune cells. An active immune system, on the other hand, produces chemical signals (cytokines) that the neurological system detects (Chaplin 2010). Bidirectional linkages between the immune system and the brain as a result lay the framework for behavioral influences on immunological function. According to the principle of classical conditioning, stressful life events and emotional states tend to reduce immunity while enhancing immunological responses (D'Acquisto 2017). These impacts appear to be involved in modifying the formation and/or progression of immunologically mediated disease processes, which are physiologically significant (Coe and Laudenslager 2007; Dantzer 2018). The discovery of brain-immune system connections and functional interactions supports the theory that immunological alterations may mediate some of the impacts of psychosocial variables on health and illness (Fig. 1.1).

The HPA axis is engaged when the hypothalamus secretes corticotrophin-releasing hormone (CRH), which is released in response to stressful physical and psychological stimuli. This hormone, in turn, induces the hypophysis to produce corticotrophin (ACTH), which stimulates the adrenal glands to secrete cortisol (Herman et al. 2016). Cortisol secretion is regulated by a negative feedback system in which cortisol suppresses the synthesis of its precursors (CRH and ACTH). This inhibition is mediated through glucocorticoid receptors in the hypothalamus and

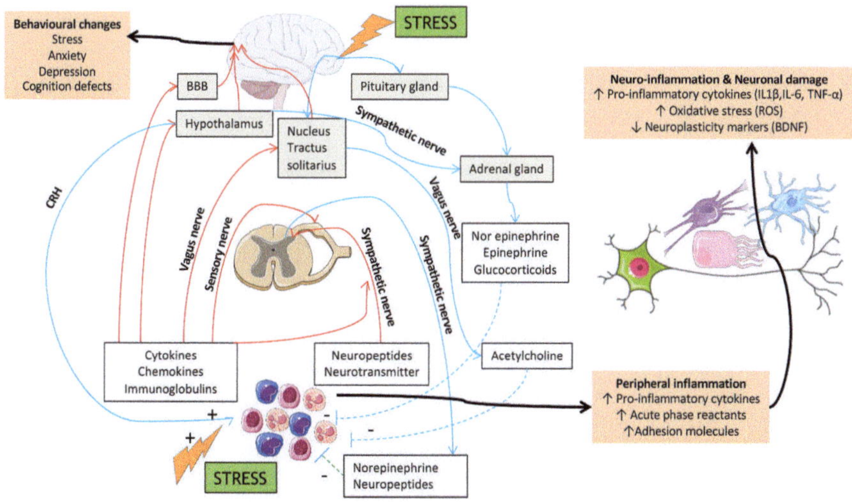

Fig. 1.1 Immunosensing afferent and immunomodulating efferent nerve pathways involved in stress response leading to inflammation, oxidative stress, neuronal damage, and behavioral changes; *BBB* blood–brain barrier, *CRH* corticotrophin-releasing hormone, *ROS* reactive oxygen species, *BDNF* brain derived neurotropic factor; red arrows: afferent nerve fibers, blue arrows: efferent nerve fibers. [The figure was created using images from the Servier Medical Art (smart.servier.com)]

pituitary, as well as glucocorticoid and mineralocorticoid receptors in the hippocampus (Gjerstad et al. 2018). Although cortisol limits the peripheral inflammatory response by acting as an immunosuppressor, other hormones in the HPA axis, such as the CRH, can stimulate inflammation. The stimulation of the peripheral inflammatory response can activate the HPA axis, hence the link is bidirectional. The results of this peripheral stimulation, which includes macrophages and lymphocytes, may penetrate the blood–brain barrier (BBB) and stimulate the microglia, triggering a neuroinflammatory response (Silverman et al. 2005). The microglia activation will trigger an inflammatory cascade by releasing cytokines and free radicals like reactive oxygen species (ROS) and reactive nitrogen species (RNS), stimulating the activation of the astroglia, which will amplify the inflammatory signals inside the central nervous system (CNS). The stimulation of the indolamine 2,3 dioxygenase (IDO) enzyme results in an excessive release of glutamate by astrocytes as well as oxidative stress mediators by activated microglia (Chen et al. 2020). The neuronal damage and neuro-inflammation negatively affect the process of neurogenesis and neuroplasticity (Tausk et al. 2008; Dantzer 2018). Allostasis refers to the maintenance of homeostasis through activating neuronal, neuroendocrine, and neuroendocrine-immune processes in response to stressful events (Sterling and Eyer 1988). Cortisol penetrates the blood–brain barrier and alters the architecture of the brain by acting on glucocorticoid and mineralocorticoid receptors, which are mostly found in the hippocampus, amygdala, prefrontal cortex, and hypothalamus, and so affects a stressed person's learning and thinking (Williams and Ghosh 2020).

However, recent studies have shown that Yoga can counteract the effects of stress by lowering cortisol levels, boosting blood flow to the brain's ventricles and other areas, activating the brain, and inducing neurogenesis and synaptogenesis, or improving neural plasticity (Tolahunase et al. 2017, 2018; Krishnakumar et al. 2015; Gothe et al. 2019). Yoga promotes neuronal proliferation and development as it upregulates several factors like neurotrophins, neureglin, BDNF, which is a marker of neuroplasticity and is cytoprotective in nature and promotes synaptogenesis. Chronic stress causes persistent activation of HPA axis and dysregulates the immune response and *Yoga* buffers this and restores homeostasis between the SNA and PNS (Gautam et al. 2020, 2021a; Basu-Ray 2021).

1.3 *Yoga* and Psychoneuroimmunology

Yoga is a centuries-old Indian tradition that promotes physical, mental, and spiritual well-being. The term "*Yoga*" comes from the Sanskrit root "*yuj*," which means "to bind, join, or yoke." *Yoga* is distinguished from other workout regimens by its reflection of the integration of the body, mind, and spirit.

1.3.1 HPA Axis Modulation by Yoga

Yoga works via parasympathetic activation, vagal stimulation, sympatho-adrenal suppression, and suppression of HPA axis (Tolahunase et al. 2021; Sengupta 2012; Raveendran et al. 2018). Yogic therapies reduce the activity of the hypothalamus's paraventricular nuclei, causing the anterior pituitary gland to release less ACTH (Arora and Bhattacharjee 2008). The production of cortisol by the adrenal glands is reduced as ACTH levels fall. Dada R and colleagues documented in several studies on complex diseases that levels of β-endorphin, dehydroepiandrosterone (DHEA), melatonin, BDNF, and sirtuin-1 (SIRT-1) increased, while cortisol and Interleukin (IL)-6 decreased, indicating a decrease in stress and inflammatory response due to regulatory changes in the brain via the HPA axis (Gautam et al. 2019, 2020, Gautam et al. 2021a, b). Stress reactions are known to be reduced when corticosteroids and catecholamines levels are low.

Yoga has also shown promising effects on immune system as a result of its ameliorative activity on HPA, which results in an optimal cortisol release which may be the cause of its buffering effect on stress-induced reduction in cellular immunity. In the presence of a stressor, *Yoga* also restores the autonomic reflex regulating systems, balancing the sympathetic and parasympathetic limbs, which likely prevents the damage of cell-mediated immunity (Gopal et al. 2011; Venkatesh et al. 2020). *Yoga* causes parasympathetic dominance via activation and release of neurotransmitters responsible for relaxation response, improved cardio-vagal tone, and decreased heart rate and reduction in blood pressure; and activation of limbic system and suppression of SNS activation, which is responsible for the induction of stress response by releasing cortisol (Pramanik et al. 2009). PNS activation takes precedence over SNS activation after a long period of *Yoga* practice, especially during stressful conditions (Ross and Thomas 2010).

1.3.2 **Yoga** *and Metabolic Profile*

A recent study from our laboratory on impact of *Yoga* on rheumatoid arthritis (RA) patients showed a substantial reduction in depressed symptoms, as seen by their time-dependent lowered BDI-II (Beck Depression Inventory-II) scores (Gautam et al. 2019). *Yoga*, a profound science and technology, positively impacts genome, epigenome, and DNA integrity. *Yoga* improves mitochondrial health by upregulation of mitochondrial DNA copy number, increase in NAD+ levels, and improvement in COX-II activity which improves mitochondrial membrane potential, thus indicating an increase in mitochondrial bioenergetics (Gautam et al. 2021b, c). This results in reduced production of free radicals. *Yoga* reduces oxidative stress and positively modifies epigenetic changes such as global methylation levels, global hydroxyl methylation levels, and histone deacetylase levels which cause dysregulation of gene expression. *Yoga* regulates soluble HLA-G levels, pro-inflammatory cytokine levels, and anti-inflammatory cytokine levels to regulate immunological homeostasis and aid in the regression of inflammatory processes (Gautam et al. 2019, 2021a, 2022). It can be an effective complementary therapy for RA patients who want to enhance their physical and emotional health. It has a proven role in RA rehabilitation, according to several research studies as it helps to reduce pain, enhance function, and create a pleasant mental state (Moonaz et al. 2015). Diaphragmatic breathing is a fundamental practice in *Pranayama* and when done effectively and mindfully leads one to a meditative state. Meditation has been shown to decrease levels of psychological stress, relieve chronic pain, improve cognitive abilities, and lower OS levels (lower cortisol levels and higher melatonin levels) (Khalsa et al. 2012; Martarelli et al. 2011). These findings demonstrate the value of *Yoga* and meditation as adjunct therapies for managing RA symptoms, extending periods of remission, and avoiding relapses in active RA patients (Gautam et al. 2019). *Yoga* maintains telomere length by upregulation of telomerase enzyme levels and activity and decrease in levels of free radicals hence reduces the rate of biological aging and also improved mind body communications (Tolahunase et al. 2017; Rathore and Abraham 2018).

1.3.3 Hormonal Homeostasis

Yoga normalizes circadian rhythm by increased levels of melatonin and serotonin and reduced cortisol levels, thus helps to maintain the hormonal milieu (Gautam et al. 2021b). *Yoga* and meditation practitioners displayed downregulation of inflammatory and apoptotic pathways as well as gene expression changes connected to genomic stability pathways such as telomere packing and telomere maintenance (Bhasin et al. 2013). According to randomized control trials and related research, *Yoga* decreases anxiety and depression symptoms while also having favorable effects on inflammation, SNS tone, and vagal activity (Bernardi et al. 2001; Pilkington et al. 2005; Vijayaraghava et al. 2015; Riley 2004; Woolery et al. 2004; Shete et al. 2017). Anxiety, social phobia, and depression are all linked to a lack of neurotransmitters including dopamine, serotonin, and norepinephrine (Schneier et al. 2000; Kim and Gorman 2005; Selden et al. 1990). After an 8-week YBLI, studies show that *Yoga* dramatically increases neuroplasticity indicators including β-endorphins BDNF, serotonin, and neuregulins and neurotrophins, as well as a reduction in depression measuring scales like the BDI-II (Yadav et al. 2012; Lee et al. 2014; Kinser et al. 2012; Brown and Gerbarg 2005).

Studies show that a 10-week group-based cognitive behavioral stress management intervention for HIV-positive men improved their mood. Dehydroepiandrosterone-sulphate (DHEA-S) and testosterone levels increased together with lower cortisol and norepinephrine levels in the 24-h urine, suggesting that these mood changes may affect the regulation of adrenal hormones (Antoni et al. 2000).

1.3.4 **Yoga** *and Immune Modulation*

Yoga has the ability to modulate the immune system and control the psycho-neuro-immune axis. Long-term changes in lymphocyte subpopulations have been seen, including transitional naive CD4 cells and suppressor/cytotoxic CD8 cells, which are associated with lower urinary noradrenaline synthesis and decreased urinary cortisol output (Antoni et al. 2000; Besedovsky et al. 2019). These changes indicate immune system reconstitution. The possible mechanisms implicated are modulation of the HPA axis, hypothalamic–pituitary–gonadal (HPG) axis, and sympathetic–adrenal–medullary (SAM) system, as well as their action on the immune system after cognitive behavioral stress management intervention (Smith and Vale 2006). Practices like *Yoga* and meditation have been proven to raise melatonin levels, which sensitize monocytes to lipopolysaccharide (LPS) activation and boost the production of interleukin (IL)-1 and interferon (IFN). In addition to modulating the immune system, it prevents tumor development and coronary atherosclerosis (Arora and Bhattacharjee 2008; Gopal et al. 2011). Melatonin, when given in vivo, counteracts cortisol's immunosuppressive effects and inhibits cortisol-induced thymic atrophy (Carrillo-Vico et al. 2013).

Cytokines and their receptors, which control pituitary development, cell proliferation, and hormone release, are expressed by the hypothalamus and anterior pituitary cells. They function as immunoneuroendocrine modulators, communicating with the HPA to link inflammation and stress (peripheral and central) (Bellavance and Rivest 2014). The pituitary cells are directly stimulated by IL-1, IL-2, IL-6, and tumor necrosis factor (TNF)-α to generate ACTH and β-endorphins. Glucocorticoid receptor expression in macrophages is upregulated by IFN-γ during immune system activation (Bellavance and Rivest 2014). Relaxation response, particularly after long-term practice of *Yoga* and meditation, has been shown to improve mitochondrial energy generation and use, therefore boosting mitochondrial integrity and resilience through increase of ATPase and insulin function, according to recent research. Relaxation-induced downregulation of NF-κB-associated upstream and downstream targets, which mitigates stress, may also increase mitochondrial resilience (Bhasin et al. 2013). The NF-κB pathway genes—whose expression was reduced following relaxation response elicitation—are known to have a key role in inflammation, metabolic syndrome, diabetes, PCOS, autoimmune diseases, trauma, and cancer. In another study by Qu et al. it was found that a *Yoga* programme had a rapid and more substantial impact on gene expression in mononuclear cells in peripheral blood than in a control group who practiced walking with soothing music (Qu et al. 2013). The capacity to handle stress in a proactive manner in everyday life may reduce the endocrine system's activity, increasing the immune system's efficacy. Hypnosis, relaxation, exercise, classic conditioning, self-disclosure, and exposure to phobic stressors have all been employed in psychoimmunological investigations to improve perceived coping and self-efficacy (Vickers et al. 2001).

Yoga causes the posterior or sympathetic portion of the hypothalamus to be inhibited, which regulates the sympathetic reactions to stressful stimuli and restores stress-related autonomic regulating reflex systems (Woodyard 2011) (Fig. 1.2). The parasympathetic system's activity may rise or remain unchanged. Fear, aggression, and rage are inhibited by yogic practices, while rewarding regions in the middle forebrain and other places are stimulated, resulting in a sense of joy and pleasure (Woodyard 2011). In another study on students who practiced *Yoga* and meditation had lower anxiety, heart rate, respiratory rate, blood pressure, and cardiac output than those who do not (Birkel and Edgren 2000). Yogic techniques are thought to reduce the activity of the hypothalamic paraventricular nuclei, causing the anterior pituitary gland to generate less ACTH. The adrenal gland's ability to produce cortisol is reduced as ACTH levels fall. Various studies have documented that *Yoga* lowers cortisol levels (Tolahunase et al. 2017, 2018; Gautam et al. 2021b). Hence lower cortisol levels result in reduction of phenylethanolamine-*N*-methyl transferase, resulting in catecholamine decline (Arora and Bhattacharjee 2008; Smith and Vale 2006). The decreased levels of corticosteroids and catecholamines are known to decrease stress responses.

The twisting movements of various yogic positions help massage and revitalize immune organs and channels. *Yoga* also provides balanced energy, which is

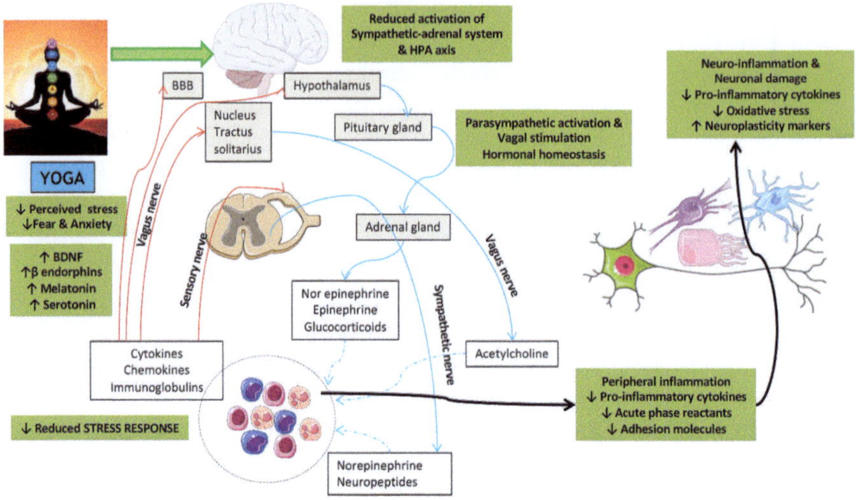

Fig. 1.2 Possible mode of action of *Yoga* in modulation of stress response and various associated changes. *BBB* blood–brain barrier, *HPA axis* hypothalamus–pituitary–adrenal axis, *BDNF* brain derived neurotropic factor; red arrows: afferent nerve fibers, blue arrows: efferent nerve fibers. [The figure was created using images from the *Servier Medical Art (*smart.servier.com*)*]

necessary for the immune system to function properly (Woodyard 2011; Pilkington et al. 2005). There are various key poses which can create benefits to improve immune functioning: (a) Both *sukhAsana* and *Pranayama* encourage deep breathing, which reduces stress hormones, calms the heart rate, and relieves nervous tension, all of which boost immunity; (b) *MatsyAsana* (fish pose) variation is an elevated stance that detoxifies the body and increases energy levels. Additionally, it can clear up nasal airways and alleviate congestion, which is beneficial if you've been sick; (c) *KurmAsana* (tortoise pose) supports the thymus gland; (d) *AdhoMukhaSvanAsana* (Downward Facing Dog pose) improves the flow of the sinuses and help flush mucus from the lungs; (e) *Viparita Karani* (Legs up the wall) forces one to invert and destress and boosts nerve connections, improves blood flow, and vitalizes your body; (f) *UstrAsana* (Camel pose) helps in relieving congestion, increase lung mobility and flush out the lungs and protect the sinuses and mucus membranes and rejuvenates the immune system. *Yoga* is a philosophical approach that aims to bring harmony to all parts of the body, mind, and spirit. *Yoga* is an art of living which has tremendous transformative power, its actions are dynamic and positively modulate the epigenome and can reverse the harmful epigenetic alterations due to unhealthy lifestyle, regular automatic negative thoughts, unhealthy highly refined processed nutrient depleted food and sedentary lifestyle (Ross and Thomas 2010; Tolahunase et al. 2018; Gautam et al. 2019; Kinser et al. 2012).

As *Yoga* improves nuclear DNA integrity and reduces oxidative damage to sperm DNA, this has impact not only on the reproductive health of the individual but can also reduce the disease burden of complex neuropsychiatric disorders like autism, bipolar disorders, spontaneous schizophrenia in the offspring.

1.4 Clinical Implications

Yoga is a mind body energy medicine and is widely believed as a profound science of wellbeing. It acts on the epigenome and thus brings about dynamic changes in gene expression and by switching on gene programs beneficial for human health and wellbeing. *Yoga* promotes neuroplasticity and can bring about changes of key neurotransmitters but also affects the brain anatomy by increasing size of hippocampus, dentate gyrus, prefrontal cortex and reduce size of amygdala. As it reduces inflammation it can delay onset of Alzheimer's and dementia and reduces its severity. *Yoga* impacts the behavior through psycho-neuro-endocrine and immunological processes. *Yoga* by improving nuclear DNA integrity in sperm can reduce disease burden of complex neuropsychiatric disorders in the offspring. *Yoga* enhances overall health and improves overall quality of life. Possessing the ability to prevent/delay the onset of complex lifestyle disorders, *Yoga* can be utilized as an adjuvant for the management of psychosomatic diseases and has tremendous rehabilitative potential.

1.5 Future Direction for Research

Yoga's role in the modulation of the psycho-neuro-immune axis can be comprehensively investigated by future studies focussing on the impact of *Yoga* on epigenetic profile, specific immune effector cells, and a complete cytokine profile.

1.6 Conclusion

This chapter has attempted to explore the beneficial effect of *Yoga* on psycho-neuro-immune axis and a possible mode of action of *Yoga* which can also be the focus of future research. *Yoga* likely modulates the epigenome and switches on gene programs beneficial for health and homeostasis. It relieves stress, enhances metabolic profile, and maintains hormonal balance by modulating the HPA axis and the ANS. *Yoga* holds the potential to modulate the immune response by reduction of pro-inflammatory cytokines, hormonal homeostasis, and elevation of anti-inflammatory and immune-modulatory cytokines. *Yoga* acts by gradually reducing the free radical levels in the blood, upregulation of telomerase activity and levels of sirtuins thereby reducing the biological age, reduction in cortisol, and upregulation in the expression of various anti-inflammatory and cell cycle control genes. It improves mitochondrial integrity and upregulates expression levels of multiple genes that code for antioxidants and reduces the rate of cellular ageing and psychological stress. *Yoga* has the power to manage brain and behavior by controlling its activity through psycho-neuro-endocrine and immunological processes, restoring, and maintaining complete health. *Yoga* promotes health and has preventive and rehabilitative potential and hence, can be used as an adjunct in the management of complex psychosomatic diseases.

References

Antoni MH, Cruess DG, Cruess S, Lutgendorf S, Kumar M, Ironson G et al (2000) Cognitive-behavioral stress management intervention effects on anxiety, 24-h urinary norepinephrine output, and T-cytotoxic/suppressor cells over time among symptomatic HIV-infected gay men. J Consult Clin Psychol 68(1):31–45

Arora S, Bhattacharjee J (2008) Modulation of immune responses in stress by *yoga*. Int J Yoga 1(2):45–55

Basu-Ray I (2021) A mechanistic model for *yoga* as a preventive and therapeutic modality. Int J Yoga 14(2):152–157. https://www.ncbi.nlm.nih.gov/pmc/articles/PMC8191226/

Bellavance MA, Rivest S (2014) The HPA—immune axis and the immunomodulatory actions of glucocorticoids in the brain. Front Immunol 5:136

Bernardi L, Sleight P, Bandinelli G, Cencetti S, Fattorini L, Wdowczyc-Szulc J et al (2001) Effect of rosary prayer and *yoga* mantras on autonomic cardiovascular rhythms: comparative study. BMJ 323(7327):1446–1449

Besedovsky L, Lange T, Haack M (2019) The sleep-immune crosstalk in health and disease. Physiol Rev 99(3):1325–1380. https://www.ncbi.nlm.nih.gov/pmc/articles/PMC6689741/

Bhasin MK, Dusek JA, Chang BH, Joseph MG, Denninger JW, Fricchione GL et al (2013) Relaxation response induces temporal transcriptome changes in energy metabolism, insulin secretion and inflammatory pathways. PLoS One 8(5):e62817

Birkel DA, Edgren L (2000) Hatha *yoga*: improved vital capacity of college students. Altern Ther Health Med 6(6):55–63

Bisht S, Banu S, Srivastava S, Pathak RU, Kumar R, Dada R et al (2020) Sperm methylome alterations following *yoga*-based lifestyle intervention in patients of primary male infertility: a pilot study. Andrologia 52(4):e13551

Brown RP, Gerbarg PL (2005) Sudarshan Kriya yogic breathing in the treatment of stress, anxiety, and depression: part I—neurophysiologic model. J Altern Complement Med 11(1):189–201

Carrillo-Vico A, Lardone PJ, Álvarez-Sánchez N, Rodríguez-Rodríguez A, Guerrero JM (2013) Melatonin: buffering the immune system. Int J Mol Sci 14(4):8638–8683. https://www.ncbi.nlm.nih.gov/pmc/articles/PMC3645767/

Chaplin DD (2010) Overview of the immune response. J Allergy Clin Immunol 125(2 Suppl 2):S3–S23. https://www.ncbi.nlm.nih.gov/pmc/articles/PMC2923430/

Chen Y, Qin C, Huang J, Tang X, Liu C, Huang K et al (2020) The role of astrocytes in oxidative stress of central nervous system: a mixed blessing. Cell Prolif 53(3):e12781. https://www.ncbi.nlm.nih.gov/pmc/articles/PMC7106951/

Coe CL, Laudenslager ML (2007) Psychosocial influences on immunity, including effects on immune maturation and senescence. Brain Behav Immun 21(8):1000–1008. https://www.ncbi.nlm.nih.gov/pmc/articles/PMC2682340/

D'Acquisto F (2017) Affective immunology: where emotions and the immune response converge. Dialogues Clin Neurosci 19(1):9–19. https://www.ncbi.nlm.nih.gov/pmc/articles/PMC5442367/

Dantzer R (2018) Neuroimmune interactions: from the brain to the immune system and vice versa. Physiol Rev 98(1):477–504. https://www.ncbi.nlm.nih.gov/pmc/articles/PMC5866360/

Gard T, Noggle JJ, Park CL, Vago DR, Wilson A (2014) Potential self-regulatory mechanisms of *yoga* for psychological health. Front Hum Neurosci 8:770

Gautam S, Dada R (2021) Molecular mechanisms underlying the effects of *yoga*. In: Handbook of research on evidence-based perspectives on the psychophysiology of *yoga* and its applications. IGI Global, pp 103–123. https://www.igi-global.com/chapter/molecular-mechanisms-underlying-the-effects-of-Yoga/www.igi-global.com/chapter/molecular-mechanisms-underlying-the-effects-of-Yoga/261146

Gautam S, Tolahunase M, Kumar U, Dada R (2019) Impact of *yoga* based mind-body intervention on systemic inflammatory markers and co-morbid depression in active rheumatoid arthritis patients: a randomized controlled trial. Restor Neurol Neurosci 37(1):41–59

Gautam S, Kumar M, Kumar U, Dada R (2020) Effect of an 8-week *yoga*-based lifestyle intervention on psycho-neuro-immune axis, disease activity, and perceived quality of life in rheumatoid arthritis patients: a randomized controlled trial. Front Psychol 11:2259. https://doi.org/10.3389/fpsyg.2020.02259

Gautam S, Kumar U, Dada R (2021a) *Yoga* and its impact on chronic inflammatory autoimmune arthritis. Front Biosci (Elite Ed) 13:77–116

Gautam S, Kumar U, Kumar M, Rana D, Dada R (2021b) *Yoga* improves mitochondrial health and reduces severity of autoimmune inflammatory arthritis: a randomized controlled trial. Mitochondrion 58:147–159

Gautam S, Saxena R, Dada T, Dada R (2021c) *Yoga*—impact on mitochondrial health: clinical consequences. Ann Neurosci 28:114–116

Gautam S, Kumar U, Mishra R, Dada R (2022) HLA-G 3′UTR polymorphisms and response to a *yoga*-based lifestyle intervention in rheumatoid arthritis: a randomized controlled trial. Indian J Med Res 155(2):253–263

Gjerstad JK, Lightman SL, Spiga F (2018) Role of glucocorticoid negative feedback in the regulation of HPA axis pulsatility. Stress 21(5):403–416. https://www.ncbi.nlm.nih.gov/pmc/articles/PMC6220752/

Goldin PR, Gross JJ (2010) Effects of mindfulness-based stress reduction (MBSR) on emotion regulation in social anxiety disorder. Emotion 10(1):83–91

Gopal A, Mondal S, Gandhi A, Arora S, Bhattacharjee J (2011) Effect of integrated *yoga* practices on immune responses in examination stress—a preliminary study. Int J Yoga 4(1):26–32. https://www.ncbi.nlm.nih.gov/pmc/articles/PMC3099098/

Gothe NP, Khan I, Hayes J, Erlenbach E, Damoiseaux JS (2019) *Yoga* effects on brain health: a systematic review of the current literature. Brain Plast 5(1):105–122. https://www.ncbi.nlm.nih.gov/pmc/articles/PMC6971819/

Halappa NG, Thirthalli J, Varambally S, Rao M, Christopher R, Nanjundaiah GB (2018) Improvement in neurocognitive functions and serum brain-derived neurotrophic factor levels in patients with depression treated with antidepressants and *yoga*. Indian J Psychiatry 60(1):32–37

Herman JP, McKlveen JM, Ghosal S, Kopp B, Wulsin A, Makinson R et al (2016) Regulation of the hypothalamic–pituitary–adrenocortical stress response. Compr Physiol 6(2):603–621. https://www.ncbi.nlm.nih.gov/pmc/articles/PMC4867107/

Khalsa SBS, Hickey-Schultz L, Cohen D, Steiner N, Cope S (2012) Evaluation of the mental health benefits of *yoga* in a secondary school: a preliminary randomized controlled trial. J Behav Health Serv Res 39(1):80–90

Kim J, Gorman J (2005) The psychobiology of anxiety. Clin Neurosci Res 4(5):335–347. http://www.sciencedirect.com/science/article/pii/S1566277205000101

Kinser PA, Goehler LE, Taylor AG (2012) How might *yoga* help depression? A neurobiological perspective. Explore (NY) 8(2):118–126

Krishnakumar D, Hamblin MR, Lakshmanan S (2015) Meditation and *yoga* can modulate brain mechanisms that affect behavior and anxiety-A modern scientific perspective. Anc Sci 2(1):13–19. https://www.ncbi.nlm.nih.gov/pmc/articles/PMC4769029/

Lee M, Moon W, Kim J (2014) Effect of *yoga* on pain, brain-derived neurotrophic factor, and serotonin in premenopausal women with chronic low back pain. Evid Based Complement Alternat Med 2014:203173

Martarelli D, Verdenelli MC, Scuri S, Cocchioni M, Silvi S, Cecchini C et al (2011) Effect of a probiotic intake on oxidant and antioxidant parameters in plasma of athletes during intense exercise training. Curr Microbiol 62(6):1689–1696

Mateos-Aparicio P, Rodríguez-Moreno A (2019) The impact of studying brain plasticity. Front Cell Neurosci 13:66. https://www.ncbi.nlm.nih.gov/pmc/articles/PMC6400842/

McEwen BS (2006) Protective and damaging effects of stress mediators: central role of the brain. Dialogues Clin Neurosci 8(4):367–381. https://www.ncbi.nlm.nih.gov/pmc/articles/PMC3181832/

Moonaz SH, Bingham CO, Wissow L, Bartlett SJ (2015) *Yoga* in sedentary adults with arthritis: effects of a randomized controlled pragmatic trial. J Rheumatol 42(7):1194–1202

Pilkington K, Kirkwood G, Rampes H, Richardson J (2005) *Yoga* for depression: the research evidence. J Affect Disord 89(1–3):13–24

Pramanik T, Sharma HO, Mishra S, Mishra A, Prajapati R, Singh S (2009) Immediate effect of slow pace bhastrika *pranayama* on blood pressure and heart rate. J Altern Complement Med 15(3):293–295

Qu S, Olafsrud SM, Meza-Zepeda LA, Saatcioglu F (2013) Rapid gene expression changes in peripheral blood lymphocytes upon practice of a comprehensive *yoga* program. PLoS One 8(4):e61910

Rathore M, Abraham J (2018) Implication of *asana*, *pranayama* and meditation on telomere stability. Int J Yoga 11(3):186–193. https://www.ncbi.nlm.nih.gov/pmc/articles/PMC6134740/

Raveendran AV, Deshpandae A, Joshi SR (2018) Therapeutic role of *yoga* in type 2 diabetes. Endocrinol Metab (Seoul) 33(3):307–317. https://www.ncbi.nlm.nih.gov/pmc/articles/PMC6145966/

Riley D (2004) Hatha *yoga* and the treatment of illness. Altern Ther Health Med 10(2):20–21

Ross A, Thomas S (2010) The health benefits of *yoga* and exercise: a review of comparison studies. J Altern Complement Med 16(1):3–12

Schneier FR, Liebowitz MR, Abi-Dargham A, Zea-Ponce Y, Lin SH, Laruelle M (2000) Low dopamine D(2) receptor binding potential in social phobia. Am J Psychiatry 157(3):457–459

Segerstrom SC, Miller GE (2004) Psychological stress and the human immune system: a meta-analytic study of 30 years of inquiry. Psychol Bull 130(4):601–630

Selden NR, Robbins TW, Everitt BJ (1990) Enhanced behavioral conditioning to context and impaired behavioral and neuroendocrine responses to conditioned stimuli following ceruleocortical noradrenergic lesions: support for an attentional hypothesis of central noradrenergic function. J Neurosci 10(2):531–539

Sengupta P (2012) Health impacts of *yoga* and *pranayama*: a state-of-the-art review. Int J Prev Med 3(7):444–458

Shete SU, Verma A, Kulkarni DD, Bhogal RS (2017) Effect of *yoga* training on inflammatory cytokines and C-reactive protein in employees of small-scale industries. J Educ Health Promot 6:76

Silverman MN, Pearce BD, Biron CA, Miller AH (2005) Immune modulation of the hypothalamic–pituitary–adrenal (HPA) axis during viral infection. Viral Immunol 18(1):41–78

Smith SM, Vale WW (2006) The role of the hypothalamic–pituitary–adrenal axis in neuroendocrine responses to stress. Dialogues Clin Neurosci 8(4):383–395. https://www.ncbi.nlm.nih.gov/pmc/articles/PMC3181830/

Sterling P, Eyer J (1988) Allostasis: a new paradigm to explain arousal pathology. In: Handbook of life stress, cognition and health. Wiley, Oxford, pp 629–649

Tausk F, Elenkov I, Moynihan J (2008) Psychoneuroimmunology. Dermatol Ther 21(1):22–31

Tolahunase M, Sagar R, Dada R (2017) Impact of *yoga* and meditation on cellular aging in apparently healthy individuals: a prospective, open-label single-arm exploratory study. Oxidative Med Cell Longev 2017:7928981

Tolahunase MR, Sagar R, Faiq M, Dada R (2018) *Yoga*- and meditation-based lifestyle intervention increases neuroplasticity and reduces severity of major depressive disorder: a randomized controlled trial. Restor Neurol Neurosci 36(3):423–442

Tolahunase MR, Gautam S, Sagar R, Kumar M, Dada R (2021) *Yoga* in major depressive disorder: molecular mechanisms and clinical utility. Front Biosci (Schol Ed) 13(1):56–81

Tsigos C, Kyrou I, Kassi E, Chrousos GP (2000) Stress: endocrine physiology and pathophysiology. In: Feingold KR, Anawalt B, Boyce A, Chrousos G, de Herder WW, Dhatariya K et al (eds) Endotext. MDText.com, Inc., South Dartmouth. http://www.ncbi.nlm.nih.gov/books/NBK278995/

Venkatesh H, Ravish H, Wilma Delphine Silvia C, Srinivas H (2020) Molecular signature of the immune response to *yoga* therapy in stress-related chronic disease conditions: an insight. Int J Yoga 13(1):9–17. https://www.ncbi.nlm.nih.gov/pmc/articles/PMC6937878/

Vickers A, Zollman C, Payne DK (2001) Hypnosis and relaxation therapies. West J Med 175(4):269–272. https://www.ncbi.nlm.nih.gov/pmc/articles/PMC1071579/

Vijayaraghava A, Doreswamy V, Narasipur OS, Kunnavil R, Srinivasamurthy N (2015) Effect of *yoga* practice on levels of inflammatory markers after moderate and strenuous exercise. J Clin Diagn Res 9(6):CC08–CC12

Villemure C, Čeko M, Cotton VA, Bushnell MC (2015) Neuroprotective effects of *yoga* practice: age-, experience-, and frequency-dependent plasticity. Front Hum Neurosci 9:281. https://www.ncbi.nlm.nih.gov/pmc/articles/PMC4428135/

Williams S, Ghosh C (2020) Neurovascular glucocorticoid receptors and glucocorticoids: implications in health, neurological disorders and drug therapy. Drug Discov Today 25(1):89–106. https://www.ncbi.nlm.nih.gov/pmc/articles/PMC6980291/

Woodyard C (2011) Exploring the therapeutic effects of *yoga* and its ability to increase quality of life. Int J Yoga 4(2):49–54. https://www.ncbi.nlm.nih.gov/pmc/articles/PMC3193654/

Woolery A, Myers H, Sternlieb B, Zeltzer L (2004) A *yoga* intervention for young adults with elevated symptoms of depression. Altern Ther Health Med 10(2):60–63

Yadav RK, Magan D, Mehta N, Sharma R, Mahapatra SC (2012) Efficacy of a short-term *yoga*-based lifestyle intervention in reducing stress and inflammation: preliminary results. J Altern Complement Med 18(7):662–667

Chapter 2
The Regulation of Neuro-immune System by Pi (Parasympathetic) and SiGMA (Sympathetic Induction Governed by Meditation and *Asana*) Rhythmicity

Rhythm Arora, Himanshu Verma, N. Banumathy, and Maryada Sharma

Contents

Abstract The sympathetic system is overactivated under neurological stress states; however, it has been reported that certain extraordinary brain centers can help to counterbalance the heightened sympathetic system by activation of the parasympathetic arm of the autonomic system. The master regulator of the autonomic system is the paraventricular nucleus (PVN) that along with the ventral striatum expresses the *VMAT2* (vesicular monoamine transporter 2) gene, also known as the GOD gene, which regulates the transport of essential brain chemicals known as monoamine neurotransmitters and has also been associated with the genetic variance in spirituality. We discuss the emerging molecular determinants of regulation of sym-

Rhythm Arora and Himanshu Verma contributed equally with all other contributors.

R. Arora · H. Verma · N. Banumathy · M. Sharma (✉)
Department of Otolaryngology and Head and Neck Surgery, Postgraduate Institute of Medical Education and Research, Chandigarh, India
e-mail: sharma.maryada@pgimer.edu.in

A. Anand (ed.), *Neuroscience of Yoga*,
https://doi.org/10.1007/978-981-97-2851-0_2

pathetic and parasympathetic arms of the autonomic nervous system with a brief discussion on the *GOD* gene, *VMAT2* that can regulate the transport of catecholamines, especially the dopamine and the norepinephrine that have an important effect on memory and cognition in the limbic system during the stress states. It is proposed that the parasympathetic arm of PVN in the hypothalamus can be activated by meditation and asana, which can improve the mental state and cognitive functions by promoting anti-inflammatory and anti-oxidative actions via RAAS inhibition through the overexpression of *VMAT2* (*GOD* gene), that participates in dopamine vesicular transport and facilitates neurotransmission to essential parts of brain for elevating moods, cognition, and memory improvement.

Keywords Sympathetic and parasympathetic · Long non-coding · Hypothalamus · Neonates · Meditation and yoga

2.1 Introduction

Yoga is an ancient discipline of science, which originated thousands of years ago in India. Yoga is the longest practiced science in India, in recent years it has spread all over the world due to its incredible health benefits. Yoga was designed to balance the mind and body connection, and one of the major aims of Yoga is to attain calmness of the mind, improve attention, increase self-confidence, decrease irritability, improve efficiency, and increase optimism in life. Many studies documented the health benefits of Yoga such as enhanced flexibility, detoxification, improved muscle tone, increased muscle strength, and lubrication of the joints, tendons, and ligaments. Previous studies (Sugama and Kakinuma 2020) reported that increased stress leads to the hyper-regulation of hypothalamic–pituitary–adrenal (HPA) and sympathetic nervous system (SNS). Due to the hyper-regulation of HPA and SNS, cortisol, renin, and catecholamines (epinephrine and norepinephrine) were released, leading to the hypervigilance state. The continuous presence of hypervigilance syndrome may cause dysregulation of the system and lead to various diseases such as autoimmune disorders, obesity, depression, cardiovascular disease, substance abuse, and diabetes (Cramer et al. 2014).

Various studies (Aggarwal et al. 2020; Wadikar 2020) revealed that regular Yoga practice would help optimize the body's sympathetic responses to stress. Yoga practice further helps restore the autonomic regulatory reflex mechanisms related to stress by inhibiting the posterior hypothalamic area. The areas responsible for aggressiveness and fear get inhibited by the Yoga practice; rather, it stimulates the centers in the median forebrain related to pleasure and reward (Wadikar 2020). Many studies found that continuous practicing of Yoga can reduce depression symptoms by significantly increasing serotonin levels and decreasing monoamine oxidase levels. From past literature (Aggarwal et al. 2020; Wadikar 2020) it can be delineated that Yoga practice improved regulation of the HPA axis and SNS, thereby reducing the risk of many health-related diseases or disorders. The authors reported that Yoga helps reduce psychological stress and alters the autonomic functions in the premenstrual phase by balancing the neuro-endocrinal axis. We discuss the emerging role of neurogenic regulation of the sympathetic system in regulating metabolic diseases like diabetes and hypertension [cardiovascular-renal (RAAS) pathways] that are deregulated during neurological stress states resulting in neuro-inflammatory diseases. We propose a potential role of alternative therapies like asanas and meditation in ameliorating the neurogenic RAAS activity (autonomic system) that has implications in modulating the twin-pandemics of diabetes mellitus (DM) and hypertension (HTN).

2.1.1 *PITTRep (Plasma-Induced Transcriptomics/epi-Transcriptomics Reprograming)*

A novel multiomics-based model to establish a link between neuro-immune deregulation under neurological diseases.

Based on our preliminary findings derived from a novel cellular stress-induced neural crest stem cell model, PITTRep (Sharma et al. 2022; Panda et al. 2022), employed for modeling neuroinflammatory diseases associated with neurosensory and sympathetic deregulation, we propose that the GPCR-coupled β1 adrenergic receptors activate the adenylate cyclase-cAMP-PKA activity by Gs arm (GPCR), which eventually results in the calcium-dependent action potential generation associated with ionotropic effects that trigger high blood pressure (BP) and heightened heart rate. This sympathetic overflow is further heightened and remains unchecked

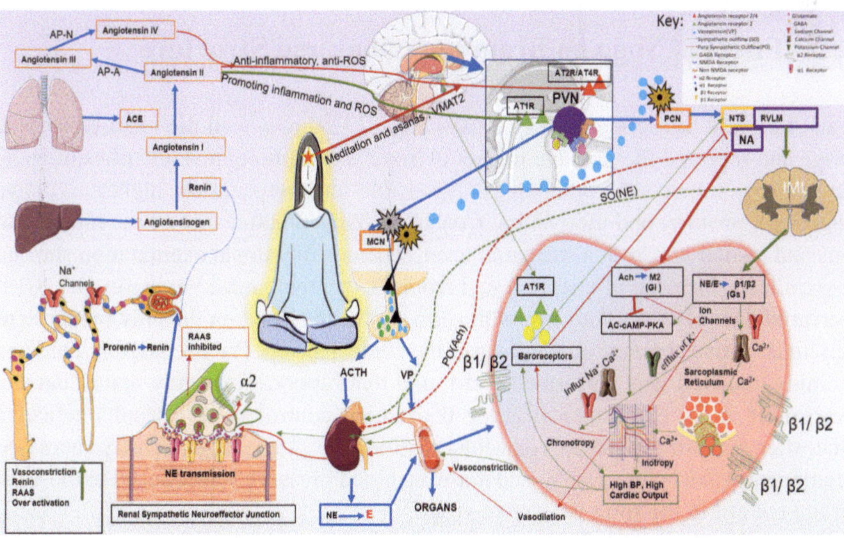

Fig. 2.1 The autonomic regulation and role of *VMAT2* (the *GOD* gene) in homeostasis

due to downregulation of alpha 2 adrenoreceptors that function to control norepinephrine activity, hence triggering the neuroinflammatory signaling. The concomitant downregulation of *VMAT2* (the *GOD* gene linked to meditation and spirituality) was strongly linked to the disruption of a homeostatic balance in the sympathetic and parasympathetic activity (Sharma et al. 2022; Panda et al. 2022) (Fig. 2.1).

PVN in the hypothalamus is the main regulator of the autonomic arm of the nervous system. Under stress states, the excitatory neurons get upregulated resulting in the overactivation of sympathetic outflow, following which the PC neurons play an important role in regulating the cardiac activity via nucleus tractus solitarius (NTS), rostral ventrolateral medulla (RVLM) in brainstem, and *iliacus median* (ILM) in spinal cord. They relay innervation to the heart following which norepinephrine reception to the heart by beta-1 adrenergic receptors activates adenylate cyclase-cAMP-PKA activity, resulting in modulation in ion channels and heart contractility. The subsequent release of excessive calcium results in increased heart contractility that is sensed by the baroreceptors that signal to NTS for normalizing the system. However, due to overactivation of sympathetic outflow, the RAAS pathway is predominantly functional, resulting in renin-mediated angiotensin upregulation, and AngII to AT1 receptors (AT1Rs) causing pathological conditions by delimiting the action of baroreceptors. This results in copious ROS production and inflammatory induction. It is proposed that the parasympathetic arm of PVN in the hypothalamus can be activated by meditation and YogAsana, which can improve the mental state and cognitive functions by promoting anti-inflammatory and anti-oxidative actions via RAAS inhibition through the overexpression of *VMAT2* (*GOD* gene), which participates in dopamine vesicular transport and facilitates neurotransmission to essential parts of the brain for elevating moods, cognition, and memory improvement.

2.2 Effect of Yoga on Brain Functions and Structure

Van-Aalst and his colleagues (van Aalst et al. 2020) reviewed the neurobiology of Yoga and reported the positive impact of Yoga at morphological and physiological levels. Yoga impacts the brain areas responsible for interoception, higher executive functions, posture, and motivation. Consistent Yoga practice affects the hippocampus and insular cortex at a structural level, whereas functional examination showed a significant increase in attention and frontal executive areas. Garner et al. (2019) performed an MRI study to assess the effect of *hatha* Yoga on the gray matter density in an active sports group and a passive control group. Prior to starting the Yoga training, an MRI was performed and found that hippocampal gray matter density was lower in the experimental group than in the control group. Results indicated that after performing *hatha* Yoga intervention, the gray matter density increased significantly. Also, the systolic and diastolic blood pressure was reduced in the Yoga group but not in the sport or passive group.

Similarly (Yang et al. 2019; Fayed et al. 2013), various studies reported positive effects of Yoga practice at the neuro-structural and neurophysiological levels. Grant and his colleagues (Grant et al. 2013) investigated the relationship between meditation and cortical thickness. The authors observed the greater cortical thickness in the experiment group for various regions, viz., attention network of cingulo-fronto-parietal region. From the literature, it can be delineated that Yoga and meditation improve attention by way of activating the various cortical and subcortical areas of the brain.

2.3 Pregnancy-Induced Stress and Anxiety

Stress, depression, and anxiety are seen to commonly emerge during the gestational period (van Heyningen et al. 2017). In the literature, multiple studies (Hou et al. 2018; Wall et al. 2018; Rosario et al. 2017) discussed pregnancy-induced stress and anxiety and its consequences. Rosario and his colleagues (Rosario et al. 2017) conducted a qualitative survey to study pregnancy-related anxiety among Tanzania's women. The results of the study revealed a high level of pregnancy-related anxiety among women of Mwanza, Tanzania. The study further revealed the rationales behind pregnancy-related anxiety including lack of knowledge, financial hardship, cultural beliefs, fear of HIV/AIDS, and poor social status. Van Heyningen et al. (2017) found a high prevalence of pregnancy-related anxiety among low-income pregnant women. Hou et al. (2018) found a correlation between the poor maternal lifestyle (i.e., prolonged usage of mobile and poor sleep) and stress during gestation period. Okun et al. 2013) observed a high prevalence of depressive symptoms among pregnant women who sleep less than 7 h. The prevalence and severity of stress and anxiety among pregnant and postpartum women increased during the COVID-19 pandemic period (Stepowicz et al. 2020). The stress and anxiety during pregnancy activate the HPA axis and the gamma-amino-butyric acid (GABAergic) neuro-steroids which are derived from progesterone (Crowley et al. 2016).

2.4 Molecular and Regulatory Aspects of HPA Axis

The limbic system in brain is the central region for maintaining the body's homeostasis amidst the states of stress, which is primarily governed by the hypothalamic and hippocampal regions that restore homeostasis by regulating the centers of energy balance and memory. The limbic neurons are, therefore, at the confluence of metabolism and cognition (Ramírez et al. 2022). Furthermore, the hypothalamic paraventricular nucleus (PVN) and arcuate nucleus (ARC) play important roles in regulating the neurological and metabolic processes such as behavior, cognition, food metabolism, and various other physiological and pathological pathways. In addition to the hypothalamus, the hippocampus is another significant region in the limbic system that is crucial to cognitive and memory function during physical and psychological stimuli.

The autonomic (sympathetic and parasympathetic) system is the master regulator for controlling the physiological and pathological pathways; therefore, a delicate balance between the sympathetic and parasympathetic branches of the autonomic system is central to balancing the states of physical and mental stress. Following the overactivation of the sympathetic system (SS) and lack of a parasympathetic (PS) branch to counterbalance it, results in the induction of pathological pathways during many neurological and metabolic conditions marked by sympathetic overactivity.

2.5 Hypothalamus: The Master Controller

A key area of the hypothalamus that is associated with regulating the sympathetic outflow is the paraventricular nucleus (PVN) that is controlled by excitatory (glutamate) and inhibitory (GABA) neurotransmitters inside the PVN. PVN is further divided into two parts: the magnocellular neurons (MCNs) that regulate neuroendocrine functions and the parvocellular neurons (PCNs) that regulate autonomic functioning. PVN's role in the cardinal control of the cardiovascular function is well appreciated, as it plays a key role in the autonomic modulation of blood pressure (BP) to environmental stressors. PCN tends to promote sympathetic outflow toward the heart, blood vessels, and kidneys whereas magnocellular neurons carry out the neuroendocrine role by vasopressin and oxytocin release to the systemic branch (Savić et al. 2022). Glutamatergic transmission in the hypothalamus plays a significant role in neuroendocrine functioning, body homeostasis, and social behavior (Royo et al. 2022). Oxytocin (OT) is released from the supraoptic nucleus via a combination of NMDA and non-NMDA glutamate receptors, and arginine vasopressin (AVP) is released from this region by stimulation of non-NMDA glutamate receptors. Additionally, the PVN's NMDA receptors have an inhibitory effect on the production of AVP and OT. AMPARs are the type of non-NMDA glutamate receptors that play an essential role in the sympathetic regulation; also the synaptic plasticity, learning, and memory are linked to the rapid trafficking of AMPARs to the postsynaptic membrane. It has been established that some psychiatric and neurodegenerative disorders have a pathophysiology that involves defects in the controlled AMPAR trafficking (Ge and Wang 2021). A recent study reported that the drug methylphenidate repaired behavioral deficits and modulated neuroplasticity as a result of alterations in the composition and distribution of AMPAR subunits when used as a treatment for Attention Deficit Hyperactivity Disorder (Contreras et al. 2022). In addition to the hypothalamus, the hippocampus is an essential part of the limbic system and helps regulate the stress, as the deregulated hippocampal pathways such as those associated with epileptic hippocampus exhibited reduced AMPAR expression and phosphorylation in a phosphatase PTEN dependent manner, indicating PTEN as a potential AMPAR antagonist to have anti-epileptic effects (Kim et al. 2020). AMPAR accessory proteins associated with AMPAR trafficking and synaptic localization are also targeted by miRNAs, indicating stringent gene regulation (Hanley 2021).

Ionotropic excitatory amino acid activities acting locally at NMDA and non-NMDA receptors are required for the cardiovascular and renal sympathetic actions in response to acute GABA-A receptor blockage in the PVN. The PVN's GABAergic inhibitory input is essential for limiting sympathetic outflow, and osmoregulatory systems have developed to maintain the extracellular compartment's tonicity within narrow bounds as a part of tight regulation of the system (Mecawi et al. 2022).

The sympathetic system overactivation results in overexpression of glutamate excitatory neurons in PVN of the hypothalamus, due to which parvocellular neurons (PCNs) regulating autonomic functioning release glutamate, which further binds to the NMDA and non-NMDA receptors, resulting in excitatory effects. Following the hypothalamus activity, the hippocampus triggers the noradrenaline release in response to nicotinic receptors (α_7nAchr receptors) activation leading to sympathetic outflow to the target organs (Ortiz et al. 2022). There are numerous projections from brainstem catecholamine neurons, including the one that reaches out to the PVN that is involved in the neuroendocrine and cardiorespiratory responses to hypoxia. Through the cAMP-protein kinase A pathway, melanocortin 3/4 receptor activation in the PVN raises sympathetic output and blood pressure.

2.6 The Impact of Stress

The PVN is also an important region for controlling autonomic circulation to the venous region, and the systemic vasoconstriction is caused by the stimulation of bradykinin-sensitive pericardial afferent fibers associated with PVN involvement (Martin et al. 2022). TNFα in PVN causes sympathetic nerve activity to intensify through mechanisms that depend on the presence of nearby ionotropic glutamate receptors but are mainly unrelated to TNFα autoinduction (Mourão et al. 2021). Due to the fact that PVN is the primary regulator of the nervous system's cardiac regulation, stress, exercise, and asana are proposed to influence the heart rate, blood pressure, and baroreception. Mutual interactions between the paraventricular and coeruleus excitatory centers tend to raise the systolic pressure and the heart rate (Cayupe et al. 2021). The sympathetic nerves innervating the heart target the β1 adrenergic receptors that boost the cardiac output and heart rate. Nitric oxide, produced by neuronal-NOS, plays a role in a neurogenic process of hyperbaric hypoxia-induced lung injury and is connected to the central nervous system's ROS via adrenergic/cholinergic pathways. In ischemic heart failure, the sympathetic excitation is facilitated by NF-kB activation in the PVN that modifies the neurotransmitter activity. The spinal ionotropic glutamate receptors and GABA receptors in the medial nucleus tractus solitarius regulate the vagal suppression and sympathetic activation that contribute to PVN-induced tachycardia.

The incremental blood pressure rise is sensed by the baroreceptors (myelinated nerve fibers) in the carotid bodies, and the chemosensory input from carotid bodies (CB) gets transmitted to the medullary centers by the carotid sinus nerve (CSN), which is activated by leptin hormone. Additionally, the impact of leptin is mediated by Trpm7 (transient receptor potential melastatin) channels in CB glomus cells (Yeung et al. 2021), which are extensively distributed on the plasma membrane of cells and function as sensors that are capable of perceiving environmental stress/stimuli (Liu et al. 2022). Baroreceptors relay sensory innervations to the nucleus tractus solitarius (NTS) via CSN that signals to the PVN for maintaining homeostasis under physiological conditions; however, in pathological conditions, the excitatory neurons are overactivated, thereby disrupting the homeostasis. The blood pressure is reported to be raised by brain-derived neurotrophic factor (BDNF) in part by reducing the inhibitory hypotensive feedback from catecholaminergic neurons in the PVN's nucleus of the solitary tract (NTS). In response to hypertensive stimuli including cellular ionic/volume (hemodynamic) stress and hyperosmolality, BDNF is reported to be elevated in the PVN, suggestive of a regulatory role in osmoregulation and stress-associated hypertension responses. BDNF acts in the PVN to elevate sympathetic outflow and blood pressure to regulate homeostatic responses (Thorsdottir et al. 2021). The RAAS pathway intermediate angiotensin II and metabolic insulin hormone play a role in the etiology of HTN due to DM-associated metabolic disturbances by amino acids, lipids, bile acids, and carnitines (Ning et al. 2021). The PVN neurons' increased production of ROS in response to glutamate is suggested to be linked to angiotensin-II-induced slow-pressor hypertension, resulting in NOX2 being mobilized to locations close to the NMDA receptor (NMDAR).

Following the activation of RAAS and increased blood pressure, the juxtaglomerular (JGA) cells in the kidney are signaled to convert prorenin to renin and its secretion into the blood to convert liver secreted angiotensinogen (AGT) to angiotensin I (AT1) by renin. Angiotensinogen expression outside of the liver or in the adrenal glands is suggested to be essential for healthy renal growth and survival (Sapouckey et al. 2020). AT1 is further converted into angiotensin II (ATII) by angiotensin-converting enzyme (ACE) from lungs or by Chymase I (Cosarderelioglu et al. 2020). The resulting AT2 further activates AT1 receptors (AT1Rs) present in various organs of the body (Cosarderelioglu et al. 2020) to relay the effects of RAAS pathway signaling. However, in pathological conditions, the RAAS sensing is altered, due to the overactivation of sympathetic outflow, by ouabain, an endogenous cardiotonic steroid which is present in PVN of the hypothalamus and has a crucial role in the regulation of blood pressure. Ouabain acts as a natriuretic hormone that promotes the enhancement of Na^+ excretion after an excessive Na^+ intake and is suggested to be an important molecular link between dietary Na^+ intake to the development of hypertension. However, further research is warranted to understand the detailed role of ouabain in the renal control of Na^+ excretion and the interaction of ouabain and other endogenous cardiotonic steroids such as marinobufagenin that are implicated in the regulation of body electrolyte homeostasis.

During pregnancy, the PVN's AT1Rs become more excitatory and the GABAergic regulation declines. During a healthy pregnancy, basal SNA rises and the ability of the baroreflex to control heart rate is compromised. This is probably owing to an increase in the SNA-supporting factor AT2 levels. It essentially counteracts significant primary vasodilation, but pregnancy-induced hypertensive diseases cause even larger elevations in SNA, which almost certainly cause hypertension (Brooks et al. 2020). In response to AT1R activation, MCN of PVN in the hypothalamus releases the vasopressin to the kidney along with the autonomic regulation by PCN that helps in the formation of aldosterone in the kidney, which results in the upregulation of gene expression in kidney principal cells that are involved in Na reabsorption to regulate the blood pressure (Cosarderelioglu et al. 2020). In parallel, the local brain RAAS is also activated and is regulated by the local brain cells like astrocytes, microglial cells, oligodendrocytes, and neurons, which are capable of secreting renin and AGT that work toward maintaining local neurogenic RAAS homeostasis. In pathological conditions, the RAAS is overexpressed, resulting in brain renin and angiotensin II formation that are linked to hypertension and heart failure. These pathological conditions can be caused due to the dysregulation caused in PVN, i.e., due to overactivation of AT1Rs, inflammation, and oxidative stress that results in increased ROS production and elevated sympathetic outflow (Cosarderelioglu et al. 2020).

Neurologic dysfunction is caused by a variety of metabolic alterations in the brain; two newer forms of neurologic disorder are hepatic encephalopathy and diabetic encephalopathy (Cheon and Song 2021). The dysregulation of glutamatergic and GABAergic neurons in the hippocampus during diabetic encephalopathy plays a significant role in neurological damage. The neurogliovascular system becomes dysfunctional as a result of hepatic encephalopathy, which also affects the blood-brain barrier's ability to operate. Astrocytes and microglia begin to malfunction as a result of an aggregation of neurotoxic chemicals and finally the lymphatic dysfunction makes it more difficult for these neurotoxins to be cleared from the body resulting in neurological deregulation (Claeys et al. 2021).

Circulating hormones regulate a vast network of brain areas including the arcuate nucleus (ARC) of the hypothalamus that regulates food intake and energy expenditure, in order to control energy balance (Mehay et al. 2021). In addition to regulating food, energy state-sensing neurons in the ARC also govern a number of physiological processes, such as blood pressure, innate immune responses, and glucose homeostasis. These studies have characterized them as fundamental regulators that adjust integrative metabolism to the organism's energy state (Jais and Brüning 2022). The persistent dysfunction of the glucoregulatory system in the mediobasal hypothalamus plays a role in the emergence of metabolic illness (Arrieta-Cruz et al. 2022).

Ang II regulates the ARC that serves as an important center for energy balance, reproduction, and arterial pressure (AP). Further, the activation of AT1R in response to Ang II causes vasoconstriction due to vasopressin release and then gradually increases sympathetic nerve activity and the heart rate (Shi et al. 2022). The fluid homeostasis and BP responses are unaffected by Muscarinic type 4 receptors (MC4R) signaling; however, Ang II in the brain that regulates resting metabolic rate (RMR) and energy expenditure is MC4R-dependent (Oliveira et al. 2022). In hypothalamic ARC cells that express Agouti-related peptide (AgRP), the AT1R localizes with the leptin receptor predominantly. AT1Rs in this subpopulation are necessary for the stimulation of RMR and thermogenic sympathetic nerve activity. Suppression of RMR has an impact on the onset and persistence of obesity. Within the ARC, the proopiomelanocortin neurons and AgRP work together to integrate RMR regulation. The hypothalamus' melanocortin system and leptin play a role in the regulation of blood pressure and energy balance (Oliveira et al. 2022).

2.7 The GOD Gene and Its Role in Spiritual Healing

Under pathological conditions, the sympathetic system is overactivated; however, it has been reported that certain extraordinary brain centers can help to counterbalance the sympathetic overactivity by activation of the parasympathetic system to suppress the sympathetic activity. The master regulator of the autonomic system, PVN along with the ventral striatum are documented to express the *VMAT2* (vesicular monoamine transporter 2) gene, also known as the God gene (Valli et al. 2021). *VMAT2* regulates the transport of essential brain chemicals known as monoamine neurotransmitters and has also been associated with the genetic variance in spirituality. As a transporter, it can regulate the transport of catecholamines, especially dopamine and norepinephrine that have important effects on memory and cognition in the limbic system. *VMAT2* also plays a very important role in regulating pathological conditions, as it has dual expression in dopaminergic neurons and pancreatic β cells; therefore, it is important for transporting dopamine into presynaptic vesicles and can selectively regulate pancreatic β cells. It was recently proposed to co-regulate diabetes and neurodegenerative conditions like Parkinson's disease (Kong et al. 2020).

2.8 Clinical Implication

2.8.1 Beneficial Effect of Yoga *Training During Pregnancy*

In ancient India, information on healthy pregnancy was usually passed on from elderly women to expected mothers. One of the significant recommendations was to remain active and perform Yoga activities during the gestational period for smooth delivery and healthy growth of the fetus. Today, various studies (Hu et al. 2021; Dal et al. 2021; Yekefallah et al. 2021) documented the positive effect of Yoga during pregnancy. Regular Yoga during pregnancy helps in reducing the risk of gestational diabetes, hypertension, back pain, pelvic pain, anxiety, stress, depression, and mood swings.

Several researchers (Hu et al. 2021; Dal et al. 2021; Rakhshani et al. 2015) further provided the evidence on benefits of Yoga in high-risk pregnancies and neonates. Rakhshani and colleagues (Rakhshani et al. 2015) advocated that Yoga helps in increasing the utero-fetal-placental circulation in high-risk pregnancy. The study revealed that fetal ultrasound measurements, viz., head circumference, abdominal circumference, femur length, heart rate, estimated fetal health, and biparietal diameter, were more significant in women practicing Yoga. Hu et al. (2021) reviewed the literature and observed that pregnant women practicing Yoga have better intrauterine fetal growth and immune functions. The presence of good fetal outcomes in women practicing Yoga may be linked to the changes at a hormonal level. The group with Yoga intervention had the lower salivary alpha-amylase, salivary cortisol, and salivary a-amylase levels and higher immunoglobulin-A level. Yekefallah et al. (2021) executed a clinical trial to explore the effects of Yoga on pregnancy, delivery, and neonatal outcomes. The study revealed beneficial effects of Yoga during and post-delivery, such as reduced episiotomy rupture, labor pain, duration of labor, and the induction of labor. Results further reported good neonatal outcomes (e.g., appropriate gestational age and adequate birth weight) in the Yoga group. A recent study (Dal et al. 2021) observed that women enrolled in Yoga program during pregnancy had increased breastfeeding self-efficacy. A previous study (Stepowicz et al. 2020) found a high level of stress and anxiety among pregnant and postpartum women during the pandemic.

2.8.2 *Effect of Yoga on Sensorimotor and Cognitive Development of the Offspring and Neonates*

Aristotle (2022) hypothesized that the development of the sensory system begins in the mother's womb and the fetus is prepared for the hurdles of extra-uterine life. He postulated epigenetic processes and compared the influence of the mother's environment with how the earth supports the development of plants. Presently growing body of literature (Arabin et al. 2021; De Asis-Cruz et al. 2020; Walsh et al. 2019; Pirasteh et al. 2018; Laplante et al. 2004; Carmichael and Shaw 2000) suggests that prenatal stress can have adverse long-term effects on the neonates' development. The long-term effects may include asthma, autism spectrum disorders, attention deficit hyperactive disorder (ADHD), attachment difficulties, allergy, affective disorders, atypical speech errors, impaired temporal processing of auditory stimuli, poor language abilities, craniofacial anomalies, and poor neurodevelopment. The above-stated changes may occur due to the direct effect of prenatal stress on the physiological functions of the fetus. Maternal stress during pregnancy leads to changes in the neuro-endocrinal system, which may lead to many sensorimotor and neurodevelopmental disorders (Coussons-Read 2013). The pregnancy-induced stress also disturbs the fetal brain functional connectivity and hence damages the fetal programming and processing skills (De Asis-Cruz et al. 2020). The altered processing skills may lead to learning disability, poor language organization, fluency disorders, dyspraxia, autistic traits, and other sensorimotor disorders in late childhood.

Humphrey et al. (2020) conducted an fMRI study to explore the effect of prenatal stress on the amygdala to medial prefrontal cortex (mPFC) structural and functional connectivity. The circuitry of the amygdala and mPFC is responsible for the regulation of emotions in later development. MRI results indicated that prenatal stress abrupt the functional connectivity and increased structural connectivity between the regulatory circuitry. As previously described, stress during pregnancy dysregulates the functioning of the HPA axis. The HPA axis is a significant and intricate neuroendocrine system. The inadequate functioning of the HPA axis leads to the long-term defect in neuropeptide, the neurotransmitter of the CNS, and glucocorticoid hormone of the peripheral nervous system. The alterations of the HPA axis can interrupt the functioning of the behavioral, neuroendocrine, autonomic, and metabolic systems in adulthood. The interruption in these stated systems may lead to schizophrenia, anxiety, depressive symptoms, and other neuropathologies in adulthood (Arabin et al. 2021).

Arabin et al. (2021) suggested facilitating the expected mother's environment to reduce the stress and for adequate neurodevelopment of the fetus. The facilitation strategies included listening to music, a healthy diet, olfactory and gustatory stimulation, and regular exercises. The studies (Albeck et al. 2006) in the past reported the beneficial effect of antenatal exercises on the cognitive functions of humans and animals. Further studies (Robinson and Bucci 2014; Dayi et al. 2012) also reported that women who performed some form of exercise during pregnancy their offspring had reduced anxiety-like behaviors and enhanced learning capability. Studies (Laplante et al. 2004) advocated Yoga practice and other leisure-time physical activities during pregnancy for better language development. They found increased vocabulary in the offspring of mothers who practiced gestational Yoga.

Animal studies also demonstrated the beneficial effect of gestational exercises on the offspring's cognitive functions such as spatial learning and memory (Dayi et al. 2012). The enhanced cognitive function in offsprings is due to the increased hippocampal neurogenesis (Dayi et al. 2012), better hippocampal brain-derived neurotrophic factor (BDNF) mRNA, and increased hippocampal leptin receptor expression (Dayi et al. 2012). As described earlier, pregnancy-induced stress dysregulates the HPA axis and hampers the neurodevelopment of fetuses and neonates. The stress-exposed fetus may potentially develop various sensorimotor disorders due to low levels of cortisol. However, appropriate rationale behind the process is still unrevealed. Practicing Yoga and exercise helps to regulate the HPA axis and increase the cortisol levels, thereby boosting the neurodevelopment and sensorimotor development in infants.

2.8.3 *The Systemic Effect of Meditation and* Asana

Meditation is becoming more popular, particularly for stress-related medical disorders, and is linked to the alleviation of stress response, immunological function, and amyloid beta A metabolism. Over the last two decades, research has generally supported the idea that meditation reduces stress, promotes wellness, and has positive benefits on overall health as well as cognitive function. Meditation has proven its best effects by curing and improving many mental disorders or metabolic conditions (Bremer et al. 2022).

In the hypothalamus, the renal SNA, arterial BP, and heart rate are all excitatory effects of PVN's NMDA glutamate receptors. The NMDA-mediated increases in sympathetic nerve outflow could be inhibited by nitric oxide in the PVN that is released by stimulation of the NMDA receptors. Under pathological conditions with elevated sympathetic tone, the negative feedback of nitric oxide on the glutamate system inside the PVN is known to play an appreciable role in maintaining the overall homeostasis and tone of sympathetic activity. The enhanced vasodilation and normalized blood pressure and cardiac rate will regulate the RAAS activity. The activation of alternative regulatory angiotensin III and angiotensin IV receptors (Mas related G protein coupled [MaSR], Angiotensin type 4 receptor [AT4R]) results in anti-inflammatory and anti-oxidative effect, therefore, restoring the blood pressure by Na reabsorption (Cosarderelioglu et al. 2020). Exercise has also been proposed to modulate the RAAS activity and reversal of hypertension-induced brain damage (Chang et al. 2022). Enalapril, an ACE inhibitor, was reported to inhibit ACE in the hypothalamic PVN, thereby reducing the ANG II-induced hypertension through the restoration of neurotransmitters and cytokines. Oxytocin lowers blood pressure and supports cardiac repair as opposed to vasopressin that significantly accelerates the course of hypertension and heart failure.

The benzodiazepine (BZD) receptor agonists are extensively used in medical practice; however, due to their role in physiological processes, a novel carbonyl group of compounds 4c was developed on the basis of the receptor model for the BZD ligand binding to the GABA receptor 1,2,4-triazole class of scaffolds that exhibit antifungal, anticancer, anticonvulsant, analgesic, and anti-inflammatory properties have also been proposed (Aggarwal et al. 2020). The excitability of presympathetic PVN neurons can also be controlled by the GABA (A)-mediated inhibitory mode. The BZDs and inhaled anesthetics target the $\alpha 1$, $\beta 2$, and $\gamma 2$ subunit-containing $GABA_{AR}$s distributed throughout the CNS. Yet, the overdose/overuse can cause anxiety or neurological conditions; therefore, the natural alternative therapies proposed for these BZDs include Yogasana and meditation Also, melatonin can have neuroprotective effects in patients with hepatic encephalopathy as it can cross the blood–brain barrier (Arjunan et al. 2022), but by practicing Yoga and meditation, endogenous melatonin levels are proposed to increase significantly.

2.8.4 Gene Expression Control of Neurological Diseases

Two categories of non-coding RNAs (ncRNAs)—long non-coding RNAs (lncRNAs) and circular RNAs (circRNAs)—can interfere with the suppression of specific messenger RNAs by microRNAs (miRNAs) (mRNAs). LncRNAs and circRNAs can contribute to inflammation, neuronal injury, microglia malfunction, astrocyte dysfunction, and other neurological diseases. Thus, neurological conditions can be positively or adversely regulated by lncRNAs and circRNAs (Chen et al. 2021). LncRNAs are non-coding transcripts over 200 nucleotides in length that are reported to affect the polarization of microglia in the CNS by controlling transcription directly or by sponging downstream miRNAs that regulate microglia gene expression. Through regulation by LncRNA, the microglia can deliver neuroprotective outcomes (Gao et al. 2022). The lncRNA, KCNQ1OT1 was downregulated by the regulation of the miR-873-5p-TRAF6-p38/NF-B pathway that was associated with neuroprotective effects (Liu et al. 2021). LncRNAs may be extremely important in traumatic brain injuries as they are widely distributed throughout CNS and take part in numerous pathological processes such as oxidative stress, inflammation, apoptosis, blood-brain barrier protection, angiogenesis, and neurogenesis (Wang et al. 2022). The expression of several lncRNAs and mRNAs changed drastically during inflammatory conditions in the hippocampus. Therefore, it is suggested that these dysregulated lncRNAs and mRNAs could be intriguing therapeutic targets to circumvent problems with long-term recovery following traumatic brain injury (Yang et al. 2021). LncRNAs and circRNAs largely function as competing candidates to the endogenous RNAs and are linked to regulation of the BBB disruption and recovery in pathological CNS conditions (Sun et al. 2022). Multiomics approaches based investigative studies are based on delineating the role of regulatory RNAs in the asana- and meditation-mediated beneficial effects of these alternative de-stressing therapies is a way forward.

2.9 Future Direction for Research

2.9.1 Linking Autonomic Responses to Inflammatory Responses

Somatosensory and autonomic neuronal regulation of the immune responses is becoming increasingly recognized and the concept of neuro-immune regulation is rapidly emerging. Recent elegant studies define the molecular mechanisms underlying the neuro-immune signaling mediated regulation of the tissue homeostasis and pathology. However, this is a field in its infancy and lot many studies are warranted for a complete understanding of the neural circuitry and how it regulates immune responses. A strong dialogue between the peripheral nervous system and the immune system is central to mounting immune responses to tissue damage and inflammatory triggers. Both the somatosensory and autonomic neurons (SS and PS) regulate immune responses primarily at the barrier tissues and in the peripheral organs. The nervous system orchestrates acute and chronic inflammatory diseases by modulating the immune cells as they express receptors for neuropeptides and neurotransmitters, the chemical mediators of neurogenic control (Gao et al. 2022).

Pyroptosis has been identified as a key pathogenic mechanism in a number of neurological illnesses and has also been linked to a number of diseases with a neuroinflammatory component. Numerous studies have demonstrated that pyroptosis and the resulting neuroinflammation of important brain regions, including the hippocampus, amygdala, and prefrontal cortex neurons, are closely related to depression. The nucleotide-binding oligomerization domain-like receptor protein 3 (NLRP3) is a key player in this process (Wan et al. 2022). Additionally, neurodegenerative disorders such as subarachnoid hemorrhage, traumatic brain injuries, and neuroinflammatory diseases are accompanied by dysregulated caspase activation frequently that play a variety of crucial but overlapping roles (Wu et al. 2022).

Inflammatory mechanisms of depression include Gasdermin D (GSDMD)-mediated pyroptosis in microglial cells, which gives a special therapeutic potential for paeoniflorin (PF) to treat depression. In the hippocampus, PF was reported to have antidepressant effects by reducing neuroinflammation by blocking the CASP-11-dependent pyroptosis signaling cascade triggered by the overactive microglia (Tian et al. 2021). To diminish the formation of GSDMD and minimize pyroptosis in microglia, CD73 inhibits the activity of NLRP3 inflammasome complexes (Xu et al. 2021). Treatment with TGF-1 reduces the proinflammatory pyroptosis of activated microglia and results in decreased levels of the crucial pyroptosis effector GSDMD, inflammasomes, and cleaved-IL-1, significantly reducing neuroinflammation (Xie et al. 2022). Also, excessive salt consumption triggers oxidative stress as well as the NLRP3-dependent inflammation upsets the balance of inhibitory and excitatory neurotransmitters in the PVN. Extracellular superoxide dismutase (Ec-SOD), a crucial antioxidant enzyme, reduces the blood pressure by removing ROS from the PVN and likely does this by reducing NLRP3-dependent inflammation and facilitating excitable neurotransmission in the PVN under salt-induced hypertension (Su et al. 2021). Through the inhibition of RAAS functioning in the hypothalamus that subsequently affects the corticotrophin-releasing hormone (CRH) expression and sympathetic nerve activity, the VD/vitamin D receptor is known to play a crucial role in energy balance restoration (Su et al. 2021). Along with this, synaptotagmin-7 therapy enhances neuronal function and prevents aging-related memory loss (Xie et al. 2022). The deregulation of PP2B-ERK1/2-SGK1-NEDD4–2-mediated GRIA1 ubiquitination causes refractory seizures and this pathway is proposed to be a viable therapeutic target in tractable epilepsy in response to AMPAR antagonists (Kim et al. 2021). In Parkinson's disease, the NEDD4 ligases are proposed as promising therapeutic targets (Conway et al. 2022). Perampanel is also proposed to prevent brain necroptosis and neuroinflammation (Yang et al. 2021). Atorvastatin treatment improved neuron death following subarachnoid hemorrhage (SAH) by increasing the survival of neurons. It also decreased the expression levels of NLRP1, cleaved caspase-1, interleukin-1 (IL-1), and IL-18 by inhibiting pyroptosis and neuroinflammation (Chen et al. 2021).

2.10 Conclusion

In the states of cellular stress, it has been demonstrated that a disequilibrium of excitatory and inhibitory neurotransmitter activities can impact cardiovascular diseases, including hypertension, which are worsened by alterations in the inhibitory GABAergic synaptic functioning. In conditions like heart failure, a dissociation of NMDARs and $GABA_{ARs}$ may help to aggravate neuronal activity and, as a result, sympatho-humoral activation. The Yogasanas are proposed to have a beneficial role under these pathological conditions marked by sympathetic overactivity. By enhancing the spiritual points and brain centers, meditation and Yogasana can enhance the activity of parasympathetic vagus innervations into the heart, and the inhibitory muscarinic acetylcholine receptors that are coupled to inhibitory GPCRs that can control the heart contractility and calcium ion channels. Multiomics investigative studies based on delineating the role of molecular players associated with the *Asana* and meditation-mediated beneficial effects of these alternative de-stressing therapies is a way forward.

References

Aggarwal A, Rao T, Palekar T, Paranjape P, Singh G (2020) Effect of *YogAsana* and *pranayama* on pain, severity and quality of life in primary dysmennorhea. Int J Med Public Health 10(1):38–42

Albeck DS, Sano K, Prewitt GE, Dalton L (2006) Mild forced treadmill exercise enhances spatial learning in the aged rat. Behav Brain Res 168(2):345–348. https://doi.org/10.1016/j.bbr.2005.11.008

Arabin B, Hellmeyer L, Maul J, Metz GAS (2021) Awareness of maternal stress, consequences for the offspring and the need for early interventions to increase stress resilience. J Perinat Med 49(8):979–989. https://doi.org/10.1515/jpm-2021-0323

Aristoteles (2022) De generatione animalium 384–22;2,4. 738b25–35

Arjunan A, Sah DK, Jung YD, Song J (2022) Hepatic encephalopathy and melatonin. Antioxidants 11(5):837

Arrieta-Cruz I, Torres-Ávila BS, Martínez-Coria H, López-Valdés HE, Gutiérrez-Juárez R (2022) Diet-induced metabolic dysfunction of hypothalamic nutrient sensing in rodents. Int J Mol Sci 23(7):3958. https://doi.org/10.3390/ijms23073958

Bremer B, Wu Q, Mora Álvarez MG, Hölzel BK, Wilhelm M, Hell E, Tavacioglu EE, Torske A, Koch K (2022) Mindfulness meditation increases default mode, salience, and central executive network connectivity. Sci Rep 12(1):1–5

Brooks VL, Fu Q, Shi Z, Heesch CM (2020) Adaptations in autonomic nervous system regulation in normal and hypertensive pregnancy. Handb Clin Neurol 171:57–84. https://doi.org/10.1016/B978-0-444-64239-4.00003-5

Carmichael SL, Shaw GM (2000) Maternal life event stress and congenital anomalies. Epidemiology 11(1):30–35. https://doi.org/10.1097/00001648-200001000-00008

Cayupe B, Morgan C, Puentes G et al (2021) Hypertension in prenatally undernourished young-adult rats is maintained by tonic reciprocal paraventricular-coerulear excitatory interactions. Molecules 26(12):3568. https://doi.org/10.3390/molecules26123568

Chang YS, Lin CL, Lee CW, Lin HC, Wu YT, Shih YH (2022) Exercise normalized the hippocampal renin–angiotensin system and restored spatial memory function, neurogenesis, and blood–brain barrier permeability in the 2K1C-hypertensive mouse. Int J Mol Sci 23(10):5531

Chen J, Zhang C, Yan T, Yang L, Wang Y, Shi Z, Li M, Chen Q (2021) Atorvastatin ameliorates early brain injury after subarachnoid hemorrhage via inhibition of pyroptosis and neuroinflammation. J Cell Physiol 236(10):6920–6931

Cheon SY, Song J (2021) The association between hepatic encephalopathy and diabetic encephalopathy: the brain–liver axis. Int J Mol Sci 22(1):463. https://doi.org/10.3390/ijms22010463

Claeys W, Van Hoecke L, Lefere S et al (2021) The neurogliovascular unit in hepatic encephalopathy. JHEP Rep 3(5):100352. https://doi.org/10.1016/j.jhepr.2021.100352

Contreras D, Piña R, Carvallo C et al (2022) Methylphenidate restores behavioral and neuroplasticity impairments in the prenatal nicotine exposure mouse model of ADHD: evidence for involvement of AMPA receptor subunit composition and synaptic spine morphology in the hippocampus. Int J Mol Sci 23(13):7099. https://doi.org/10.3390/ijms23137099

Conway JA, Kinsman G, Kramer ER (2022) The role of NEDD4 E3 ubiquitin–protein ligases in Parkinson's disease. Genes 13(3):513

Cosardereliogiu C, Nidadavolu LS, George CJ et al (2020) Brain renin–angiotensin system at the intersect of physical and cognitive frailty. Front Neurosci 14:586314. https://doi.org/10.3389/fnins.2020.586314

Coussons-Read ME (2013) Effects of prenatal stress on pregnancy and human development: mechanisms and pathways. Obstet Med 6(2):52–57. https://doi.org/10.1177/1753495X12473751

Cramer H, Lauche R, Haller H, Steckhan N, Michalsen A (2014) Effects of *yoga* on cardiovascular disease risk factors: a systematic review and meta-analysis. Int J Cardiol 173(2):170–183. https://doi.org/10.1016/j.ijcard.2014.02.017

Crowley SK, O'Buckley TK, Schiller CE, Stuebe A, Morrow AL, Girdler SS (2016) Blunted neuroactive steroid and HPA axis responses to stress are associated with reduced sleep quality and negative affect in pregnancy: a pilot study. Psychopharmacology 233(7):1299–1310. https://doi.org/10.1007/s00213-016-4217-x

Dal NA, Gümüşsoy S, Sarıdoğan E, Özdemir E (2021) The effect of pregnancy *yoga* on breastfeeding self-efficacy and prenatal attachment: a semi-experimental study. Int Med 3(3):80–85

Dayi A, Agilkaya S, Ozbal S et al (2012) Maternal aerobic exercise during pregnancy can increase spatial learning by affecting leptin expression on offspring's early and late period in life depending on gender. Sci World J 2012:429803. https://doi.org/10.1100/2012/429803

De Asis-Cruz J, Krishnamurthy D, Zhao L et al (2020) Association of prenatal maternal anxiety with fetal regional brain connectivity. JAMA Netw Open 3(12):e2022349. https://doi.org/10.1001/jamanetworkopen.2020.22349

Fayed N, Lopez Del Hoyo Y, Andres E et al (2013) Brain changes in long-term zen meditators using proton magnetic resonance spectroscopy and diffusion tensor imaging: a controlled study. PLoS One 8(3):e58476. https://doi.org/10.1371/journal.pone.0058476

Gao X, Tan H, Fan G, Li P, Su W, Guo W, Wan T (2022) LncRNA, an emerging approach for neurological diseases treatment by regulating microglia polarization. Front Neurosci 16:903472

Garner M, Reith W, Krick C (2019) 10-Week hatha *yoga* increases right hippocampal density compared to active and passive control groups: a controlled structural cMRI study. J Neuroimaging Psychiatry Neurol 4(1):1–11

Ge Y, Wang YT (2021) GluA1-homomeric AMPA receptor in synaptic plasticity and neurological diseases. Neuropharmacology 197:108708. https://doi.org/10.1016/j.neuropharm.2021.108708

Grant JA, Duerden EG, Courtemanche J, Cherkasova M, Duncan GH, Rainville P (2013) Cortical thickness, mental absorption and meditative practice: possible implications for disorders of attention. Biol Psychol 92(2):275–281. https://doi.org/10.1016/j.biopsycho.2012.09.007

Hanley JG (2021) Regulation of AMPAR expression by microRNAs. Neuropharmacology 197:108723. https://doi.org/10.1016/j.neuropharm.2021.108723

Hou Q, Li S, Jiang C et al (2018) The associations between maternal lifestyles and antenatal stress and anxiety in Chinese pregnant women: a cross-sectional study. Sci Rep 8(1):10771. https://doi.org/10.1038/s41598-018-28974-x

Hu SJ, Xu T, Wang XW (2021) *Yoga* as an exercise prescription for the pregnancy or postpartum period: recent advances and perspective. Yangtze Med 5:157–170. https://doi.org/10.4236/ym.2021.53016

Humphreys KL, Camacho MC, Roth MC, Estes EC (2020) Prenatal stress exposure and multimodal assessment of amygdala-medial prefrontal cortex connectivity in infants. Dev Cogn Neurosci 46:100877. https://doi.org/10.1016/j.dcn.2020.100877

Jais A, Brüning JC (2022) Arcuate nucleus-dependent regulation of metabolism—pathways to obesity and diabetes mellitus. Endocr Rev 43(2):314–328

Kim JE, Park H, Lee JE, Kim TH, Kang TC (2020) PTEN is required for the anti-epileptic effects of AMPA receptor antagonists in chronic epileptic rats. Int J Mol Sci 21(16):5643. https://doi.org/10.3390/ijms21165643

Kim JE, Lee DS, Park H, Kim TH, Kang TC (2021) AMPA receptor antagonists facilitate NEDD4-2-mediated GRIA1 ubiquitination by regulating PP2B-ERK1/2-SGK1 pathway in chronic epilepsy rats. Biomedicines 9(8):1069. https://doi.org/10.3390/biomedicines9081069

Kong Y, Zhou H, Feng H, Zhuang J, Wen T, Zhang C, Sun B, Wang J, Guan Y (2020) Elucidating the relationship between diabetes mellitus and Parkinson's disease using 18F-FP-(+)-DTBZ, a positron-emission tomography probe for vesicular monoamine transporter 2. Front Neurosci 14:682

Laplante DP, Barr RG, Brunet A et al (2004) Stress during pregnancy affects general intellectual and language functioning in human toddlers. Pediatr Res 56(3):400–410. https://doi.org/10.1203/01.P0000136281.34035.44

Liu N, Sun H, Li X, Cao W, Peng A, Dong S, Yu Z (2021) Downregulation of lncRNA KCNQ1OT1 relieves traumatic brain injury induced neurological deficits via promoting "M2" microglia polarization. Brain Res Bull 171:91–102

Liu Y, Lyu Y, Wang H (2022) TRP channels as molecular targets to relieve endocrine-related diseases. Front Mol Biosci 9:895814. https://doi.org/10.3389/fmolb.2022.895814

Martin D, Reihe C, Drummer S, Roessler K, Boomer S, Nelson M (2022) Venoconstrictor responses to activation of bradykinin-sensitive pericardial afferents involve the region of the hypothalamic paraventricular nucleus. Physiol Rep 10(6):e15221. https://doi.org/10.14814/phy2.15221

Mecawi AS, Varanda WA, da Silva MP (2022) Osmoregulation and the hypothalamic supraoptic nucleus: from genes to functions. Front Physiol 13:887779. https://doi.org/10.3389/fphys.2022.887779

Mehay D, Silberman Y, Arnold AC (2021) The arcuate nucleus of the hypothalamus and metabolic regulation: an emerging role for renin–angiotensin pathways. Int J Mol Sci 22(13):7050. https://doi.org/10.3390/ijms22137050

Mourão AA, Shimoura CG, Andrade MA, Truong TT, Pedrino GR, Toney GM (2021) Local ionotropic glutamate receptors are required to trigger and sustain ramping of sympathetic nerve activity by hypothalamic paraventricular nucleus TNFα. Am J Physiol Heart Circ Physiol 321(3):H580–H591. https://doi.org/10.1152/ajpheart.00322.2021

Ning Z, Song Z, Wang C et al (2021) How perturbated metabolites in diabetes mellitus affect the pathogenesis of hypertension? Front Physiol 12:705588. https://doi.org/10.3389/fphys.2021.705588

Okun ML, Kline CE, Roberts JM, Wettlaufer B, Glover K, Hall M (2013) Prevalence of sleep deficiency in early gestation and its associations with stress and depressive symptoms. J Womens Health (Larchmt) 22(12):1028–1037. https://doi.org/10.1089/jwh.2013.4331

Oliveira V, Riedl RA, Claflin KE, Mathieu NM, Ritter ML, Balapattabi K, Wackman KK, Reho JJ, Brozoski DT, Morgan DA, Cui H (2022) Melanocortin MC4R receptor is required for energy expenditure but not blood pressure effects of angiotensin II within the mouse brain. Physiol Genomics 54(6):196–205

Ortiz V, Costa Campos R, Fofo H, Fernandez SP, Barik J (2022) Nicotinic receptors promote susceptibility to social stress in female mice linked with neuroadaptations within VTA dopamine neurons. Neuropsychopharmacology 47(9):1587–1596. https://doi.org/10.1038/s41386-022-01314-4

Panda NK, Sharma M, Ramavat AS, Patro SK, Nayak GR, Kaushal K (2022) Novel cellular stress models with implications in understanding and treating ENT pathologies. In: Biomedical translational research: from disease diagnosis to treatment. Springer Nature Singapore, Singapore, pp 139–180

Pirasteh E, Akbari M, Pourbakht A, Hosseini A, Bahadorifar M et al (2018) The effect of prenatal stress on auditory brainstem response in rat pups. Health Scope 7(3):e55019. https://doi.org/10.5812/jhealthscope.55019

Rakhshani A, Nagarathna R, Mhaskar R, Mhaskar A, Thomas A, Gunasheela S (2015) Effects of *yoga* on utero-fetal-placental circulation in high-risk pregnancy: a randomized controlled trial. Adv Prev Med 2015:373041. https://doi.org/10.1155/2015/373041

Ramírez S, Haddad-Tóvolli R, Radosevic M et al (2022) Hypothalamic pregnenolone mediates recognition memory in the context of metabolic disorders. Cell Metab 34(2):269–284.e9. https://doi.org/10.1016/j.cmet.2021.12.023

Robinson AM, Bucci DJ (2014) Physical exercise during pregnancy improves object recognition memory in adult offspring. Neuroscience 256:53–60. https://doi.org/10.1016/j.neuroscience.2013.10.012

Rosario MK, Premji SS, Nyanza EC, Bouchal SR, Este D (2017) A qualitative study of pregnancy-related anxiety among women in Tanzania. BMJ Open 7(8):e016072. https://doi.org/10.1136/bmjopen-2017-016072

Royo M, Escolano BA, Madrigal MP, Jurado S (2022) AMPA receptor function in hypothalamic synapses. Front Synaptic Neurosci 14:833449. https://doi.org/10.3389/fnsyn.2022.833449

Sapouckey SA, Morselli LL, Deng G et al (2020) Exploration of cardiometabolic and developmental significance of angiotensinogen expression by cells expressing the leptin receptor or agouti-related peptide. Am J Physiol Regul Integr Comp Physiol 318(5):R855–R869. https://doi.org/10.1152/ajpregu.00297.2019

Savić B, Murphy D, Japundžić-Žigon N (2022) The paraventricular nucleus of the hypothalamus in control of blood pressure and blood pressure variability. Front Physiol 13:858941. https://doi.org/10.3389/fphys.2022.858941

Sharma M, Vanam HP, Panda NK, Patro SK, Arora R, Bhadada SK, Rudramurthy SM, Singh MP, Koppula PR (2022) Deciphering the neurosensory olfactory pathway and associated neo-immunometabolic vulnerabilities implicated in COVID-associated mucormycosis (CAM) and COVID-19 in a diabetes backdrop—a novel perspective. Diabetology 3(1):193–235

Shi Z, Stornetta DS, Stornetta RL, Brooks VL (2022) Arcuate angiotensin II increases arterial pressure via coordinated increases in sympathetic nerve activity and vasopressin secretion. Eneuro 9(1):ENEURO.0404

Stepowicz A, Wencka B, Bieńkiewicz J, Horzelski W, Grzesiak M (2020) Stress and anxiety levels in pregnant and post-partum women during the COVID-19 pandemic. Int J Environ Res Public Health 17(24):9450. https://doi.org/10.3390/ijerph17249450

Su H, Liu N, Zhang Y, Kong J (2021) Vitamin D/VDR regulates peripheral energy homeostasis via central renin–angiotensin system. J Adv Res 33:69–80

Sugama S, Kakinuma Y (2020) Stress and brain immunity: microglial homeostasis through hypothalamus–pituitary–adrenal gland axis and sympathetic nervous system. Brain Behav Immun Health 7:100111. https://doi.org/10.1016/j.bbih.2020.100111

Sun P, Hamblin MH, Yin KJ (2022) Non-coding RNAs in the regulation of blood–brain barrier functions in central nervous system disorders. Fluids Barriers CNS 19(1):1–39

Thorsdottir D, Einwag Z, Erdos B (2021) BDNF shifts excitatory–inhibitory balance in the paraventricular nucleus of the hypothalamus to elevate blood pressure. J Neurophysiol 126(4):1209–1220. https://doi.org/10.1152/jn.00247.2021

Tian DD, Wang M, Liu A, Gao MR, Qiu C, Yu W, Wang WJ, Zhang K, Yang L, Jia YY, Yang CB (2021) Antidepressant effect of paeoniflorin is through inhibiting pyroptosis CASP-11/GSDMD pathway. Mol Neurobiol 58(2):761–776

Valli M, Cho SS, Uribe C, Masellis M, Chen R, Mihaescu A, Strafella AP (2021) VMAT2 availability in Parkinson's disease with probable REM sleep behaviour disorder. Mol Brain 14(1):1–9

van Aalst J, Ceccarini J, Demyttenaere K, Sunaert S, Van Laere K (2020) What has neuroimaging taught us on the neurobiology of *yoga*? A review. Front Integr Neurosci 14:34. Published 2020 Jul 8. https://doi.org/10.3389/fnint.2020.00034

van Heyningen T, Honikman S, Myer L, Onah MN, Field S, Tomlinson M (2017) Prevalence and predictors of anxiety disorders amongst low-income pregnant women in urban South Africa: a cross-sectional study. Arch Womens Ment Health 20(6):765–775. https://doi.org/10.1007/s00737-017-0768-z

Wadikar SS (2020) Stress, hypertension and *yoga*. In: Effects of stress on human health. IntechOpen. https://doi.org/10.5772/intechopen.88147

Wall V, Premji SS, Letourneau N, McCaffrey G, Nyanza EC (2018) Factors associated with pregnancy-related anxiety in Tanzanian women: a cross sectional study. BMJ Open 8(6):e020056. https://doi.org/10.1136/bmjopen-2017-020056

Walsh K, McCormack CA, Webster R et al (2019) Maternal prenatal stress phenotypes associate with fetal neurodevelopment and birth outcomes. Proc Natl Acad Sci USA 116(48):23996–24005. https://doi.org/10.1073/pnas.1905890116

Wan T, Li X, Fu M, Gao X, Li P, Guo W (2022) NLRP3-dependent pyroptosis: a candidate therapeutic target for depression. Front Cell Neurosci 16:863426

Wang JP, Li C, Ding WC, Peng G, Xiao GL, Chen R, Cheng Q (2022) Research progress on the inflammatory effects of long non-coding RNA in traumatic brain injury. Front Mol Neurosci 15:835012

Wu Y, Liu Y, Zhou C, Wu Y, Sun J, Gao X, Huang Y (2022) Biological effects and mechanisms of caspases in early brain injury after subarachnoid hemorrhage. Oxidative Med Cell Longev 2022:3345637

Xie Y, Chen X, Li Y, Chen S, Liu S, Yu Z, Wang W (2022) Transforming growth factor-β1 protects against LPC-induced cognitive deficit by attenuating pyroptosis of microglia via NF-κB/ERK1/2 pathways. J Neuroinflammation 19(1):1–20

Xu S, Wang J, Zhong J, Shao M, Jiang J, Song J, Zhu W, Zhang F, Xu H, Xu G, Zhang Y (2021) CD73 alleviates GSDMD-mediated microglia pyroptosis in spinal cord injury through PI3K/AKT/Foxo1 signaling. Clin Transl Med 11(1):e269

Yang CC, Barrós-Loscertales A, Li M et al (2019) Alterations in brain structure and amplitude of low-frequency after 8 weeks of mindfulness meditation training in meditation-naïve subjects. Sci Rep 9(1):10977. https://doi.org/10.1038/s41598-019-47470-4

Yang L, Wang Y, Zhang C, Cheng H (2021) Perampanel, an AMPAR antagonist, alleviates experimental intracerebral hemorrhage-induced brain injury via necroptosis and neuroinflammation. Mol Med Rep 24(2):1–2

Yekefallah L, Namdar P, Dehghankar L, Golestaneh F, Taheri S, Mohammadkhaniha F (2021) The effect of *yoga* on the delivery and neonatal outcomes in nulliparous pregnant women in Iran: a clinical trial study. BMC Pregnancy Childbirth 21(1):351. https://doi.org/10.1186/s12884-021-03794-6

Yeung BH, Griffiths K, Berger L et al (2021) Leptin induces epigenetic regulation of transient receptor potential melastatin 7 in rat adrenal pheochromocytoma cells. Am J Respir Cell Mol Biol 65(2):214–221. https://doi.org/10.1165/rcmb.2020-0374OC

Chapter 3
Neurobiology of *Yoga* and Mindfulness and Its Impact on Neuronal Function, Wellbeing Addiction, and Palliative Care

Kavita Bakshi and Vinod Srivastava

Contents

Abstract Understanding the neurophysiology of *Yoga* and mindfulness, its action, and its impact on addiction and palliative care is the overarching focus of the chapter. *Yoga* and mindfulness are ancient behavioral practices that conjoin different body organ systems with the mind to have positive health outcomes. The literature has substantive evidence about *Yoga* and mindfulness as integrative approaches to health and well-being, influencing cognitive, affective, psychiatric problems/stress, immune-endocrine-neurology axis. Thus, this chapter also aims to identify the scope and application of the psycho-neuro-immuno-endocrinological approach in highlighting the neurophysiology of *Yoga* and mindfulness and its effects on general well-being, mental health, palliative care, and addiction rehabilitation. Possible mechanisms behind the effectiveness of *Yoga* and mindfulness are mentioned such as neurophysiological processes, neuro visceral integration modal, polyvagal theory, effect on cellular level and epigenetics, and mindfulness as a communication between intention, attention, and attitude. A module of *Yoga* which can be practiced

K. Bakshi
Department of Allied and Applied Sciences, University of Patanjali, Haridwar, Uttarakhand, India

V. Srivastava (✉)
College of Health and Behavioral Sciences, Fort Hays State University, Hays, KS, USA
e-mail: vksrivastava@fhsu.edu

A. Anand (ed.), *Neuroscience of Yoga*,
https://doi.org/10.1007/978-981-97-2851-0_3

by all to keep good health and that is also endorsed by government of India is provided for reference.

Keywords *Yoga* · Mindfulness · PNEI · Wellness · Palliative care · Addiction · Mechanisms

3.1 Introduction to *Yoga* and Mindfulness

With the increasing burden of chronic diseases worldwide, a new approach to health and well-being has emerged called "lifestyle modification" (Hallsworth and Adams 2019), which resonates with ancient Indian lifestyles of *Yoga* and ayurvedic medicine for healthy and quality living. "*Yoga*" and mindfulness date back to 5000 B.C (Deshpande 2018; Prokopová 2010), but of late, the scientific community has emphasized the importance of integrative health approaches and has been producing scientific evidence to strengthen integrative approaches for preventing and treating various diseases and leading a fulfilling life. The word *Yoga* comes from the Sanskrit language, which means *to unite*, and represents the conjunction of mind and body for optimal health. The literature review indicates the effectiveness of *Yoga* in preventing cardiovascular diseases, metabolic disorders, chronic painful conditions, and helps in their management as well. Furthermore, *Yoga* has turned out to be advantageous for various neurological diseases and has shown positive effects on Parkinson's disease, stroke, Guillain-Barre syndrome, hemiparesis, dementia, epilepsy, migraines, and Alzheimer's disease (Prabhakaran and Chandrasekaran 2020; Sreedevi et al. 2017; Hilton et al. 2017; Mishra et al. 2012; Paikkatt et al. 2015; Mooventhan and Nivethitha 2017). Although it is not very clearly understood at present how *Yoga* helps to improve these conditions, the growing evidence of the association between environmental and epigenetic factors with neuroimmunology, endocrinology, and nervous systems has drawn scientist's attention to *Yoga*, mindfulness, and lifestyle modification (Butzer 2017; Wimberly 2020). This chapter attempts to present a coherent view of *Yoga*, mindfulness, and neurophysiology and how they affect neurophysiological functioning in addition, maintaining good mental-physical health, normalizing the life of terminally ill patients through palliative care, and some new approaches to explain the mechanism behind these.

3.1.1 What Is Yoga*?*

A mind-body practice that harmonizes one's physical health and mind. According to *Patanjali Yoga* Sutra, it is "control over modifications of the mind." This control over modifications or patterns of thoughts leads to a balanced and equanimous mind creating a mindful action towards present situations. Thus, *Yoga* makes one opt for better life choices and good health. *Yoga* has eight limbs that include ethical principles, physical practices, and mind training to live a purpose-driven, self-disciplined, and ethically sound life (Khanna and Greeson 2013). There are various schools of *Yoga*, i.e., *Bhakti Yoga*, *Gyan Yoga*, *Karma Yoga*, *Mantra Yoga*, *Hatha Yoga*, and *Raj Yoga*. Although it has been a spiritual practice for the past few decades, *Yoga* as a therapy has started to gain recognition. In *Yoga* therapy, a proper module including physical practice (*Asana*), breathing practices (*Pranayama*), and mental pieces of training (meditation techniques) are incorporated. One of these modules of *Yoga* used today for its therapeutic purposes is "mindfulness."

3.1.2 What Is Mindfulness?

Mindfulness has its origin in the Buddhist tradition (Hsu et al. 2008). It has been refined from the *Vipassana* meditation technique, which emphasizes living in the present moment without judgment. *Vipassana* implies the practice of seeing things as they are. It is an awareness skill where the practitioner tries to be mindful of one's thoughts and circumstances and accept impermanence. One learns not to get carried away with thoughts or situations and remains unattached, enabling the meditator to modify perspective and bring positive change in behavioral patterns (Tapper 2018).

Now, various programs have been developed specifically for certain diseases using mindfulness. For example, there are several models, such as the Mindfulness-Based Stress Reduction (MBSR) model to alleviate stress, Mindfulness-Based Cognitive Therapy (MBCT) for psychiatric issues, Mindfulness-Based Relapse Prevention (MBRP), and Mindfulness Oriented Recovery Enhancement (MORE) for addiction and relapse prevention (Garland and Howard 2018). Several Mindfulness-Based Interventions (MBI) have been noted to affect chronic painful conditions positively (Yuan et al. 2020).

3.2 How *Yoga* and Mindfulness Affect Neurophysiology?

3.2.1 Brain Structure and Function

The human brain can change itself in response to the environment called neuroplasticity.

Insula: Structural changes in the insular cortex have been found to occur post-meditation (Yuan et al. 2020). The insula is responsible for sensing what is happening within the body, i.e., physiological changes. It has been proposed that the posterior insula receives homeostatic stimuli and sensations, and the anterior insula refines them into higher level awareness (Craig 2011).

Cingulate Gyrus: Meditation induces changes in the volume of grey matter. A report suggests an increase in volume in adults, while a study on adolescents shows a decrease in volume (Yuan et al. 2020; Babu et al. 2020).

The cingulate gyrus is responsible for the autonomic nervous function and is part of the limbic system that regulates behavior and emotions.

Cortex: Whole cortex increases in volume with mindfulness intervention. Cortical thickness has been shown to increase in normal and diseased individuals doing mindfulness meditation. The cortex is the largest area for neuronal connection and regulates language skills, memory, thinking, attention, perception, and consciousness (Yang et al. 2019).

Hippocampus: Volume increases in the hippocampus with mindfulness practice. Neuronal circuitry improves according to experience with mindfulness (Sevinc 2019).

It stores long-term memory.

3.2.2 Physiological Changes with Yoga *and Mindfulness*

HPA Axis: It is the communication channel between the hypothalamus, anterior pituitary, and adrenal cortex. This channel regulates the cortisol secretion in the body, i.e., a stress hormone (Fig. 3.1). Hypothalamus and pituitary are situated in the brain, while adrenal glands rest on the kidneys in the abdominal cavity.

The HPA axis plays an essential role in response to stress. Cortisol released by the adrenal cortex regulates:

- Stress responses
- Immunological response

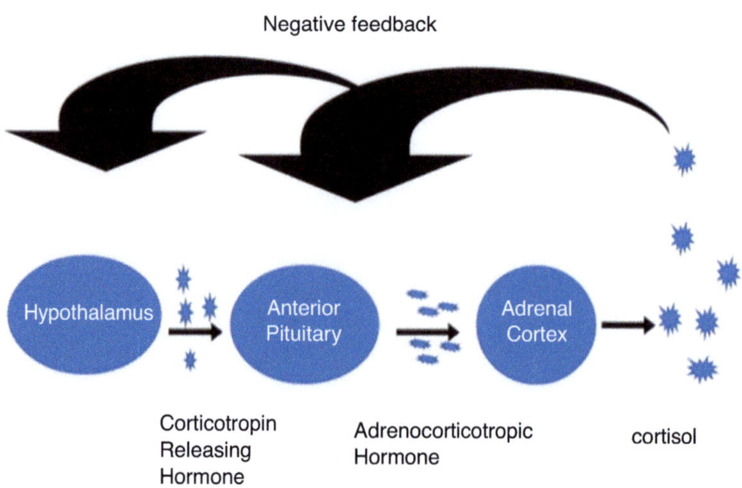

Fig. 3.1 HPA axis—hypothalamus pituitary adrenal axis

- Glucose metabolism
- Inflammatory response

When cortisol remains high for a longer duration, it ages the body faster. The brain's executive function gets affected, and neurodegeneration speeds up, leading to functional impairment and depression-like symptoms (Shindel et al. 2020). Studies have confirmed that *Yoga* enhances body functions and improved depression and quality of life (D'Silva et al. 2012). Thus, the cortisol hormone plays a significant role in stress management, and below is some evidence of how *Yoga* and mindfulness can help in reducing cortisol levels:

A 10-week *Yoga* protocol interventional quasi-experimental study on veterans with Post Traumatic Depression demonstrated better cognitive function and reduced cortisol output (Cross et al. 2016). There was a statistically significant improvement in depression levels after the *Yoga* protocol intervention amongst all the 27 veterans included in the study. In a randomized control trial done on 108 nursing students, mindfulness training showed reduced perceived stress, $F = 0.16$, $p < 0.001$ and serum cortisol level $F = 17.10$, $p < 0.001$, with a weekly 30-min intervention of mindfulness meditation for 5 weeks (Alhawatmeh et al. 2022).

Various neurotransmitters are affected by *Yoga* and mindfulness, like Gamma Amino Butyric Acid (GABA). It is the primary inhibitory neurotransmitter that is synthesized from glutamic acid decarboxylase (GAD) from glutamate (a metabolite abundantly found in the brain that forms with nitrogen and oxygen metabolism) (Koch and Magnusson 2009). GABA is present in the brain as well as in particular food. Taking GABA orally helps improve sleep and stress conditions (Hepsomali et al. 2020). Glutamate, the main excitatory neurotransmitter in the brain (Danbolt 2014), assists in growth, development, and neuroplasticity and supports the cognitive function of the brain. Stable neuronal activity in the forebrain is maintained by well-balanced excitatory (glutamatergic) and inhibitory (GABAergic) inputs (Sears and Hewett 2021).

Dopamine is a neurotransmitter that makes one feel good and satisfied. This neurochemical boosts mood, motivation, and focus and regulates motor coordination, learning, and emotional reactions. Dopamine receptors are expressed in the peripheral and central nervous systems (Klein et al. 2019). The mood regulator serotonin gets synthesized from tryptophan, an essential amino acid (David and Gardier 2016). Most serotonin is present in the gut other than in the brain. It helps in sleep, sexual function, and bone health. Norepinephrine, a hormone, and neurotransmitter, is responsible for flight and fight response to the stimulus. This increases alertness, anxiety, and focused attention and memory (Prokopová 2010). A pilot study on healthy individuals of a 90-min *Yoga* intervention once a week for 12 weeks improved serotonin levels and reduced norepinephrine levels in the *Yoga* group in comparison to the control group (Lim and Cheong 2015). Also, acetylcholine (ACh) which helps increase cognition, sleep, and memory by bringing a change in neuron excitability plays a role in synaptic transmission and plasticity. It is a neuromodulator and fast-acting neurotransmitter (Zannone et al. 2018) and helps mainly in learning, memory, and neuromuscular coordination.

3.2.2.1 Neurotrophic Factors

Neurogenesis involves the growth of new nerve cells, migration, and differentiation (Bath and Lee 2010). Brain-derived neurotrophic factor (BDNF) is also called abrineurin. BDNF improves neuroprotection and nerve cell regeneration. Reduced BDNF level has been associated with diseases like parkinsonism. Physical activity tends to improve the BDNF level, and its therapeutic use is under study for neurodegenerative diseases (Palasz et al. 2020).

Studies have generated significant evidence that several changes occur at the molecular level through *Yoga* and mindfulness interventions, as mentioned above. These changes are not disconnected from other interconnected brain structures and neural circuitry. It is important to comprehend how these changes caused by *Yoga* and mindfulness affect psychology, immunology, and neurology through various systems of mind and body, which are discussed below.

3.3 Concept of Psycho-Neuro-Endocrine-Immunology (PNEI)

This is a new field of study in which health is seen from a psychological, neurological, endocrinological, and immune system standpoint and their correlation with respect to physical health. This approach uses a bio-psycho-social perspective of human existence and various mind-body practices, which creates the vantage point of PNEI studies. There is significant evidence in the literature that many abnormalities in immune responses can lead to the development of various diseases, such as those in the domain of psychiatry (Haroon et al. 2012), dermatology (Lotti and França 2017), neurology (Haroon et al. 2012) and gastroenterology (Labanski et al. 2020), and autoimmunity (Bechter et al. 2019). However, endocrinology is another important discipline closely related to the above-mentioned linkage. The linkage of various neuroendocrine axes may well explain the pathology of diseases and their complications.

The Psycho-Neuro-Endocrine-Immunology (PNEI) framework provides a tool to study disease processes in all their complexity. From the PNEI viewpoint, psychology plays a significant role in pathology. Claude Bernard distinguished that all living beings need to keep the internal environment balanced despite the ever-changing outer environment showing its effect on living entities. Cannon coined the term "Homeostasis" for it (Cooper 2008). Selye added that stress disrupts homeostasis by affecting the hypothalamus and adenohypophysis—the brain's area that maintains homeostasis. The stimulus giving stress is a stressor, and an individual's response towards it is the stress response. It is observed that severe long-term stress responses can damage tissues and cause diseases with increased inflammation (Schneiderman et al. 2005). In particular, the stress response has diversity in physiological responses leading to burnout, depression, or post-traumatic stress disorder.

No single behavioral or bio-marker would suffice for diagnosing stress-related disorders and helping discover effective treatments (McEwen 2019).

Human brains respond to stress by invoking a coping response depending on perceived stress. Central Nervous System (CNS) starts to produce complex integrated means of response. The physiology of fight, flight, and freeze responses generates an increased autonomic and hormonal response with increased muscular exertion. In contrast, when an active fight or flight response is not required, a vigilance response activates the sympathetic response by inhibiting movement and shutting the blood supply to the periphery (Schneiderman et al. 2005). There is a hypothalamus–pituitary–axis response to stressful events along with the release of stress hormones, epinephrine, catecholamine, and adrenocorticotropin, which indicate the involvement of the endocrine stress response.

During stress, the immune system gets activated—macrophages and natural killer cells responsible for innate immunity start entering the bloodstream in a rush. There occurs leukocytosis, which is an increased number of white blood cells (responsible for immunity) in the bloodstream. These cells deposit into the skin from the bloodstream to counter microbes entering the body if there are physical injuries (Schneiderman et al. 2005). If acute stress persists for a longer duration stress response process becomes maladaptive. For instance, if the long-term stress-induced blood pressure remains high for a sustained period, this may cause heart muscles to get hypertrophied. Chronically elevated and changing blood pressure levels can cause damage to the vascular system by damaging arteries and plaque formation. A sustained immune response to stress will suppress immunity by altering the cytokines, which are the molecules responsible for communication between various immune cells. Furthermore, alteration in the immune system leads to autoimmunity, which is enhanced with chronic stress as cortisol gets released and pro-inflammatory cytokines remain at a higher level. Thus, psychological issues like stress significantly impact through PNEI mechanism and play a vital role in developing diseases in humans. While various epidemiological studies have proposed that the development of disorders occurs due to the interaction of different genetic factors, and developmental and environmental factors like infections, newer studies suggest PNEI plays a part as well (González-Díaz et al. 2017; Barouki et al. 2012). Utilizing the PNEI framework, certain behavioral interventions that impact the PNEI systems of the individual are coming into the limelight. Amongst these interventions, *Yoga* has turned out to top the chart.

3.3.1 Background of Psycho-Neuro-Endocrine-Immunology Based Meditation

Antonia Carosella and Francesco Bottaccioli developed PNEI-based Meditation (PNEIMED) in the year 1998. The approach includes the current scientific concept along with Buddhist philosophy and meditation techniques to work in real-life situations. PNEIMED throws light on the mind-body concept of looking at a human being as the constituents of a psycho-neuro-endocrine-immunology network. This amalgamates science with philosophy and the ancient mind-controlling practice of meditation. There are studies that show brief interventions of PNEIMED, as brief as a 4-day study, can decrease cortisol levels (Bottaccioli et al. 2014, 2020).

3.4 Importance of PNEI and Its Application

3.4.1 Mental Health

According to WHO, mental health is a state of well-being in which an individual realizes his or her abilities, can cope with the everyday stresses of life, can work productively, and can make a contribution to their community. In contrast, mental disorders, such as depression, anxiety disorder, conduct disorder, bipolar disorder, and psychosis, represent struggles to a person's mental health that are often characterized by an amalgamation of troubling thoughts, emotions, behavior, and relationships with others. Depression is one of the costliest and most common leading causes of disability due to health-related reasons in the world (Roman and Irwin 2020). Pathology involved in depression is associated with the immune system responses through inflammatory factors. Studies have found elevated levels of cytokines, mostly CRP (CSF), IL-6, IL-1, and TNF-α in the body (Roman and Irwin 2020; Haroon et al. 2016; Ting et al. 2020). Factors known to reduce stress also impact the level of these molecules in the body, like exercise and antidepressants. *Yoga* is effective in reducing the level of these cytokines; according to research done on a randomized sample of patients having major depressive disorders in the American population, there was a reduction in the levels of IL-6 by 0.8 μL, TNF-α by 0.2 μL, and CRP level by 0.2 μL, in a 10-week duration *Yoga* intervention compared with healthy living workshop (Paolucci et al. 2018; Köhler et al. 2018; Nugent et al. 2022).

3.4.1.1 Impact of *Yoga* on Mental Health

Palmer et al. reported a systematic review on *Yoga* intervention that reduced symptoms of anxiety and depression in children and adolescents. Twenty-seven studies were included, and 70% of these studies endorsed improvements (Zannone et al. 2018). Another systematic review and meta-analysis published in 2021, which included RCTs done on patients with mental health disorders, stated that *Yoga* decreased depressive symptoms in 578 patients compared to 502 patients of the control group enrolled in 19 different studies (Brinsley et al. 2021). Evidence in the literature is strong that *Yoga* moderated the effect of depressive symptoms on human subjects.

3.4.1.2 Impact of Mindfulness-Based Intervention on Mental Health

Goldberg et al. reviewed 142 RCTs with 12,005 participants. They found that mindfulness interventions are equally effective and sometimes superior to comparative treatments in many cases. Thus, they supported mindfulness-based interventions as an evidence-based treatment approach for schizophrenia, depression, anxiety, and addictions (Goldberg et al. 2018).

3.4.2 General Well Being

According to the Global Wellness Institute, "Wellness is the active pursuit of activities, choices, and lifestyle that lead to a state of holistic health." An internationally accepted definition is not available; however, a proposed definition by Simons and Baldwin is: "A state of positive feelings and meeting full potential in the world." It can be measured subjectively and objectively using a salutogenic approach (Western and Tomaszewski 2016). Being healthy is necessary to live well. Health becomes more visible through the largest organ, the skin. There is a strong correlation between gut health and skin health. There are commonalities between the two organs like both are highly vascularized, richly innervated, and heavily colonized by various microbes. Skin and gut can be considered as an amalgamation of immune and endocrine systems, which play a crucial role in maintaining homeostasis for the individual's survival by interacting at various pathological and physiological levels (O'Neill et al. 2016). PNEI regulation pathways, like Gut–Brain Axis and Gut Skin Axis, have cross-links at multi-levels and get modulated through various cell signaling pathways. All these cross-talks maintain PNEI homeostasis. There are pieces of evidence on how gut microbiota influences skin health. In a double-blind RCT done in patients with Acne vulgaris, there was an improvement in skin condition and urine bacterial extracellular vesicles by taking *Lactobacillus plantarum* CJLP55 in diet, which has already shown improvement in skin condition through in vivo experiments (Kim et al. 2021). In an animal model for atopic dermatitis, Lactobacillus strain ingestion showed suppression of atopic dermatitis lesions. Authors of the experiment suggested this happened due to a change in the balance of T-helper cells, TH1 or TH2 ratio, or by an increase in IL-10 production, which are anti-inflammatory cytokines (Won et al. 2011). These experiments show how the immune system, skin, and gut are interrelated through various PNEI axes.

The heart also acts as a PNEI organ. From embryology, heartbeats start even before brain formation. The heart has an extensive neuronal network. It gets regulated by CNS and intrathoracic nerve centers connecting nervous intra- and extra-thoracic ganglia, spinal cord, and cortical nerve centers (Dal Lin et al. 2018). The heart secretes cardiokines for homeostasis and cardiac remodeling from myocardial cells, endothelial cells, fibroblasts, and heart immune cells. It has also been found that mental stress-related diseases like depression, post-traumatic stress disorder, and anxiety predispose to cardiovascular events like myocardial infarction and heart attack even in the absence of dyslipidemia or atherosclerosis by increasing the release of catecholamines which cause sterile inflammation in heart muscles (Dal Lin et al. 2018).

3.4.2.1 Impact of *Yoga* on Well-Being

Yoga improves the individual's lifestyle choices in terms of healthy eating habits, taking responsibility for health, living an active life, maintaining healthy social relations, spiritual growth, and managing daily life stresses effectively (Lim and Hyun 2021). Even during the time of lockdown due to the COVID-19 outbreak, which put immense mental and economic pressure on the population, those who practiced *Yoga* showed significantly better mental health and general well-being in comparison to non-practitioners and those who practiced other spiritual practices, according to an online cross-sectional study done by Sahni et al. (2021).

3.4.3 Palliative Care

According to WHO, "Palliative care is the active total care of patients whose disease is not responsive to curative treatment. Control of pain, other symptoms, and psychological, social, and spiritual problems are paramount. The goal of palliative care is to achieve the best possible quality of life for patients and their families. Many aspects of palliative care are also applicable earlier in the course of the illness, in conjunction with anticancer treatment" (World Health Organization 2007).

3.4.3.1 Impact of *Yoga* on Palliative Care

After cancer detection, most patients develop stress and fatigue-related issues that impact sleep quality. Various *Yoga* practices have been found to benefit cancer survivors. According to a review published by Cochrane Collaborations, *Yoga* improved quality of life, sleep, and fatigue levels in 2166 women detected with breast cancer who were subjects in 24 different studies (Cramer et al. 2017). In another study on cancer survivors, an 8-week *Yoga* program helped reduce cancer-related fatigue and improved quality of life and depression severity. It also implied that a better attendance rate led to more symptom reduction, thus establishing a stronger correlation between *Yoga* therapy and cancer coping (Selman et al. 2015).

3.4.4 Addiction

Substance addiction is a neuropsychiatric disorder in which there is a repeated craving to continue taking a substance or drug despite having harmful consequences. There is behavioral or non-substance addiction as well, which includes gambling, addiction to food or mobile use, or net surfing. Most people have alcohol, tobacco, heroin, cannabis, and other drug addictions. Addiction depends upon genetic makeup and the psychological and social context in which drug use happens (Sahni et al. 2021; Uhl et al. 2019; Greene 2021; Carbia et al. 2020), in contrast to previous assumptions of addiction resulting from weak willpower or coping abilities of substance users (Zou et al. 2017).

Dopamine plays the most significant role in addiction. Addictive drugs increase dopamine in the nucleus accumbens, the locus for reward circuitry. Starting to use drugs is a voluntary process of feeling high or rewarded; however, it becomes compulsive in later stages. The alteration of psychological and physiological compulsions occurs following a change in neuronal circuitry. In starting, the basolateral amygdala plays a role, followed by the nucleus accumbens having higher activity when behavior becomes compulsive, and the dorsolateral striatum plays a vital role in showing dopamine activity (Uhl et al. 2019).

Role of *Yoga* in Addiction: *Yoga* serves as a preventive measure to keep substance use at bay in adolescents. Long-term follow-up of adolescents practicing *Yoga* at school showed better emotional regulation and prevented them from getting into drug abuse later (Butzer 2017). A qualitative study with an intervention of 1.5 h of *Yoga* once a week in returning citizens from jail showed either *Yoga*-maintained nonuse or reduced substance abuse in HIV-infected individuals (Wimberly 2020). Through an RCT done on women addicts undergoing heroin detox treatment in a Chinese hospital, *Yoga* improved mood and quality of life compared to the control group (Zhuang et al. 2013). A 3-arm study done on active smokers that compared mindfulness-based yogic breathing (M.B.) versus cognitive strategy (C.S.) versus no treatment showed two times better smoking cessation in M.B. than C.S. than no treatment group in 24 h. M.B. also reduced the acute negative effects of breathing. *Sadaf* et al. advocated M.B. as a safe, easy, and economical strategy to quit smoking (Lotfalian et al. 2020). All these significant effects on drug or alcohol abusers may be due to the disruption of pathways responsible for addictive behavior.

3.4.4.1 Role of Mindfulness in Addiction

Mindfulness has a negative relationship with substance use. Mindfulness helps to regulate emotions and protects from harmful addiction behavior. In a cross-sectional study done on opioid misuse youth, it was established that participants were less mindful and had reduced emotional regulation (Wilson et al. 2017). Mindfulness-based interventions help develop more self-control and attention by working on brain areas responsible for it, i.e., the anterior cerebral cortex (Tang et al. 2015), which gets dysregulated in addicts (Goldstein and Volkow 2011). Integrated body-mind training based on mindfulness meditation showed that smokers had better self-control, emotional regulation, and stress reactivity. This study showed that activity in the anterior cingulate cortex and adjacent prefrontal cortex helped give up addiction (Tang et al. 2016). In a residential rehabilitation center for drug abuse for criminals, Mindfulness-Based Relapse Prevention (MBRP) was compared with the relapse prevention technique for 15 weeks. The women participant showed reduced frequency of drug use, medical issues, and legal issues in the MBRP group compared to the relapse prevention group (Witkiewitz et al. 2014). These studies show that mindfulness-based intervention reduces relapse and prevents addiction behavior.

3.4.5 Possible Mechanisms Behind Yoga *and Mindfulness*

3.4.5.1 Popular Theories

There are various mechanisms and postulated theories on how *Yoga* and mindfulness work. Yogic practices function by integrating top-down neurocognitive and bottom-up neurophysiological mechanisms. Top-down mechanisms such as regulating attention and focusing on intention decrease stress and reduce sympathetic and HPA axis activity, thereby reducing inflammation and overactive immune system response. These trigger how a person responds to challenges in life. Bottom-up processes influence various organ systems like cardiovascular, musculoskeletal, and central nervous systems along with HPA and sympathetic nervous system activity, bringing about changes in the immune system and improving emotional health. These bidirectional mechanisms play a vital role in the development of chronic diseases and are responsible for the individual's well-being (Taylor et al. 2010).

There is a possible neurovisceral integration (NVI) model working on how *Yoga* impacts health at various levels of being. The parasympathetic or vagal tone has been associated with other organ systems and has been found to impact psychological and behavioral aspects along with the physical body. Mostly vagal tone is measured using Heart Rate Variability (HRV). Various findings implicate that HRV is associated with better emotional regulation and higher cognitive functions like attention and working memory. Further, HRV is inversely proportional to the occurrence of all-cause mortality, reduced inflammatory markers in the blood, glucose regulation, and better HPA function, thus, influencing health outcomes (Zhang et al. 2021).

The polyvagal theory is another emerging mechanism that suggests how *Yoga* works. It links the autonomic nervous system and prosocial behavior and recommends that nervous pathways involved in prosocial behavior impact good health. It is also a connecting link between the top-down and bottom-up approaches suggested above. There are three polyvagal neural platforms: first is the ventral vagal complex (VVC) that mediates social communication. The nucleus ambiguous is the motor component of the Ventral Vagal Complex. It regulates the muscles of the head, neck, and face with bronchi and heart, which are involved in facial and verbal expression; thereby, social engagement is managed. The second platform is the defensive strategy of mobilization associated with the sympathetic nervous system, which is a fight and flight response.

When ventral vagal complex, i.e., social communication strategies, fails, this response gets activated, and huge physiological changes occur. There is increased heart rate, respiratory rate, catecholamine release in the bloodstream, and peripheral blood flow for muscles to work correctly in case the need to escape arises. Here sympathetic nervous system plays the role. Furthermore, the second platform arouses fear and anger to achieve a protective and safe environment, while the first helps develop social connectedness. The dorsal vagal complex is the third platform, and it supplies motor innervation to organs below the diaphragm. It is an evolutionary and very primal response to stressful conditions. It puts the person into a passive state with reduced metabolism and oxygen demand to reserve energy for survival. There is hypotonia in muscle, decreased cardiac output, reflexive urination, and defecation; sometimes, there may also be loss of consciousness. This third response is the shutdown or freeze response and occurs when life is at stake. All three exist simultaneously and produce complex physiological, emotional, and behavioral patterns in the individual (Sullivan et al. 2018). Better vagal control leads to better emotional regulation, interpretation, and improved responses to situations, while *Yoga* helps attain all of that (Breit et al. 2018; Tyagi and Cohen 2016).

Mindfulness helps change perspective and self-concept by focusing on the present moment without reacting and judging. Neural circuits of the cortical midline (including the prefrontal cortex and posterior cingulate) give a narrative focus and experiential focus helps in insular activation. Narrative focus helps in explaining mental events, while experiential focus improves connecting to the present situation through thoughts or body. By doing mindfulness training, there is an increase in connectivity in the insula and reduced medial frontal activation. All this leads to improved awareness of the individual (Gibson 2019).

According to *Shapiro* et al. there are interacting processes of intention, attention, and attitude that comprise mindfulness (Fig. 3.2). First, the meditator develops an intention to regulate, explore and afterwards liberates himself by paying attention to the present moment with a compassionate attitude. This leads to detachment from his emotions, and he/she imbibes that it is just a transient mental phenomenon. This process leads to a change in perspective (Shapiro et al. 2006).

Mindfulness meditation works on multiple high-level cognitive functions, and large-scale networks get affected by it. According to *Lutz* et al. there are three main neural networks related to mindfulness functioning. First is the Central Executive Network (CEN), which is responsible for the top-down regulation of attention and

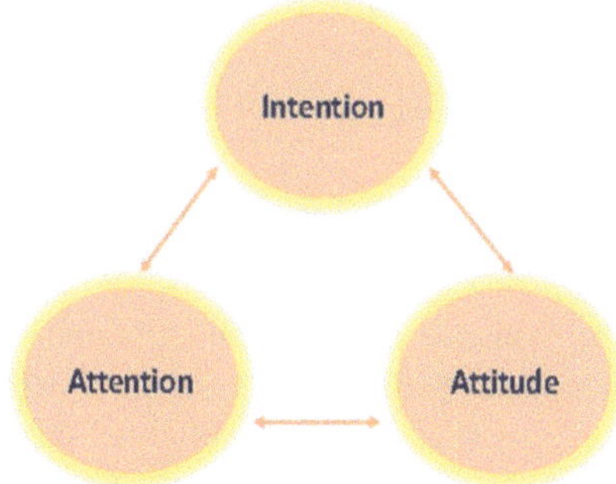

Fig. 3.2 Mindfulness mediation comprises intention to liberate, attention to the present moment, and an attitude of compassion which occur simultaneously and lead to a shift in perspective

has nodes in the prefrontal and parietal cortex. Salience Network (S.N.) is the second one, which regulates bottom-up attention and includes the anterior insula, amygdala, anterior cingulate cortex, and dopaminergic midbrain regions. The third one is Default Mode Network (DMN) which includes a cortical region around the midline mainly involving the posterior cingulate cortex and medial prefrontal cortex. During mindfulness meditation, these networks become more active (Wielgosz et al. 2017).

3.4.6 Cellular and Epigenetic Effects of Yoga and Mindfulness

Yoga helps re-establish immunological tolerance by working on the cellular and molecular levels. It helps in autoimmune and inflammatory disorders as a complementary therapy (Gautam et al. 2019). In healthy adults, *Yoga* induces angiogenesis and neurogenesis through the regulation of Brain-Derived Neurotropic Factor (Neurogenesis), Vascular Endothelial Growth Factor (Angiogenesis), and Amyloid β molecules (a marker for neurodegenerative diseases), even within 1 month of *Yoga* practice according to a pilot study done on 64 individuals (Sharma et al. 2020). A study on meditators found an upregulation of 220 genes involved in immune response, especially those involved in cell interferon signaling pathways. Also, there was a downregulation of cell signaling pathways involved in oxidative stress, cell cycle, and detoxification. Thus, this study revealed how meditation could impact cellular and genetic levels (Koka et al. 2021). In a study on meditators who practiced meditation for 8 h for 1 day, their methylome of peripheral blood mononuclear cells was examined for DNA methylation profile. Post-intervention, there were 61 differently methylated sites mostly associated with immune response and aging. Suggesting meditation in a short while can impact epigenome at sites related to immune response (Chaix et al. 2021).

Yoga and mindfulness affect PNEI by normalizing the altered neurochemistry and physiology, bringing positive molecular and anatomical changes, and even impacting epigenetics. *Yoga* and mindfulness's role in managing chronic ailments, improving quality of life, and rehabilitation post-drug addiction has been proven through modern research techniques. *Yoga* and mindfulness are the oldest behavioral healthcare practices ever recorded that originated in India as behavioral health treatment modalities to lead a healthy and accomplished life.

3.5 Clinical Implications

The evidence in the literature supports how *Yoga* and mindfulness benefit the diseased individual; however, the mechanism behind all these effects is unclear. Previous studies have highlighted the effect of these practices on cognition, memory, psychiatry, immunology, endocrinology, and psychology (Cross et al. 2016;

Zannone et al. 2018; Chatterjee and Mondal 2014; Pascoe et al. 2020). Also, the PNEI approach, another new field, supports the bio-psycho-social perspective of human existence, like *Yoga* and mindfulness. The PNEI approach clarifies the pathology behind various psychiatric, dermatological, neurological, and autoimmune diseases (Haroon et al. 2012; Lotti and França 2017; Labanski et al. 2020; Bechter et al. 2019).

Mental health issues can have debilitating effects on people. On a positive note, the evidence of *Yoga* and mindfulness showing great improvement in mental disorders and stress is growing and holds significance in preventing and treating such disorders (Brinsley et al. 2021). In a randomized controlled trial on patients with major depressive disorders, the *Yoga* intervention of 12 weeks reduced the depressive symptoms and also increased Gaba Amino Butyric Acid (GABA). Low levels of GABA are associated with major depressive disorders (Streeter et al. 2020). In another trial, *Yoga* intervention reduced stress-related molecules and inflammatory markers—plasma cortisol decreased, beta-endorphins increased, IL-6 decreased, and TNF-alpha increased within 10 days of *Yoga* intervention in patients having chronic inflammatory diseases (Yadav et al. 2012). Similarly, mindfulness has been useful in reducing stress and improving mental and physical health by reducing symptoms of depression and anxiety and improving cognition and social behavior in individuals (Poissant et al. 2019). Mindfulness *Yoga* program has been utilized in patients with neurological disorders like parkinsonism, and it turned out to be on par with the conventional treatment of "stretch and resistance training" with additional mental benefits of reduced depression symptoms (Kwok et al. 2019). Thus, by impacting neurotransmitters and immunomodulators, *Yoga* and mindfulness benefit mental health involving the PNEI mechanism.

During palliative care, the focus is on improving the quality of life rather than extending the length of life to live. In the last days, palliative care helps to live life with dignity, equanimity, and peace. Medicine may not be promising, but *Yoga* and mindfulness, through an integrated approach, can bring the required changes in life. Mindfulness and *Yoga* enhance life's spiritual perspective and may be helpful in cognitive reframing and restricting to be at peace with life outcomes (de Latorraca et al. 2017).

The nervous, endocrine, and immune systems are interconnected functionally, and there is a bidirectional coordinated channel that gets affected by psychology, stress, infection, or inflammation (Chesnokova and Melmed 2002). Addiction also disrupts this coordination by impacting PNEI. It is known that the HPA axis plays a role in addiction (Pascoe et al. 2020). There is more release of glucocorticoids in the relapse phase, but it is reduced if the patient remains abstinent. A 1-year observational study showed that a lower cerebral spinal fluid concentration of cortisol indicates long-term abstinence from alcohol (Walter et al. 2006). There is a direct relationship between neuroendocrine changes and alcohol abuse; leptin, beta-endorphin, and low corticotrophin-releasing hormone indirectly regulate the HPA axis and enhance craving for alcohol, thereby increasing the chances of relapse of alcohol abuse. Alcohol abuse leads to changes in gut microbiota and alters the gut–brain axis. This leads to chronic inflammation and changes in fronto-limbic circuitry, leading to emotional disbalance and further accelerating addiction (Carbia et al. 2020). There are changes in the immune system even in acute alcohol consumption, pro-inflammatory cytokines, TNF-alpha, and IL-8 levels alter, and there is impaired immune response (Hillmer et al. 2019).

Yoga and mindfulness help interrupt the pathways responsible for addictive behavior. The anterior cingulate cortex and prefrontal cortex showed better activity in people trying to quit smoking (Tang et al. 2016). According to *Goldberg* et al. mindfulness-based interventions were superior to active controls and no treatments while it was on par with education-based therapies (Goldberg et al. 2018). While comparing smoking cessation rates through mindfulness-based interventions to active controls or no treatment, Jackson et al. did not find clear-cut benefits; they included 21 studies, of which 15 were done in the USA (Jackson et al. 2022). But mindfulness and *Yoga* have shown better results when looking at substance abuse. According to Wen Li et al. mindfulness treatments increased post-treatment cessation rates. There were reduced frequency and intensity of craving for substances and reduced stress. Here 42 studies were included having 34 RCTs that showed significant positive results in favor of mindfulness-based interventions (Li et al. 2017).

3.6 Future Directions for Research

The positive aspects of mindfulness and *Yoga* intervention are better health, well-being, and improved quality of life. It has shown good application in preventing diseases, addiction, and palliative care. However, there are certain limitations to the studies done on mindfulness and *Yoga*. Most did not look for harm or adverse effects, and some were biased. A meta-analysis on poor mental health quality in tertiary education students (college and university students) also highlighted concerns regarding ignoring the adverse effects of interventions. When compared to active-controlled, it showed moderate effectiveness (Breedvelt et al. 2019).

Applications of various *Yoga* and mindfulness in palliative care and addiction are worth noting, but research on them lacks rigor, reducing its reliability. There are issues of bias, and a lack of active comparison groups decreases research credibility (Colgrove et al. 2019; Hilcove et al. 2021). Some studies are not blinded, and it is difficult to maintain research treatment fidelity (Cohen et al. 2014). More studies involving RCT design with detailed descriptions of interventions to reduce bias are needed.

Additionally, studies with biomarker assessment rather than qualitative and subjective assessment are needed to improve the research's internal and external validity and strengthen evidence-based practice. Comparison amongst different *Yoga* approaches is required to establish which *Yoga* module is beneficial for a particular condition (Simon et al. 2021). Confounding factors need to be focused upon as this may lead to biased results, for example, too much faith in gurus/*Yoga* teachers, religious beliefs, and socioeconomic status (Bokulich et al. 2017). Comparative studies amongst different population sets for similar techniques need to be done (Zhang et al. 2021). More large sample long-term follow-up studies will help to know the effectiveness of *Yoga* and Mindfulness in different population sets. The IDY-Protocol serves as a good example of an evidence-based *Yoga* practice to keep good health has been provided in Table 3.1 (Pascoe et al. 2020).

Table 3.1 Practices of common *Yoga* protocol developed by experts and heads of eminent *Yoga* institutions of India, which can be practiced for health

S. no.	Practice	Name of practice	Benefits
1.	Prayer		Develops focus and enhances benefits of practice
2.	Loosening practices	Neck bending Trunk movements Knee movements	Improves microcirculation in the body
3.	*Yogasana* (each *asana* for 1 min)	Standing postures: *Tadasana* (palm tree pose)	Stretches the whole body and lengthens the spine, and corrects bad posture
		Vrikshasana (tree pose)	Improves body balance, alertness, and focus
		Padhastasana (hand to feet pose)	Improves spinal flexion, digestion, and menstrual cramps
		Ardha Chakrasana (half wheel pose)	Improves spinal extension and improves cervical spondylosis
		Trikonasana (triangle pose)	Stretches sides and legs and increases lung capacity
		Sitting postures: *Bhadrasana* (auspicious pose)	Stabilizes the mind and improves menstrual cramps
		Vajrasana (thunderbolt pose)	Good for digestion and strengthens the legs
		Ardha Ushtra asana (half camel pose)	Strengthens the upper back, relieves constipation, and increases circulation in the chest area
		Sasakasana (hare pose)	Reduces stress and anxiety and stimulates abdominal organs
		Uttana Mandukasana (stretched up frog pose)	Improves neck pain and lung capacity
		Vakrasana (spinal twist pose)	Improves spinal twists and the function of the pancreas
		Prone postures: *Makarasana* (crocodile pose)	Relaxes mind and body
		Bhujangasana (cobra pose)	Reduces abdominal fat and bronchial problems
		Shalabhasana (locust pose)	Relieves sciatica and strengthens legs and lungs
		Supine postures: *Setubandhasana* (bridge pose)	Relieves depression, anxiety, and constipation
		Uttanapadasana (raised feet pose)	Strengthens abdomen and pelvic floor

(continued)

Table 3.1 (continued)

S. no.	Practice	Name of practice	Benefits
		Ardha halasana (half plough pose)	Relieves constipation and improves hypertension condition
		Pavana Muktasana (wind releasing pose)	Reduces flatulence and bloating and lengthens back muscles
		Shavasana (corpse pose)	Relaxes psychophysiological system
4.	Cleansing practice	*Kapalbhati*	Purifies frontal air sinuses, reduces cough and other respiratory disorders, strengthens the nervous system, and tones the abdomen
5.	Breathing practice	*Nadishodhan* or *Anulom Vilom pranayama* (alternate nostril breathing)	Induces tranquility, improves concentration, and reduces anxiety
		Shitali *pranayama*	It has a cooling effect on the body and mind, beneficial in hypertension and skin and eye disorders
		Bhramari pranayama	Relieves stress and improves concentration
6.	Meditation	*Dhyana*	Eliminates negative emotions and improves memory clarity of thoughts and willpower
7.	Resolution	*Sankalpa*	To commit to keeping the mind balanced at all times, to fulfill duty towards the world for the promotion of health, peace, and harmony
8.	Closing prayer	*Shanti path*	For the good health of all

3.7 Conclusion

Lifestyle modifications by incorporating *Yoga* and mindfulness in daily life prevent diseases. This incorporation also serves as a treatment for managing cardiovascular, metabolic, chronic pain conditions, and neurological disorders. *Yoga* and mindfulness practices have an impact on brain structure and function, especially on the Insula (senses physiological), cingulate Gyrus (regulates behavior and emotion along with ANS function), Cortex (higher mental functions), and Hippocampus (stores long-term memory). The physiological changes associated with *Yoga* involve HPA Axis (cortisol secretion), Gaba Amino Butyric Acid (sleep and stress improvement), Glutamate (excitatory hormone), Dopamine (boosts satisfaction), Serotonin (boosts mood), Acetylcholine (coordination), Norepinephrine (increases alertness), and neurotropic factors like BDNF (neuroprotection and nerve repair).

The bio-psycho-social perspective of human existence infers that adverse psychology can impact the body's physiology and develop diseases. Modulating the immune response and endocrine system stress can create pathology and disrupt homeostasis. In the long term, this leads to increased inflammation and damaged tissues. The HPA axis responds to stress by releasing epinephrine, catecholamine, and adreno-corticotropin. The immune system gets hyperactive in the presence of stress and macrophages, and natural killer cells enter in blood. Thus, PNEI affects mental health (with an increased level of cytokines like IL6, IL1, TNF-α, and CRP), general well-being (affecting the gut–brain axis and neuronal networks of the heart), chronic ailments like musculoskeletal pain, and even cancer (by impacting the quality of life through psychosocial and physical issues), and addiction pathology (altering neuronal circuitry) (Walter et al. 2006; Hilcove et al. 2021).

References

Alhawatmeh HN, Rababa M, Alfaqih M, Albataineh R, Hweidi I, Awwad AA (2022) The benefits of mindfulness meditation on trait mindfulness, perceived stress, cortisol, and C-reactive protein in nursing students: a randomized controlled trial. Adv Med Educ Pract 13:47–58

Babu MGR, Kadavigere R, Koteshwara P, Sathian B, Rai KS (2020) *RajYoga* meditation induces grey matter volume changes in regions that process reward and happiness. Sci Rep 10(1):1–11. https://doi.org/10.1038/s41598-020-73221-x

Barouki R, Gluckman PD, Grandjean P, Hanson M, Heindel JJ (2012) Developmental origins of non-communicable disease: implications for research and public health. Environ Health A Glob Access Sci Source 11(1):1–9

Bath KG, Lee FS (2010) Neurotrophic factor control of adult SVZ neurogenesis. Dev Neurobiol 70:339–349

Bechter K, Brown D, Najjar S (2019) Editorial: Recent advances in psychiatry from psycho-neuro-immunology research: autoimmune encephalitis, autoimmune encephalopathy, and mild encephalitis. Front Psychiatry 10:1–4

Bokulich NA, Subramanian S, Faith JJ, Gevers D, Gordon JI, Knight R, Mills DA, Caporaso JG (2017) 乳鼠心肌提取 HHS public access. Physiol Behav 176(1):139–148

Bottaccioli F, Carosella A, Cardone R, Mambelli M, Cemin M, D'Errico MM et al (2014) Brief training of psychoneuroendocrino immunology-based meditation (PNEIMED) reduces stress symptom ratings and improves control on salivary cortisol secretion under basal and stimulated conditions. Explor J Sci Health 10(3):170–179. https://doi.org/10.1016/j.explore.2014.02.002

Bottaccioli AG, Bottaccioli F, Carosella A, Cofini V, Muzi P, Bologna M (2020) Psychoneuroendocrinoimmunology-based meditation (PNEIMED) training reduces salivary cortisol under basal and stressful conditions in healthy university students: results of a randomized controlled study. Explore 16(3):189–198

Breedvelt JJF, Amanvermez Y, Harrer M, Karyotaki E (2019) The effects of meditation, *yoga*, and mindfulness on depression, anxiety, and stress in tertiary education students: a meta-analysis. Front Psychiatry 10:1–15

Breit S, Kupferberg A, Rogler G, Hasler G (2018) Vagus nerve as modulator of the brain–gut axis in psychiatric and inflammatory disorders. Front Psychiatry 9:44

Brinsley J, Schuch F, Lederman O, Girard D, Smout M, Immink MA et al (2021) Effects of *yoga* on depressive symptoms in people with mental disorders: a systematic review and meta-analysis. Br J Sports Med 55:992–1000

Butzer B (2017) HHS public access. J Youth Adolesc 176(3):139–148

Carbia C, Lannoy S, Maurage P, López-Caneda E, O'Riordan KJ, Dinan TG et al (2020) A biological framework for emotional dysregulation in alcohol misuse: from gut to brain. Mol Psychiatry 26(4):1098–1118

Chaix R, Fagny M, Lemee L, Regnault B, Davidson RJ, Lutz A et al (2021) HHS Public Access 84:36–44

Chatterjee S, Mondal S (2014) Effect of regular yogic training on growth hormone and dehydroepiandrosterone sulfate as an endocrine marker of aging. Evid Based Complement Alternat Med 2014:240581

Chesnokova V, Melmed S (2002) Minireview: neuro-immuno-endocrine modulation of the hypothalamic–pituitary–adrenal (HPA) axis by gp130 signaling molecules. Endocrinology 143(5):1571–1574

Cohen L, Spelman A, Engle R, Arun B, Wei Q, Harrison C et al (2014) Randomized trial of Tibetan *yoga* in breast cancer patients undergoing chemotherapy. J Clin Oncol 32(15_suppl):9539

Colgrove YM, Gravino-Dunn NS, Dinyer SC, Sis EA, Heier AC, Sharma NK (2019) Physical and physiological effects of *Yoga* for an underserved population with chronic low back pain. Int J Yoga 12:252–264

Cooper SJ (2008) From Claude Bernard to Walter Cannon. Emergence of the concept of homeostasis. Appetite 51(3):419–427

Craig ADB (2011) Significance of the insula for the evolution of human awareness of feelings from the body. Ann N Y Acad Sci 1225(1):72–82

Cramer H, Lauche R, Klose P, Lange S, Langhorst J, Dobos GJ (2017) *Yoga* for improving health-related quality of life, mental health and cancer-related symptoms in women diagnosed with breast cancer. Cochrane Database Syst Rev 2017(1):CD010802

Cross SJ, Linker KE, Leslie FM (2016) 乳鼠心肌提取 HHS public access. Physiol Behav 176(1):100–106

Dal Lin C, Tona F, Osto E (2018) The heart as a psychoneuroendocrine and immunoregulatory organ. Adv Exp Med Biol 1065:225–239

Danbolt ZZN (2014) Glutamate as a neurotransmitter in the healthy brain. J Neural Transm 121:799–817

David DJ, Gardier AM (2016) The pharmacological basis of the serotonin system: application to antidepressant response. Encephale 42(3):255–263

de Latorraca COC, Martimbianco ALC, Pacheco RL, Riera R (2017) Mindfulness for palliative care patients. Systematic review. Int J Clin Pract 71(12):e13034

Deshpande A (2018) *Yoga* for palliative care. Integr Med Res 7(3):211–213. https://doi.org/10.1016/j.imr.2018.04.001

D'Silva S, Poscablo C, Habousha R, Kogan M, Kligler B (2012) Mind-body medicine therapies for a range of depression severity: a systematic review. Psychosomatics 53(5):407–423. https://doi.org/10.1016/j.psym.2012.04.006

Garland EL, Howard MO (2018) Mindfulness-based treatment of addiction: current state of the field and envisioning the next wave of research. Addict Sci Clin Pract 13(1):1–14. https://doi.org/10.1186/s13722-018-0115-3

Gautam S, Tolahunase M, Kumar U, Dada R (2019) Impact of *yoga* based mind-body intervention on systemic inflammatory markers and co-morbid depression in active rheumatoid arthritis patients: a randomized controlled trial. Restor Neurol Neurosci 37(1):41–59

Gibson J (2019) Mindfulness, interoception, and the body: a contemporary perspective. Front Psychol 10:475917

Goldberg SB, Tucker RP, Greene PA, Davidson RJ, Wampold BE, Kearney DJ et al (2018) Mindfulness-based interventions for psychiatric disorders: a systematic review and meta-analysis. Clin Psychol Rev 59:52–60. https://doi.org/10.1016/j.cpr.2017.10.011

Goldstein RZ, Volkow ND (2011) Dysfunction of the prefrontal cortex in addiction: neuroimaging findings and clinical implications. Nat Rev Neurosci 12(11):652–669

González-Díaz SN, Arias-Cruz A, Elizondo-Villarreal B, Monge-Ortega OP (2017) Psychoneuroimmunoendocrinology: clinical implications. World Allergy Organ J 10(1):1–13

Greene DS (2021) *Yoga*: a holistic approach to addiction treatment and recovery. OBM Integr Complement Med 6(4):1–1

Hallsworth K, Adams LA (2019) Lifestyle modification in NAFLD/NASH: facts and figures. JHEP Rep 1(6):468–479

Haroon E, Raison CL, Miller AH (2012) Psychoneuroimmunology meets neuropsychopharmacology: translational implications of the impact of inflammation on behavior. Neuropsychopharmacology 37(1):137–162. https://doi.org/10.1038/npp.2011.205

Haroon E, Fleischer CC, Felger JC, Chen X, Woolwine BJ, Patel T et al (2016) Conceptual convergence: increased inflammation is associated with increased basal ganglia glutamate in patients with major depression. Mol Psychiatry 21(10):1351–1357

Hepsomali P, Groeger JA, Nishihira J, Scholey A (2020) Effects of oral gamma-aminobutyric acid (GABA) administration on stress and sleep in humans: a systematic review. Front Neurosci 14:559962

Hilcove K, Marceau C, Thekdi P, Larkey L, Brewer MA, Jones K (2021) Holistic nursing in practice: mindfulness-based *yoga* as an intervention to manage stress and burnout. J Holist Nurs 39(1):29–42

Hillmer AT, Nadim H, Devine L, Jatlow P, O'Malley SS (2019) Acute alcohol consumption alters the peripheral cytokines IL-8 and TNF-α. Alcohol 85:95–99

Hilton L, Hempel S, Ewing BA, Apaydin E, Xenakis L, Newberry S et al (2017) Mindfulness meditation for chronic pain: systematic review and meta-analysis. Ann Behav Med 51(2):199–213. https://academic.oup.com/abm/article/51/2/199-213/4564147

Hsu SH, Grow J, Marlatt GA (2008) Mindfulness and addiction. Recent Dev Alcohol 18:229–250

Jackson S, Brown J, Norris E, Livingstone-Banks J, Hayes E, Lindson N (2022) Mindfulness for smoking cessation. Cochrane Database Syst Rev 2022(4):CD013696

Khanna S, Greeson JM (2013) A narrative review of *yoga* and mindfulness as complementary therapies for addiction. Complement Ther Med 21(3):244–252

Kim MJ, Kim KP, Choi E, Yim JH, Choi C, Yun HS et al (2021) Effects of *Lactobacillus plantarum* CJLP55 on clinical improvement, skin condition and urine bacterial extracellular vesicles in patients with acne vulgaris: a randomized, double-blind, placebo-controlled study. Nutrients 13(4):1368

Klein MO, Battagello DS, Cardoso AR, Hauser DN, Bittencourt JC, Correa RG (2019) Dopamine: functions, signaling, and association with neurological diseases. Cell Mol Neurobiol 39(1):31–59

Koch U, Magnusson AK (2009) Unconventional GABA release: mechanisms and function. Curr Opin Neurobiol 19(3):305–310

Köhler CA, Freitas TH, Stubbs B, Maes M, Solmi M, Veronese N et al (2018) Peripheral alterations in cytokine and chemokine levels after antidepressant drug treatment for major depressive disorder: systematic review and meta-analysis. Mol Neurobiol 55(5):4195–4206

Koka M, Chandran B, Pawale D (2021) Large-scale genomic study reveals robust activation of the immune system following advanced inner engineering meditation retreat. Proc Natl Acad Sci 118(51):e2110455118

Kwok JYY, Kwan JCY, Auyeung M, Mok VCT, Lau CKY, Choi KC et al (2019) Effects of mindfulness *yoga* vs stretching and resistance training exercises on anxiety and depression for people with Parkinson disease: a randomized clinical trial. JAMA Neurol 76(7):755–763

Labanski A, Langhorst J, Engler H, Elsenbruch S (2020) Stress and the brain–gut axis in functional and chronic-inflammatory gastrointestinal diseases: a transdisciplinary challenge. Psychoneuroendocrinology 111:104501. https://doi.org/10.1016/j.psyneuen.2019.104501

Li W, Howard MO, Garland EL, McGovern P, Lazar M (2017) Mindfulness treatment for substance misuse: a systematic review and meta-analysis. J Subst Abuse Treat 75:62–96. https://doi.org/10.1016/j.jsat.2017.01.008

Lim SA, Cheong KJ (2015) Regular *yoga* practice improves antioxidant status, immune function, and stress hormone releases in young healthy people: a randomized, double-blind, controlled pilot study. J Altern Complement Med 21(9):530–538

Lim EJ, Hyun EJ (2021) The impacts of pilates and *yoga* on health-promoting behaviors and subjective health status. Int J Environ Res Public Health 18(7):3802

Lotfalian S, Spears CA, Juliano LM (2020) The effects of mindfulness-based yogic breathing on craving, affect, and smoking behavior. Psychol Addict Behav 34(2):351–359

Lotti T, França K (2017) Psycho-neuro-endocrine-immunology: a psychobiological concept. Wiley
McEwen BS (2019) From serendipity to clinical relevance: how clinical psychology and neuroscience converged to illuminate psychoneuroendocrinology. Psychoneuroendocrinology 105:36–43. https://doi.org/10.1016/j.psyneuen.2018.09.011
Mishra SK, Singh P, Bunch S, Zhang R (2012) The therapeutic value of *yoga* in neurological disorders. Ann Indian Acad Neurol 15(4):247–254
Mooventhan A, Nivethitha L (2017) Evidence based effects of *yoga* in neurological disorders. J Clin Neurosci 43:61–67. https://doi.org/10.1016/j.jocn.2017.05.012
Nugent NR, Brick L, Armey MF, Tyrka AR, Kathryn K, Uebelacker LA et al (2022) Trial of *yoga* for depression. Behav Med 47(1):21–30
O'Neill CA, Monteleone G, McLaughlin JT, Paus R (2016) The gut–skin axis in health and disease: a paradigm with therapeutic implications. BioEssays 38(11):1167–1176. https://doi.org/10.1002/bies.201600008
Paikkatt B, Singh AR, Singh PK, Jahan M, Ranjan JK (2015) Efficacy of *yoga* therapy for the management of psychopathology of patients having chronic schizophrenia. Indian J Psychiatry 57(4):355
Palasz E, Wysocka A, Gasiorowska A, Chalimoniuk M, Niewiadomski W, Niewiadomska G (2020) BDNF as a promising therapeutic agent in Parkinson's disease. Int J Mol Sci 21(3):1170
Paolucci EM, Loukov D, Bowdish DME, Heisz JJ (2018) Exercise reduces depression and inflammation but intensity matters. Biol Psychol 133:79–84
Pascoe MC, Thompson DR, Ski CF (2020) Meditation and endocrine health and well-being. Trends Endocrinol Metab 31(7):469–477. https://doi.org/10.1016/j.tem.2020.01.012
Poissant H, Mendrek A, Talbot N, Khoury B, Nolan J (2019) Behavioural and cognitive impacts of mindfulness-based interventions on adults with attention-deficit hyperactivity disorder: a systematic review. Behav Neurol 2019:5682050
Prabhakaran D, Chandrasekaran AM (2020) *Yoga* for the prevention of cardiovascular disease. Nat Rev Cardiol 17(9):536–537. https://doi.org/10.1038/s41569-020-0412-x
Prokopová I (2010) Noradrenaline and behavior. Ceskoslov Fysiol 59(2):51–58. https://pubmed.ncbi.nlm.nih.gov/21254660/
Roman M, Irwin MR (2020) Novel neuroimmunologic therapeutics in depression: a clinical perspective on what we know so far. Brain Behav Immun 83:7–21. https://doi.org/10.1016/j.bbi.2019.09.016
Sahni PS, Singh K, Sharma N, Garg R (2021) *Yoga* an effective strategy for selfmanagement of stress-related problems and well-being during COVID19 lockdown: a cross-sectional study. PLoS One 16:1–22. https://doi.org/10.1371/journal.pone.0245214
Schneiderman N, Ironson G, Siegel SD (2005) Stress and health: psychological, behavioral, and biological determinants. Annu Rev Clin Psychol 1:607–628
Sears SMS, Hewett SJ (2021) Influence of glutamate and GABA transport on brain excitatory/inhibitory balance. Exp Biol Med 246(9):1069–1083
Selman L, McDermott K, Donesky DA, Citron T, Howie-Esquivel J (2015) Appropriateness and acceptability of a tele-*yoga* intervention for people with heart failure and chronic obstructive pulmonary disease: qualitative findings from a controlled pilot study. BMC Complement Altern Med 15(1):1–13
Sevinc G (2019) Strengthened hippocampal circuits underlie enhanced retrieval of extinguished fear memories following mindfulness training. Biol Psychiatry 86(9):693–702. https://reader.elsevier.com/reader/sd/pii/S0006322319314076?token=3BD3ACF170E7E1EEADAF1779B8AECF821F767CC870624639BD7C4FCC6ED3B727A52C4E3899FE395F2430B20A6957968F&originRegion=eu-west-1&originCreation=20211120093329
Shapiro SL, Carlson LE, Astin JA, Freedman B (2006) Mechanism of mindfulness. J Clin Psychol 62(3):373–386. https://doi.org/10.1002/jclp.20237
Sharma K, Pannu V, Sayal N, Bhalla A, Anand A (2020) Effects of one month of common *yoga* protocol practice appear to be mediated by the angiogenic and neurogenic pathway: a pilot study. Explore 17:451. https://doi.org/10.1016/j.explore.2020.09.007

Shindel C, Holland JM, Gallagher-Thompson D (2020) The link between activities of daily living and cortisol in late-life depression. Clin Gerontol 43(4):430–440. https://doi.org/10.1080/07317115.2018.1561581

Simon NM, Hofmann SG, Rosenfield D, Hoeppner SS, Hoge EA, Bui E et al (2021) Efficacy of *Yoga* vs cognitive behavioral therapy vs stress education for the treatment of generalized anxiety disorder: a randomized clinical trial. JAMA Psychiatry 78(1):1

Sreedevi A, Gopalakrishnan UA, Ramaiyer SK (2017) A randomized controlled trial of the effect of *yoga* and peer support on glycemic outcomes in women with type 2 diabetes mellitus: a feasibility study. BMC Complement Altern Med 17(1):100

Streeter CC, Gerbarg PL, Brown RP, Scott TM, Nielsen GH, Owen L et al (2020) Thalamic gamma aminobutyric acid level changes in major depressive disorder after a 12-week Iyengar *yoga* and coherent breathing intervention. J Altern Complement Med 26(3):190–197

Sullivan MB, Erb M, Schmalzl L, Moonaz S, Taylor JN, Porges SW (2018) *Yoga* therapy and polyvagal theory: the convergence of traditional wisdom and contemporary neuroscience for self-regulation and resilience. Front Hum Neurosci 27:12

Tang YY, Hölzel BK, Posner MI (2015) The neuroscience of mindfulness meditation. Nat Rev Neurosci 16(4):213–225. https://doi.org/10.1038/nrn3916

Tang YY, Tang R, Posner MI (2016) Mindfulness meditation improves emotion regulation and reduces drug abuse. Drug Alcohol Depend 163:S13–S18. https://doi.org/10.1016/j.drugalcdep.2015.11.041

Tapper K (2018) Mindfulness and craving: effects and mechanisms. Clin Psychol Rev 59:101–117. https://doi.org/10.1016/j.cpr.2017.11.003

Taylor AG, Goehler LE, Galper DI, Innes KE, Bourguignon C (2010) Top-down and bottom-up mechanisms in mind-body medicine: development of an integrative framework for psychophysiological research. Explor J Sci Health 6(1):29–41

Ting EYC, Yang AC, Tsai SJ (2020) Role of interleukin-6 in depressive disorder. Int J Mol Sci 21(6):1–22

Tyagi A, Cohen M (2016) *Yoga* and heart rate variability: a comprehensive review of the literature. Int J Yoga 9(2):97

Uhl GR, Koob GF, Cable J (2019) The neurobiology of addiction. Ann N Y Acad Sci 1451(1):5–28

Walter M, Gerhard U, Gerlach M, Weijers HG, Boening J, Wiesbeck GA (2006) Cortisol concentrations, stress-coping styles after withdrawal and long-term abstinence in alcohol dependence. Addict Biol 11(2):157–162

Western M, Tomaszewski W (2016) Subjective well-being, objective well-being and inequality in Australia. PLoS One 11(10):1–20

Wielgosz J, Goldberg SB, Kral TR, Dunne JD, Davidson RJ (2017) Mindfuness meditation and psychopathology. Physiol Behav 176(12):139–148

Wilson JD, Vo H, Matson P, Adger H, Barnett G, Fishman M (2017) Trait mindfulness and progression to injection use in youth with opioid. Addiction 52(11):1486–1493. https://doi.org/10.1080/10826084.2017.1289225

Wimberly AS (2020) How yoga impacts the substance use of people living with HIV who are in reentry from prison or jail: a qualitative study. Complement Ther Med 410:1–19

Witkiewitz K, Warner K, Sully B, Barricks A, Stauffer C, Thompson BL et al (2014) Randomized trial comparing mindfulness-based relapse prevention with relapse prevention for women offenders at a residential addiction treatment center. Subst Use Misuse 49(5):536–546. https://doi.org/10.3109/10826084.2013.856922

Won TJ, Kim B, Lim YT, Song DS, Park SY, Park ES et al (2011) Oral administration of lactobacillus strains from kimchi inhibits atopic dermatitis in NC/Nga mice. J Appl Microbiol 110(5):1195–1202

World Health Organization (2007) Cancer control palliative care. Chest 123(1 Suppl):284S–311S

Yadav RK, Magan D, Mehta N, Sharma R, Mahapatra SC (2012) Efficacy of a short-term *yoga*-based lifestyle intervention in reducing stress and inflammation: preliminary results. J Altern Complement Med 18(7):662–667

Yang CC, Barrós-Loscertales A, Li M, Pinazo D, Borchardt V, Ávila C et al (2019) Alterations in brain structure and amplitude of low-frequency after 8 weeks of mindfulness meditation training in meditation-Naïve subjects. Sci Rep 9(1):1–10

Yuan JP, Connolly CG, Henje E, Sugrue LP, Yang TT, Xu D et al (2020) Gray matter changes in adolescents participating in a meditation training. Front Hum Neurosci 14:1–9

Zannone S, Brzosko Z, Paulsen O, Clopath C (2018) Acetylcholine-modulated plasticity in reward-driven navigation: a computational study. Sci Rep 8(1):1–20

Zhang D, Lee EKP, Mak ECW, Ho CY, Wong SYS (2021) Mindfulness-based interventions: an overall review. Br Med Bull 138(1):41–57

Zhuang SM, An SH, Zhao Y (2013) *Yoga* effects on mood and quality of life in Chinese women undergoing heroin detoxification: a randomized controlled trial. Nurs Res 62(4):260–268

Zou Z, Wang H, Uquillas O, Wang X (2017) Definition of substance and non-substance addiction. Springer

Chapter 4
Computer Vision-Based Systems and Alternative Therapies

Sumit Kalra and Saurabh Srivastava

Contents

Abstract Computer vision is a fascinating field of study which includes the application of algorithms over real-world visual data, such as photos and videos, in order to provide valuable suggestions for tasks that have been traditionally done by human beings only. Although not perfect, these algorithms often perform satisfactory solutions to many real-world problems and can execute ubiquitously on easily available devices. It makes them a great candidate for use across a wide range of use cases.

In this chapter, we will highlight the application of computer vision-based solutions in the field of yogic therapies. In particular, we will discuss their utility for aiding *Yoga* practitioners in the improvement of their postures. It can help the beginners in the arena to perform the *Asana* correctly, and at the same time, provide the *Yoga* trainers with crucial feedback about the progress of their disciples.

S. Kalra (✉)
Department of Computer Science and Engineering, Indian Institute of Technology Jodhpur, Jodhpur, India
e-mail: sumitk@iitj.ac.in

S. Srivastava
Department of Computer Science and Engineering, Indian Institute of Technology (ISM) Dhanbad, Dhanbad, India

A. Anand (ed.), *Neuroscience of Yoga*,
https://doi.org/10.1007/978-981-97-2851-0_4

Keywords *Yoga* · Computer vision · Internet of things · Artificial intelligence · Deep learning

4.1 Introduction

Artificial intelligence (AI) is an interdisciplinary field of study that deals with designing and creating intelligent systems. While defining intelligence is tricky, there is a general consensus among the practitioners that it involves building systems that can process information from the real world and produce actionable insights that may improve the quality of our lives. Computer vision is a branch of AI that processes visual information such as photos and videos to make valuable suggestions based on previously processed data. Humans have a definitive advantage when it comes to vision—a lifetime of contextual training. The amount of visual data that the human brain processes, along with the plethora of contextual knowledge that accompanies it, is arduous to feed to a machine for training purposes. It enables a human brain to perform tasks such as discerning an object's type, their distance from the observer, and whether they are moving or still, rather rudimentary.

Practitioners working in the field of computer vision are attempting to get these tasks done by a computer. The primary motivation to do so is that computers, when adequately trained, can perform these tasks faster, using cameras, data, and algorithms, than a human being who does it using the retinas, optic nerves, and the visual cortex. The benefits of such systems are many. For instance, these systems are deployed to evaluate hundreds of products or processes every minute to detect faults or issues that may not be easy for the human eye to detect.

A typical placement of the computers equipped with such systems may be seen in the manufacturing and automotive industries, with an ever-growing demand in other domains (Szeliski 2022). In this article, we will discuss how practitioners of *Yoga* may be benefitted from the use of such systems.

Effects of *Yoga* have been studied in the context of various diseases and researchers have published evidence-based studies. It includes neurological disorder, multiple sclerosis, stroke, epilepsy, Parkinson's disease, dementia, cerebrovascular disease, Alzheimer disease, neuropathy, myelopathy, Guillain-Barre syndrome, and many more (Stephens 2017; Mishra et al. 2012; Mooventhan and Nivethitha 2017; Meyer et al. 2012). These studies concluded that practice of *Yoga* plays an important role in treating a variety of diseases at physical, mental, and emotional level (Miller 1996).

In a randomized, controlled trial, the efficacy of *Hatha Yoga* in reducing Carpal Tunnel Syndrome (CTS) symptoms was examined. Forty-two CTS individuals were given 8 weeks of Hatha *Yoga* sessions by *Garnfinkel* et al. whereas a control group received a wrist splint to enhance their existing therapy (Garfinkel et al. 1998). In conclusion, grip strength (which grew from 162 to 187 mmHg) and pain reduction (which fell from 5.0 to 2.9 mm) were statistically improved significantly in the *Yoga* groups' individuals, but not in the control subjects. Phalen's sign also significantly improved after attending the *Yoga* class (12 improved versus 2 in the control group). The research showed that a *Yoga*-based program was an effective tool in such conditions.

We draw the conclusion that *Yoga* has potential for treating the signs and symptoms of a number of neurological and mental diseases based on the material that has been published to yet. Six of the seven published randomized controlled trials on *Yoga*'s ability to treat neurological problems indicated that it had a meaningful impact. There have been 13 published randomized controlled studies of *Yoga* for the treatment of psychiatric disorders and symptoms, and 10 of them found *Yoga* to be a significant treatment, as measured by substantial effects of the *Yoga* group alone on one or more outcome measures.

Despite the diverse protocols of *Yoga* employed, the results imply that *Yoga* may be an useful adjunct intervention for major depression, as seven of the nine randomized controlled trials found substantial favorable benefits of the *Yoga* intervention alone. Clinicians might propose *Yoga* as an adjunct treatment option for patients with serious depression in light of these findings.

4.2 How Do Computer Vision-Based Systems Work?

So how does a machine make a decision or suggestion over some visual input? The key is data—lots of it, to be precise. Take the example of detecting if an automotive tyre is faulty or not. A system that can use computer vision algorithms must be fed many images of tyres. These images must be divided into categories. At least one such category should contain images of tyres with no anomalies. The system, then, attempts to "learn" the distinctions between faulty and non-faulty tyres from the fed examples.

The salient technologies that are used to accomplish this task include a Convolutional Neural Network (CNN) (Fig. 4.1) and a Deep Learning Neural Network (DNN) (Fig. 4.2) (Yadav et al. 2019). These algorithms enable what is

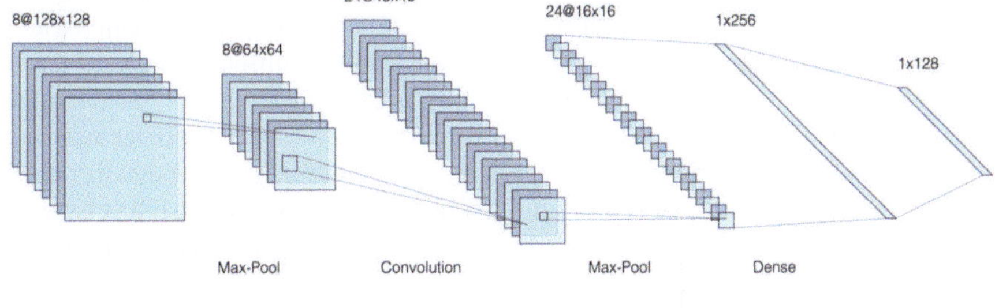

Fig. 4.1 CNN (convolutional neural network) architecture example

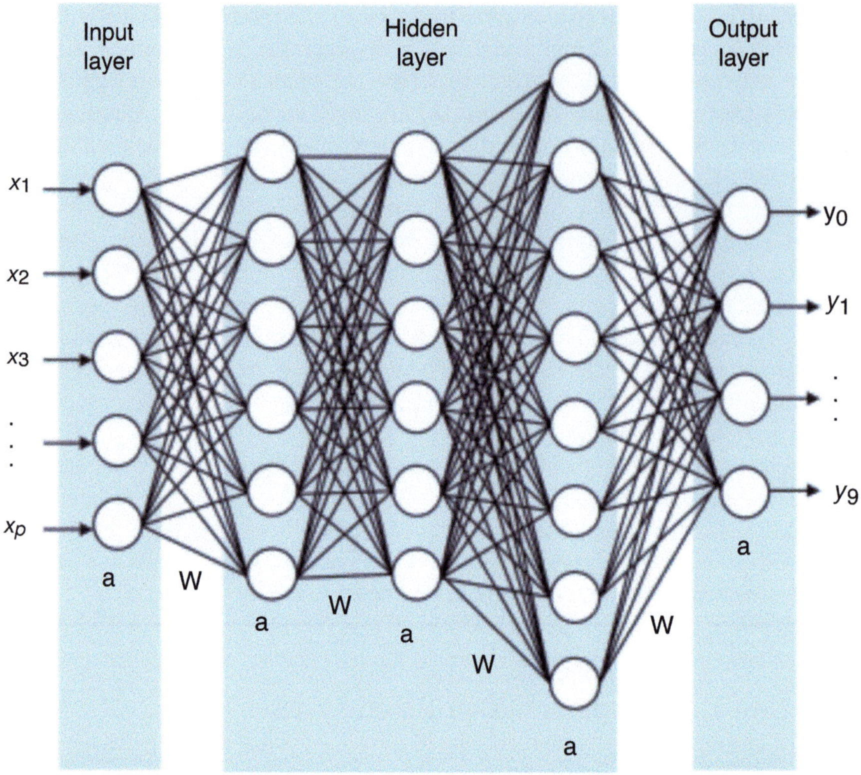

Fig. 4.2 DNN (deep learning neural network) architecture example

categorized as Machine Learning, a loosely defined term that encompasses techniques which may be used to "train" a system by providing it with an ample number of observation-inference pairs. The learning process usually does not require human intervention, except for a few "minor tweaks," commonly referred to as hyperparameters for the algorithm.

Typically, the input of any machine learning approach in computer vision is a two-dimensional array (or a matrix) of values, each representing one pixel in an image (Agrawal et al. 2020). This collection of pixels is tagged or labeled to explain what it represents in the real world. For instance, one photo (read one matrix) may be tagged as a "Cat," while another photo (read another matrix) may be labeled as a "Dog." By performing a series of convolutions over small subsets of the matrix and building up on the outputs of previous iterations, a CNN does what a typical human being does. In initial iterations of the convolution operation, it starts recognizing edges and basic shapes, and as the process proceeds, it "fills in more details" in its inferences. Due to this iterative, bottom-up understanding, a CNN attempts to learn the distinguishing factors at the macro-level. This learning can then be applied to predict the label for a new photo.

Videos are essentially a collection of "related" frames or images. While a CNN is a good pick to categorize images, a Recurrent Neural Network (RNN) is a common choice for understanding how visuals in a sequence of frames are connected. It is because RNNs can extract insights from temporal data (data that can be pieced on a line representing time). Since frames of a video have a similar relationship with each other, RNNs can provide condensed inferences across a large number of frames, i.e., the video.

4.3 Examples of Computer Vision Tasks

A social network firm has to deal with a large number of images that the users upload. These images must be categorized into at least two categories—those that do not violate the firm's policies and those who do. Image classification is a task that can do so without a human-in-loop.

Object detection involves detecting and localizing the occurrence of specific objects or patterns in an image or video. A manufacturing product line may benefit from a system that could perform the same if it detects damages or faults in a product early so that the cost of later stages could be reduced.

An autonomous vehicle must detect and track pedestrians to make safe maneuvers. Thus, not only does it require the detection of objects, but Object Tracking is another important goal that the vehicle manufacturer must achieve. This task involves processing sequenced photos or real-time video streams to follow the movement of a detected entity (Sharma et al. 2022; Yadav et al. 2019).

Images are often accompanied by metadata or labels as part of their descriptions. Content-based image retrieval involves searching and retrieving images for given search criteria. Extracting these labels automatically with the help of computer vision algorithms, instead of relying on manually provided metadata, may improve the retrieval accuracy of such operations significantly.

In this chapter, we have a look at how modern science and technology can help *Yoga* practitioners in mastering the art by using computer vision-based approaches. But before that, we will briefly touch upon the different directions of research that have been taken up recently.

Most of the AI-based research in the field focuses on the problem of *Yoga* pose grading. The core idea is to be able to evaluate the postures of a human being, attempting to perform a specific *Yoga* pose, and comment on the deviation from the standard practice. Overall, the AI-based approaches to grade *Yoga* poses can be divided into three categories. The first set includes those methods which are dependent on wearable devices. In such approaches, sensors are attached to certain joints of the *Yoga* practitioner to capture the movements and poses (Gupta and Gupta 2021). It requires the practitioner to wear devices, which may be fairly inconvenient in the long run. The second category of approaches requires specific devices, such as the Kinect device by Microsoft, to capture the contour of the body of the practitioner and proceed towards grading the pose. Such methods require specialized hardware, which may not be available everywhere. The third class of approaches, which seems the most pragmatic for building frugal *Yoga* pose grading systems, are those which involve the use of community hardware along with computer vision algorithms. We will discuss one such approach here in a little more detail.

4.4 *Yoga* Pose Grading with Contrastive Examples

The methods that use contrastive images work on contrasting standard *Yoga* pose photographs (say, when the *Yoga* pose is performed by an expert) with those taken while a relatively new practitioner performs that *Yoga* pose. We will describe, in brief, one such method proposed by Tian et al. (Wu et al. 2022). The method seeks one photo each from the apprentice and the trainer. The human skeleton points are extracted from these images and fed as input to the pose feature encoder. The quality of the apprentice's posture is evaluated by comparing the similarity between the two inputs. The proposed framework includes a model training framework and a model inference procedure. The model training framework may be divided into three phases: (a) creation of contrastive instances, (b) extraction of skeletal features, and (c) encoding of the features using contrastive skeleton feature representations as shown in Fig. 4.3.

The availability of distinct data is essential in defining the contrastive prediction tasks for learning effective distinct representations. Also, the overall idea is to come up with discriminative feature embedding representations in which similar attributes fall nearby while unique characteristics are projected as far away from each other as feasible. Two types of triplets are proposed for the training purpose as shown in Fig. 4.4. First is the coarse triplet example, which is created by picking a

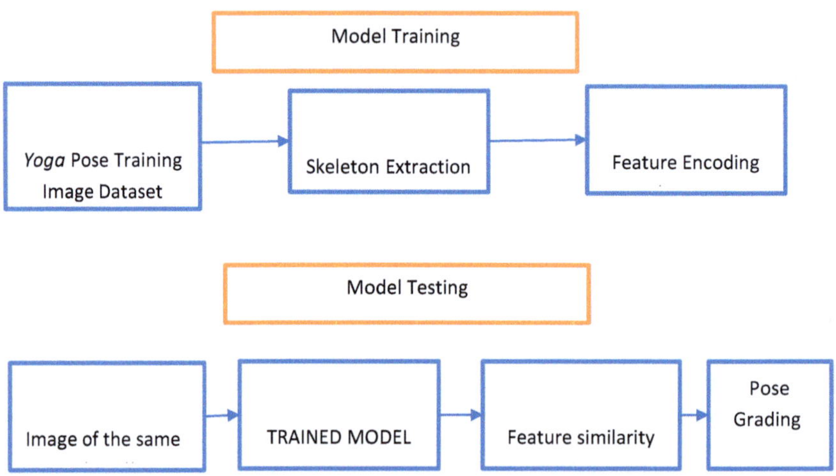

Fig. 4.3 A conceptual overview of the proposed *Yoga* pose grading framework

Fig. 4.4 Triplets with (**a**) anchor, (**b**) positive example, and (**c**) negative example for *Tadasana*

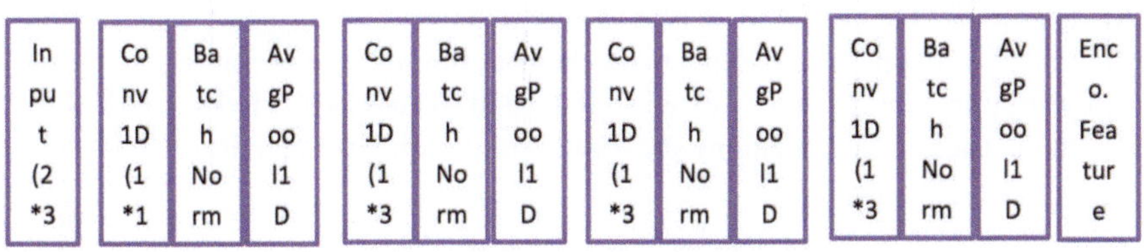

Fig. 4.5 Detailed network architecture of the pose feature encoder

standard image of a particular *Yoga* pose (the anchor), along with another image of the same pose (the positive example) and an image of some other pose (the negative example). The second is the fine triplet example, where a standard image of a particular *Yoga* pose (the anchor) is clubbed with two images of the *Yoga* pose, but while one is of decent quality, the other can be regarded as a poor form of the pose (meaning the former is a "better" approximation to the anchor, as against the latter). Some *Yoga* postures are incredibly complicated to analyze from a single viewpoint. Thus, analyzing important skeletal points from the target photographs is better. The framework only considers key skeletal points for analysis rather than processing the complete image (which may be tricky due to the presence of backgrounds and other objects).

Chen et al. performed these operations on top of the Mediapipe platform using the architecture shown in Fig. 4.5. It provides a state-of-the-art machine learning model called BlazePose for extracting skeletal points from the given input. BlazePose can detect and track multiple vital points on the body and provide their locations as coordinates in a two-dimensional Euclidean space in the form of (x, y)

Table 4.1 Accuracy and classification report in pose grade approach

Approach	Accuracy	Precision	Recall	F1
Coarse contrastive examples only	0.7760	0.6961	0.9795	0.8138
Both coarse and fine contrastive examples	0.8321	0.8819	0.7669	0.8204

pairs. The values for x and y are in the range (0, 1), which is done by normalizing the coordinates values against the image's width and height, respectively. One posture image, a total of 33 crucial points of the body, is detected by Mediapipe. Since each point has two coordinate values, a picture has (2, 33) coordinates that are used for feature encoding. The approach is to attempt to create different representations by maximizing the agreement among comparable pose images. The number of training examples for the coarse and fine triplets in every batch used for training the model is kept the same (Table 4.1).

Some companies are trying hard to help people using yogic therapies to prevent disorders. Zenia, a Belarus-based firm, is one of these companies that uses AI to assist users in identifying the most precise ways to execute *Asana*. Zenia promises to be the world's first virtual *Yoga* assistant, guiding users through their routines using machine learning and movement detection technologies. It is a digital *Yoga* coach that helps improve postures and makes *Yoga* more accessible. A representative dataset of 2,00,000 photos serves as neural network training and is an important component of the program. Zenia takes the user through each posture step by step, demonstrating how to stand, extend, and bend. The voice-activated orders also inform the user of the sensations she should have during the activity.

4.5 e-*YogaGuru*: An E-tutor for *Yoga*

Being able to grade a *Yoga* pose is only the first step towards building a solution that may benefit a *Yoga* practitioner. The ultimate goal is to help new practitioners with achieving high levels of precision while they practice the *Yoga* poses, with little or no manual help. Next, we discuss a computer-vision based *Yoga* tutor called e-*YogaGuru* (Kale et al. 2021). One major highlight of the tutor is that not only does it comment on the correctness of the posture, but it also analyzes, through a continuous stream, the holding time required for different *Yoga* poses.

An aspect of e-*YogaGuru* that makes it unique with respect to the previous efforts is its stress on variability among the practitioners, while they attempt to perform an *Asana*. This variability not only includes the differences in the angles between the joint positions, but also, the speed at which they perform the transitions. This allows them to also verify and provide feedback about an essential aspect of an *Asana*—the hold time. The approach uses "just-enough" machine learning that is required to build a working prototype, while relying on a well-thought-out process to achieve its goals.

The machine learning approaches help in transforming captured frames of a practitioner to skeletal outputs, which are further analyzed to identify angles between important joints of the practitioner's body. With the help of a state transition model, and a differencing scheme to detect the changes between poses, the system can identify the different poses in an *Asana*, as well as the transition from one pose to another. The hold time can be evaluated by calibrating the number of frames getting classified as the same pose (the higher the number of frames in a pose, the higher is the hold time). The dataset contains ample samples collected from people with varying physical features to accommodate the variance in the angles between the joints based on different anthropometry.

The use of the meticulously crafted state transition model for each *Asana* also helps in providing valuable feedback to the practitioner. For instance, an unwanted transition in the state leads to a special "error state," thus, allowing the system to signal the precise pose or transition which was imperfect at the time of execution.

The starting point for any solution of this kind is a well-crafted knowledge-base, containing enough data samples to be able to perform any kind of learning. For e-*YogaGuru*, the researchers prepared a dataset that was curated out of *Yoga* poses performed by 25 individuals with different anthropocentric parameters such as the body mass index, gender, and age, in order to create a dataset with an ample amount of anthropomorphic variations to build a generalized tutor. The dataset contained a total of 1700 *Yoga* sequences, out of which, 875 were tagged as "ground truth," i.e., as the standard performances of the *Yoga* pose in context, while the rest, i.e., 875 left-over instances, were tagged as incorrect or partially correct versions of the attempted poses. Kinect 1.0 sensors were used during the dataset creation phase, in order to capture detailed positions of the joints at different intervals. In total, video sequences of more than 29 h with 3,150,000 frames were used for the purpose of building the tutor. A flowchart explaining the working of e-*YogaGuru* is shown below.

4.6 Clinical Implication

In this chapter, we discussed how computer vision techniques can be useful to detect the yogic postures while doing *Yogasana* and a grading system to automatically grade postures performed by any person. One of the computer vision-based *Yoga* pose grading approaches has been discussed in detail. The approach was able to automatically grade the *Yoga* pose image via the learned contrastive skeleton feature representations. The proposed approach produced more accurate pose grading, as verified in experimental results using the two benchmark datasets.

Another system mentioned in this chapter is e-*YogaGuru* that presents a novel speed invariant, interactive, vision-based system for amendment in human action while performing *Yogasana*. The developed novel state transition-based approach includes angle features and thus addresses the issues raised due to human anthropometry and variance in execution speed.

The state transition model-based approach assures validation of the correct sequence. It facilitates the calculation of hold time by considering the number of posture frames for a particular posture. Furthermore, it also suggests amendment in the hold-time and sequence accordingly, if required, thus presenting an automated and intelligent system for monitoring *Yogasana*. The e-*YogaGuru* system is validated using a set of seven *Yogasana* with varying difficulty levels. Two forward bending *Yogasana* (under "hard" difficulty level) involve forward bending movements, which naturally occlude the front view of the practitioner. But the problem in this posture is that a side view of the posture is mandatory in such cases to analyze the bending angles and validate the postures. A practitioner need not change the position of the sensor or even his/her position but needs to turn around at 90° to capture the bending angles and facilitate its validation.

4.7 Future Direction for Research

In the future, the research should focus on the dataset of *Yoga* sequences and mudra posture, with an audio guidance system. *Yoga* posture detection can be further implemented along with the Internet of *Yoga* Things (IoYT) framework. Identifying the postures in other *Yoga*-like practices such as ballet, climbing, and Tai-Chi is required. In future work, it is required to consider the multi-users in single-frame scenarios such as classrooms and practice halls (Yadav et al. 2022). To improve the wearable system, we need to integrate features like blood pressure monitoring and body temperature through other wearable devices, which can be connected to an existing application program.

If more people work toward computer vision, we can positively change the world. In the future, there will be possibly more startups for computer vision-based yogic therapies like Zenia because there is a strong need to balance health and work-life properly.

4.8 Conclusion

All the mentioned approaches focus on the main-ultimate posture of the *Asana* and ignore the sequence of pre-postures and post-postures. It is highly recommended that the yogi adheres to the correct posture sequence and the hold time for the best results of the exercises. The correct sequence and appropriate hold time are reported to enhance the blood flow, build strength, and stabilize the mind and body. It plays a significant role in attaining desired benefits of the exercises. However, the state-of-the-art approaches completely ignore consideration of these vital parameters.

References

Agrawal Y, Shah Y, Sharma A (2020) Implementation of machine learning technique for identification of *yoga* poses. In: 2020 IEEE 9th international conference on communication systems and network technologies (CSNT). IEEE, pp 40–43

Garfinkel MS, Singhal A, Katz WA, Allan DA, Reshetar R, Schumacher HR Jr (1998) *Yoga*-based intervention for carpal tunnel syndrome: a randomized trial. JAMA 280(18):1601–1603. https://doi.org/10.1001/jama.280.18.1601

Gupta A, Gupta HP (2021) *Yoga*Help: leveraging motion sensors for learning correct execution of *yoga* with feedback. IEEE Trans Artif Intell 2(4):362–371

Kale G, Patil V, Munot M (2021) A novel and intelligent vision-based tutor for Yogāsana: e-*Yoga*Guru. Mach Vis Appl 32(1):1–7

Meyer HB, Katsman A, Sones AC, Auerbach DE, Ames D, Rubin RT (2012) *Yoga* as an ancillary treatment for neurological and psychiatric disorders: a review. J Neuropsychiatry Clin Neurosci 24(2):152–164

Miller BS (1996) *Yoga*: discipline of freedom: the *yoga* sutra attributed to *Patanjali*. University of California Press

Mishra SK, Singh P, Bunch SJ, Zhang R (2012) The therapeutic value of *yoga* in neurological disorders. Ann Indian Acad Neurol 15(4):247

Mooventhan A, Nivethitha L (2017) Evidence based effects of *yoga* in neurological disorders. J Clin Neurosci 43:61–67

Sharma A, Shah Y, Agrawal Y, Jain P (2022) Real-time recognition of *Yoga* Poses using computer vision for smart health care. arXiv preprint arXiv:2201.07594

Stephens I (2017) Medical *yoga* therapy. Children 4(2):12

Szeliski R (2022) Computer vision: algorithms and applications. Springer Nature

Wu Y, Lin Q, Yang M, Liu J, Tian J, Kapil D, Vanderbloemen L (2022) A computer vision-based *yoga* pose grading approach using contrastive skeleton feature representations. Healthcare 10(1):36

Yadav SK, Singh A, Gupta A, Raheja JL (2019) Real-time *yoga* recognition using deep learning. Neural Comput Appl 31(12):9349–9361

Yadav SK, Agarwal A, Kumar A, Tiwari K, Pandey HM, Akbar SA (2022) YogNet: a two-stream network for real time multi person *yoga* action recognition and posture correction. Knowl Based Syst 250:109097

Chapter 5
Endocrinology and *Yoga*

Rajat Gupta and Rama Walia

Contents

Abstract The ultra-modernization has come at a cost of pandemic of many lifestyle disorders including type 2 diabetes mellitus. *Yoga* is one of the most ancient and effective practice, harmonizing body and mind. It has a great potential in the prevention and management of lifestyle disorders and its resultant complications. Plethora of scientific work supports the role of *Yoga* in management of metabolic syndrome including type 2 diabetes mellitus. This review discusses available literature regarding *Yoga* and various endocrine disorders especially type 2 diabetes. Most of the studies suggest positive outcome regarding various parameters of glycaemia, body weight, and lipids. Some data has shown benefit of *Yoga* with regard to oxidative and inflammatory markers and adipokines. Some data discussed here has shown positive correlation of *Yoga* with thyroid disorders and polycystic ovarian syndrome. However, large-scale prospective randomized controlled studies are required so as to further establish the role of *Yoga* in metabolic syndrome and other endocrine disorders.

R. Gupta
Department of Endocrinology, Alchemist Hospital, Panchkula, Haryana, India

R. Walia (✉)
Department of Endocrinology, Postgraduate Institute of Medical Education and Research, Chandigarh, India
e-mail: walia.rama@pgimer.edu.in

A. Anand (ed.), *Neuroscience of Yoga*,
https://doi.org/10.1007/978-981-97-2851-0_5

Keywords *Yoga* · Type 2 diabetes · Metabolic syndrome · Polycystic ovarian syndrome · Thyroid hormones

5.1 Introduction

Yoga is a group of physical, mental, and spiritual practices that aim to control and still the mind, recognizing a detached witness-consciousness untouched by the mind (*Chitta)* and mundane suffering (*Duḥkha*). *Yoga* originated in India around 5000 B. C. E. Endocrinology is a medical discipline dealing with hormones; the chemical signals secreted by endocrine organs into the blood stream that act on distant tissue. Endocrinology encompasses varied organ systems including the hypothalamus, pituitary, thyroid gland, adrenal cortex and adrenal medulla, testes and ovaries, bone and mineral metabolism, pancreas and carbohydrate metabolism, and few others. External and internal environmental signals modulate the synthesis and secretion of the hormones through the central nervous system, hypothalamus, and various other pathways. *Yoga* through synchronizing mind, body, and soul can potentially provide benefits for various endocrine disorders. This chapter reviews the published data relating to *Yoga* and endocrinology disorders, mainly related to diabetes and metabolic syndrome.

5.2 *Yoga* and Diabetes

Diabetes is the most common endocrine disorder. The estimates in 2019 showed that 77 million adults have type 2 diabetes in India, and is expected to rise to about 134 million by 2045 (International Diabetes Federation (IDF) 2019). Diabetes prevention program research group showed reduction in incidence of type 2 diabetes by 58% in obese pre-diabetic individuals with lifestyle intervention; which included 150 min per week of moderate intensity exercise such as brisk walking and caloric reduction so as to achieve a sustained weight loss of about 7%. The average follow up was 2.8 years (Knowler et al. 2002). Role of *Yoga*-based lifestyle interventions has been studied variously. *Hegde* et al randomized 123 type 2 diabetic patients into standard of care or standard of care with *Yoga* practices for 3 months. *Yoga* practices included *tadasana, padahastasana, vrikshasana, trikonasana, parshvothanasana, vajrasana, vakrasana, gomukasana, paschimotasana, uttanapadasanta, pawanamuktasana, bhujangasana, shalabasana, dhanurasana, viparitakarani, sitkari and bhramari Pranayama, anuloma viloma*, and *shavasana* poses. *Yoga* group showed statistically significant improvements in body mass index (-0.5 ± 0.1 kg/m^2 vs. 0.3 ± 0.2 kg/m^2), fasting plasma glucose (-0.8 ± 0.3 mmol/L vs. 0.4 ± 0.1 mmol/L), post prandial plasma glucose (-1.1 ± 0.4 mmol/L vs. 0.3 ± 0.4 mmol/L), and HbA1c (-0.1 ± 0.2% vs. 0.5 ± 0.3%) in 60 type 2 diabetes individuals at the end of 3 months in comparison to the control group on standard of care (Hegde et al. 2011).

In a small pilot study done by similar group involving similar *Yoga* practices (75–90 min; five times per week), significant improvement was seen in 14 prediabetic individuals in terms of body mass index, waist circumference, systolic blood pressure, and fasting plasma glucose levels at the end of 3 months. No significant improvement was observed in post prandial plasma glucose levels, glycosylated haemoglobin, waist-to-hip ratio, or measured antioxidants (Hegde et al. 2013).

In one of the largest study, *Grontved* et al followed two cohorts of 99,316 US women for 8 years from the Nurses' Health Study and Nurses' Health Study II, who were free of diabetes, cancer, and cardiovascular diseases at baseline. Participants self-reported time spent on resistance exercise, lower intensity muscular conditioning exercises (*Yoga*, stretching, toning), and aerobic activity. Pooled relative risk for type 2 diabetes for women performing 1–29, 30–59, 60–150, and >150 min/week of total muscle-strengthening and conditioning activities was 0.83, 0.93, 0.75, and 0.60 compared to women reporting no muscle-strengthening and conditioning activities ($p < 0.001$ for trend). Furthermore, resistance exercise and lower intensity muscular conditioning exercises were each independently associated with lower risk of type 2 diabetes in pooled analyses. Women who engaged in at least 150 min/week of aerobic activity and at least 60 min/week of muscle-strengthening activities had substantial risk reduction compared with inactive women (relative risk—0.33) (Grontved et al. 2014).

Another study enrolled 1292 subjects with type 2 diabetes and pre-diabetes from five different states of India for *Yoga*-based lifestyle intervention for 10 days. Mean fasting plasma glucose reduced significantly to 121 mg/dL from a baseline of 133 mg/dL (Venugopal et al. 2017).

In a prospective randomized control trial done by *Kacker* et al., 102 prediabetic adults were randomized to *Yoga* group or wait list for a period of 3 months. The integrated approach to *Yoga* therapy included Prayer, Omkar recitation, *Yoga* postures (*Asana*), breathing techniques (*Pranayama*), *Shavasana*, and dietary intervention. These sessions were under the guidance of certified *Yoga* trainer and lasted for about 45 min, 6 days a week. At the end of trial intervention group had significant reduction in BMI and waist to hip ratio compared to their pre-trial levels. Also fasting plasma glucose reduced significantly from 116 to 108 mg/dL. HbA1c levels reduced significantly from 6.42 to 5.67%. Lipid parameters also improved significantly (Kacker et al. 2019).

In a multicentre retrospective study *Arumugam* et al evaluated the effects of adjunct *Yoga*-treatment in achieving composite cardiovascular goals for type 2 diabetes in rural Indian settings. Composite cardiovascular scores were based on the combination of individual ADA goals; haemoglobin A1c < 7%, blood pressure (BP) < 140/90 mmHg, stringent BP (<130/80 mmHg), and LDL-C < 100 mg/day. The study cohort comprised two treatment groups ($n = 73$ each): non-*Yoga* group (standard of care) and *Yoga* group (adjunct *Yoga*-treatment). Compared to standard of care, adjunct *Yoga*-treatment was found to significantly facilitate the attainment of ADA composite score by eight-fold; haemoglobin A1c, ~two-fold; LDL-C, ~two-fold; BP < 140/90 mmHg and <130/80 mmHg by ~8-and ~6-fold respectively (Arumugam et al. 2020).

Viswanathan et al studied the effect of *Yoga* intervention on the biochemical, oxidative stress markers and inflammatory markers, and sleep quality among subjects with type 2 diabetes. Non-*Yoga* group ($n = 150$) was advised on simple physical exercises whereas *Yoga* group ($n = 150$) did *YogAsana* (50 min for 5 days in a week). At the end of 3 months, *Yoga* group had significant reduction in fasting blood glucose levels (140–128 mg/dL), post prandial blood glucose levels (215–197 mg/dL), and HbA1c (7.5–7.2%). Inflammatory markers and oxidative stress markers (Interleukin 6, tumor necrosis factor α, and Thio-Barbituric Acid Reactive Substances) also reduced significantly in *Yoga* group as compared to non-*Yoga* group. There was marked improvement in the levels of Adiponectin, prostacyclin synthase, and sleep quality among subjects practising *Yoga* (Viswanathan et al. 2021).

Cui et al evaluated the effects of *Yoga* in adults with type 2 diabetes in a meta-analysis including 12 randomized controlled trials and total of 864 subjects. Fasting blood glucose and haemoglobin A1c was lower by 23.72 mg/dL (95% CI −37.7 to −9.65; $P = 0.001$) and 0.47% (95% CI −0.87 to −0.07; $P = 0.02$) respectively in *Yoga* group in comparison to controls. The weighted mean differences were −17.38 mg/dL (95% CI −27.88 to −6.89; $P = 0.001$) for postprandial blood glucose, −18.50 mg/dL (95% CI −29.88 to −7.11; $P = 0.001$) for total cholesterol, 4.30 mg/dL (95% CI 3.25–5.36; $P < 0.00001$) for high-density lipoprotein cholesterol, −12.95 mg/dL (95% CI −18.84 to −7.06; $P < 0.0001$) for low-density lipoprotein cholesterol, and −12.57 mg/dL (95% CI −29.91 to 4.76; $P = 0.16$) for triglycerides (Cui et al. 2017).

The regular *Yoga* practice improves cardiac autonomic functions independent of glycemic control and reduces the risk of cardiovascular events. In a systemic review and meta-analysis, including 44 RCTs with a total of 3168 participants, regular *Yoga* practitioners showed mean reduction in systolic blood pressure (5.85 mmHg) and diastolic blood pressure (4.12 mmHg), basal heart rate (6.59 beats per minute), basal respiratory rate (0.93 breaths/min), waist circumference (1.95 cm), waist/hip ratio (0.02), total cholesterol (13.09 mg/dL), VLDL (5.70 mg/dL) triglycerides (20.97 mg/dL), HbA1c (0.45%), and insulin resistance (0.19) as compared to regular care. Serum HDL increase by 2.94 mg/dL (Cramer et al. 2014).

5.3 Mechanism of Benefits of *Yoga* in Diabetes and Metabolic Syndrome

Yoga incorporates specific postures (*Asana*), meditation (*Dhyana*), breathing practices (*Pranayama*). *Yoga* practice leads to improvement in physical fitness and mental relaxation. Most of the studies groups evaluating the benefit of *Yoga* in diabetes and metabolic syndrome have practised *Tadasana, Trikonasana, VajrAsana, Padmasana, Ardhamatsyendrasana, Paschimottanasana, Bhujangasana, Dhanurasana, Halasana, Naukasana, Shavasana, Surya Namaskar*. Psychological stress and depression have bidirectional effect in the control of diabetes (Bystritsky et al. 2014).

Stress results in disorganized stimulation of hypothalamic-pituitary-adrenal and sympathetic axes and suppression of parasympathetic axis. Overall this results in increased level of cortisol, epinephrine, norepinephrine, growth hormone, glucagon, and pro-inflammatory cytokines. This in turn leads to insulin resistance and increased blood glucose levels, hypertension, and increased risk of cardiovascular events. *Kosuri* et al have shown reduction in body mass index, reduction in anxiety scores, and improvement in general well-being in 35 subjects of type 2 diabetes with 40 days of *Yoga* practice (Kosuri and Sridhar 2009).

Regular practice of *Yoga* lowers basal TNF-α and IL-6 levels. It also reduces the extent of increase of TNF-α and IL-6 to a physical challenge of moderate exercise and strenuous exercise (Vijayaraghava et al. 2015).

Yoga also results in improvement in physical exercise, behavioural changes, and dietary practices in addition to its effects on relaxation and stress management (Aswathy et al. 2013). *Yoga* practice is associated with increase in β-endorphin, serotonin, and dopamine levels leading to joy, euphoria, and decrease in negative thoughts. Chronic *Yoga* practise is associated with lateral hypothalamus stimulation and ecstatic feelings. The practice of *Yoga* postures is associated with increased brain GABA levels and hence leads to reduction in anxiety and stress disorders (Newberg and Iversen 2003). Oxidative stress is reduced by *Yoga* as shown by decrease in serum malondialdehyde, interleukin 6, and leptin levels and increase in glutathione and vitamin C and adiponectin levels (Hegde et al. 2011; Kiecolt-Glaser et al. 2012).

Six months of *Yoga* resulted in improvement in glycaemic control by increasing percentage insulin binding receptor in type 2 diabetic patients but with no significant change in cortisol and thyroid hormones (Gordon et al. 2008).

One hundred type 2 diabetes patient showed a decrease in total cholesterol, triglycerides, and LDL, with an improvement in HDL with *Yoga* for a period of 3 months (Shantakumari et al. 2013). *Yoga* practice has been shown to reduce the concentration of adipose cells in the visceral region, thus attenuating or minimizing excess free fatty acid (FFA) released from adipose cells. This improves muscle cell, leucocyte, and adipocyte-mediated glucose-uptake by affecting the receptor composition of their plasma membrane, thereby reducing circulating glucose levels and increasing insulin sensitivity.

5.4 *Yoga* and Thyroid Hormones

Mohan et al reviewed the literature in relation to benefit of *Yoga* in thyroid disorders. They searched the literature using PUBMED and GOOGLE. Out of 41 studies selected, 11 met the criteria for the final review. Various *Asanas* used in different studies were *swastikasana, vajrasana, suptavajrasana, tadasana, trikonasana, parshavkonasana, paschimottanasana, purvotanasana, janushirshasana, pavanmuktasana, bhujangasana, shalabhasana, dhanurasana, vakrasana, padottanasana, shavasana*. *Pranayama* included were mainly *anuloma-vilom, suryabhedi, sheetali, bharamari.* Authors stated that although yogic practice may regularize thyroid hormone functioning and improve total health, but the studies included had serious methodological limitations. Many studies included did not elaborate the Yogic practices. In some studies, *Yoga* was not the only intervention and a combination of *Yoga* and diet and/or conventional medicine was used (Mohan and Pal 2022).

5.5 *Yoga* and Polycystic Ovarian Syndrome (PCOS)

Polycystic ovarian syndrome is the most common form of chronic anovulation associated with androgen excess. The estimated prevalence is about 5–10% in reproductive age women. As per Rotterdam criteria two out of three features make the diagnosis of PCOS, oligo-ovulation/anovulation, clinical or biochemical hyperandrogenism, and polycystic ovaries morphology by ultrasonography (with exclusion of related disorders). PCOS is a metabolic disorder related to poor life style and obesity. In a systemic review *Thakur* et al evaluated 74 research papers and found 16 to be relevant for the review. They concluded that holistic *Yoga* program was effective in reducing serum anti mullerian hormone, luteinizing hormone, and testosterone levels. It also improved menstrual frequency and modified Ferriman-Gallwey score for hirsutism (Thakur et al. 2021; Nidhi et al. 2013).

5.6 Clinical Implications

Scientific research supports the role of *Yoga* in addressing metabolic syndrome and type 2 diabetes. Studies demonstrate improvements in various parameters such as glycaemia, body weight, lipids, oxidative and inflammatory markers, and even conditions like thyroid disorders and polycystic ovarian syndrome. Implementing *Yoga*-based interventions alongside conventional therapies may enhance management of endocrine disorders. However, further large-scale randomized controlled trials are needed to firmly establish *Yoga*'s role in addressing these conditions.

5.7 Future Direction for Research

Future research should target extensive prospective randomized controlled studies to deepen our understanding of *Yoga*'s therapeutic potential. Future studies should explore the specific mechanisms through which *Yoga* impacts metabolic syndrome, type 2 diabetes, and endocrine disorders. Research should also consider the broader implications of *Yoga* on overall well-being and quality of life, as well as its potential in preventing lifestyle-related disorders.

5.8 Conclusion

Published data on *Yoga*-based lifestyle interventions suggests benefit of *Yoga* in prevention and treatment of diabetes and other metabolic syndrome-based disease spectrum. Possible benefits are through various pathways including psycho-neuro-endocrine and immune mechanisms. The published studies however are of short duration, have included smaller number of subjects and without proper controls. Further studies are necessary to investigate the long-term effects of *Yoga* in management of metabolic syndrome and type 2 diabetes mellitus.

References

Arumugam G, Nagarathna R, Majumdar V et al (2020) *Yoga* based lifestyle treatment and composite treatment goals in type 2 diabetes in a rural south Indian setup—a retrospective study. Sci Rep 10:6402

Aswathy S, Unnikrishnan AG, Kalra S (2013) Effective management of type 2 DM in India: looking at low-cost adjunctive therapy. Indian J Endocrinol Metab 17:149–152

Bystritsky A, Danial J, Kronemyer D (2014) Interactions between diabetes and anxiety and depression: implications for treatment. Endocrinol Metab Clin North Am 43:269–283

Cramer H, Lauche R, Haller H, Steckhan N, Michalsen A, Dobos G (2014) Effects of *Yoga* on cardiovascular disease risk factors: a systematic review and meta-analysis. Int J Cardiol 173:170–183

Cui J, Yan JH, Yan LM, Pan L, Le JJ, Guo YZ (2017) Effects of *Yoga* in adults with type 2 diabetes mellitus: a meta-analysis. J Diabetes Investig 8:201–209

Gordon L, Morrison EY, McGrowder D, Penas YF, Zamoraz EM, Garwood D et al (2008) Effect of *Yoga* and traditional physical exercise on hormones and percentage insulin binding receptor in patients with type 2 diabetes. Am J Biochem Biotechnol 4:35–42

Grontved A, Pan A, Mekary RA, Stampfer M, Willett WC, Manson JE et al (2014) Muscle-strengthening and conditioning activities and risk of type 2 diabetes: a prospective study in two cohorts of US women. PLoS Med 11(1):e1001587

Hegde SV, Adhikari P, Kotian S, Pinto VJ, D'Souza S, D'Souza V (2011) Effect of 3- month *Yoga* on oxidative stress in type 2 diabetes with or without complications: a controlled clinical trial. Diabetes Care 34(10):2208–2210

Hegde SV, Adhikari P, Shetty S, Manjrekar P, D'Souza V (2013) Effect of community based *Yoga* intervention on oxidative stress and glycemic parameters in prediabetes: a randomized controlled trial. Complement Ther Med 21(6):571–576

International Diabetes Federation (IDF) (2019) IDF diabetes atlas, 9th edn. IDF, Brussels

Kacker S, Saboo N, Sharma S, Sorout J (2019) Quasi prospective comparative study on effect of *Yoga* among prediabetics on progression of cardiovascular risk factors. Int J Yoga 12(2):114–119

Kiecolt-Glaser JK, Christian LM, Andridge R, Hwang BS, Malarkey WB, Belury MA et al (2012) Adiponectin, leptin, and *Yoga* practice. Physiol Behav 107:809–813

Knowler WC, Barrett-Connor E, Fowler SE, Hamman RF, Lachin JM, Walker EA, et al.; Diabetes Prevention Program Research Group (2002) Reduction in the incidence of type 2 diabetes with lifestyle intervention or metformin. N Engl J Meed 346:393–403

Kosuri M, Sridhar GR (2009) *Yoga* practice in diabetes improves physical and psychological outcomes. Metab Syndr Relat Disord 7:515–517

Mohan S, Pal R (2022) Effect of yogic practices on thyroid hormones: a review. BLDE Univ J Health Sci 7(1):7–13

Newberg AB, Iversen J (2003) The neural basis of the complex mental task of meditation: neurotransmitter and neurochemical considerations. Med Hypotheses 61:282–291

Nidhi R, Padmalatha V, Nagarathna R, Amritanshu R (2013) Effects of a holistic *Yoga* program on endocrine parameters in adolescents with polycystic ovarian syndrome: a randomized controlled trial. J Altern Complement Med 19(2):153–160

Shantakumari N, Sequeira S, El deeb R. (2013) Effects of a *Yoga* intervention on lipid profiles of diabetes patients with dyslipidemia. Indian Heart J 65(2):127–131

Thakur D, Saurabh Singh DS, Tripathi DM, Lufang D (2021) Effect of *Yoga* on polycystic ovarian syndrome: a systematic review. J Bodyw Mov Ther 27:281–286

Venugopal V, Rathi A, Raghuram N (2017) Effect of short-term *Yoga*-based lifestyle intervention on plasma glucose levels in individuals with diabetes and prediabetes in the community. Diabetes Metab Syndr 11(2):S597–S599

Vijayaraghava A, Doreswamy V, Narasipur OS, Kunnavil R, Srinivasamurthy N (2015) Effect of *Yoga* practice on levels of inflammatory markers after moderate and strenuous exercise. J Clin Diagn Res 9(6):8–12

Viswanathan V, Sivakumar S, Prathiba AS et al (2021) Effect of *Yoga* intervention on biochemical, oxidative stress markers, inflammatory markers and sleep quality among subjects with type 2 diabetes in South India: results from the SATYAM project. Diabetes Res Clin Pract 172:108644

Chapter 6
Neurophysiological Effects of *Yoga* Intervention on Pregnancy

Pooja Nadholta and Akshay Anand

Contents

Abstract There is continuous adaptation of the maternal physiology during pregnancy. In order to accommodate the increased demand of a growing fetus, there are obvious physical and physiological discomforts. In few decades, prenatal *Yoga* has gained popularity due to its possible benefits for both mother and fetus. The acceptance of *Yoga* as a lifestyle and behavioral modification intervention has drastically

P. Nadholta
Neuroscience Research Lab, Department of Neurology, Postgraduate Institute of Medical Education and Research, Chandigarh, India

A. Anand (✉)
Neuroscience Research Lab, Department of Neurology, Postgraduate Institute of Medical Education and Research, Chandigarh, India

CCRYN-Collaborative Centre for Mind Body Intervention, PGIMER, Chandigarh, India
e-mail: anand.akshay@pgimer.edu.in

A. Anand (ed.), *Neuroscience of Yoga*,
https://doi.org/10.1007/978-981-97-2851-0_6

increased as research and awareness have advanced. *Yoga* has been found to have a positive effect on the overall health and neurophysiology of pregnancy. Studies have shown that *Yoga* can reduce stress, anxiety, and depression in pregnant women. It can also improve sleep quality, reduce fatigue, and improve overall well-being. Although process of delineating the mechanism of action of *Yoga* is still in early phase, there is still bulk of evidence to support the benefits of *Yoga* at physiological, psychological, molecular, and cellular dimensions. Continuous harmonious adaptation of *Yoga* as a lifestyle is one of the secrets of healthy pregnancy outcomes. With the aid of scientific evidence, this chapter explains how physiology changes during pregnancy and how *Yoga* can contribute to neurophysiological health in pregnancy.

Keywords Pregnancy · Prental stress · Neurophysiology · Allostasis · Prenatal Yoga

6.1 Introduction

The process of pregnancy involves hormonal, metabolic, respiratory, and cardiovascular changes, necessary for the normal fetal development and to meet the increased metabolic demands of women during this time. A woman's body undergoes many changes during pregnancy that are new to her physically, physiologically, and psychologically, and to survive through it, she must adapt to those changes throughout the gestational period. A flexible anatomy allows physical adaptations to some extent, but physiological, psychological, and hormonal changes require special attention and care. Increased plasma volume, requirement of iron (for hemoglobin, fetus, and the production of enzymes), vitamin B12, clotting factors, cardiac output, respiratory requirement, and nutrient requirement, while decrease in platelets level, hemoglobin, and red blood cells are some of the very common changes that occur during pregnancy (Soma-Pillay et al. 2016). Additionally, the increased blood flow to the brain during pregnancy can lead to increased neural activity, which can affect the way the brain functions and the response of the mother toward the circumstances. These are necessary changes accompanying pregnancy and can be managed and improved with the palliative care and lifestyle modification during pregnancy. Unmanageable physical discomfort, fatigue, insomnia, stress, anxiety, depression, and other unnecessary changes, which are not common in uncomplicated pregnancy, can be avoided and managed.

Studies indicate that *Yoga* during pregnancy is a great way to stay fit and healthy as well as to avoid unpleasant experience and unnecessary discomfort. First, *Yoga-Asana* can improve the mother's posture, flexibility, and uterine blood flow, which can reduce the physical discomfort, increase sleep quality, and increase umbilical cord blood flow. An increase in blood flow to the umbilical cord indicates that the fetus is getting enough oxygen and nutrients which are essential for the fetal growth and development.

Additionally, studies report that hormonal balance can also be achieved through the practice of *Yoga*. Reduced cortisol levels in pregnant women after the practice of *Yoga* have been reported, which help in coping the stress and hence improve sleep quality, reduce anxiety, and improve overall well-being. Therefore, practicing *Yoga* during pregnancy can be beneficial for both the mother and the fetus. There are many forms and components of *Yoga*; however, a prenatal *Yoga* protocol must be designed carefully keeping in mind the physiology of pregnancy. It should include poses that are safe for pregnant women and help in pelvic opening, whole body stretching and loosening, deep breathing practices without breath retention. It should also include breathing exercises, relaxation techniques, and meditation for sure. It is important to avoid poses that involve lying on stomach, as well as any poses that involve deep twisting or stretching.

This chapter includes the physiological changes that transpire during pregnancy, how these changes affect women and fetus, complications which arise due to maladaptation of these changes, and pre-eminently the neurophysiological effect of *Yoga* or mind-body practice on pregnancy to make it unsophisticated.

6.2 Physiology of Pregnancy

Pregnancy is a period of continuous adaptation in maternal physiology influenced by mechanical, physiological, psychological, and hormonal changes in the body to accommodate developing fetus. Each organ functions differently during pregnancy which alters the overall physiology and psychology of women. Placenta, a temporary endocrine organ, develops in ninth to tenth gestation week, which helps in mediating adaptation in maternal physiology. Its profound effect on pituitary stimulates the endocrine system to release hormones such as β-hCG, prolactin, lactogen, and massive amount of progesterone and estrogen in maternal circulation which supports maternal changes of pregnancy. These hormonal fluctuations drive brain changes (neuroendocrine effect) which lead to the physiological changes (neurophysiological effect) and psychological changes. These modifications in women's body are required to support the normal development of fetus.

Maternal physiology alters differently in early, mid, and late pregnancy, and these changes are supported by hormones which help in appropriate adjustment of maternal physiology. Hormonal changes succor the physiological adaptations which include cardiac output (by 30–50%) to deliver blood supply to uteroplacental circulation (demand increase with development of fetus), increased plasma volume as compared to RBCs (Hb drops due to dilution), respiratory discomfort due to enlargement of uterus, and pigmentation. Physiological changes during pregnancy are associated with all the organs and organ systems, viz., respiratory, cardiovascular, hematological, renal, neurological, gastrointestinal, musculoskeletal, and endocrine, to meet the increased metabolic demand of parturient and fetus (Talbot and Maclennan 2016) (Fig. 6.1).

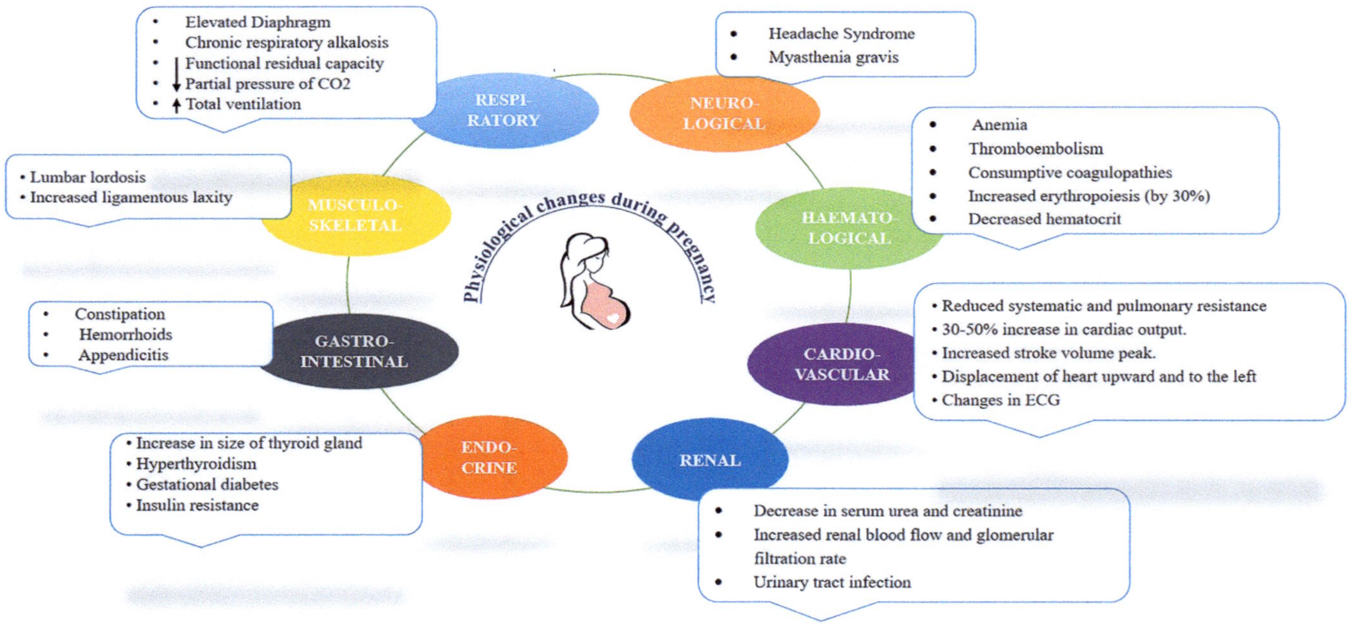

Fig. 6.1 Altered physiology of various organ systems during pregnancy

All changes listed above make pregnancy a disparate phase as compared to other phases of life. Initially, these changes are challenging and come with lots of mental and physical stress to which women adapt with time, and if the adaptation doesn't happen, stress becomes allostatic load which becomes a threat to both the mother and fetus. Therefore, to prevent the conversion of acute challenges into stress and then into allostatic load, there is a requisite need of conventional approaches which can manage these challenges.

6.3 Allostasis and Allostatic Load

Brain is the key regulator and mediator of stress and associated processes. Autonomic nervous system (ANS) along with neuroendocrine and immune system are the mediators of adaptations to daily life challenges, which is called allostasis, i.e., "maintaining stability through change." As these mediators produce hormones and cytokines to cope with a variety of situations, they are capable of producing adaptive effects in the short term. However, if these mediators are not turned off when they are no longer required, they can cause damage to the system (McEwen 2003). During stress, ANS, hypothalamic-pituitary-adrenal axis (HPA), immune, cardiovascular, and metabolic systems collectively act as protecting system through allostasis. Chronic stress can increase the chances of ill medical conditions by activating all these systems recursively, which affects the cognitive, emotional, and behavioral responses of individuals (McEwen 1998). Instant response to any stressor involves the activation of sympathetic nervous system (SNS), while longer-term stress response is mediated by the activation of HPA axis.

During pregnancy, chronic unregulated stress, which ultimately causes allostatic load, imbalances the hormones and neuropeptides which snag the physiology of pregnancy. It is considered as a contributor for various adverse pregnancy outcomes like preterm births, small gestational age, gestational diabetes, gestational hypertension, and poor sleep quality (Hux et al. 2014, 2017). These are manifestations of unregulated stress and anxiety until or unless genetic or due to pre-existing conditions. They can be managed with a healthy lifestyle during pregnancy, which includes increased psychological resilience and better adaptation to pregnancy physiology.

6.4 Neurophysiological Changes in Pregnancy

Neurophysiology is mainly associated with the chemical and electrical activity of individual nerve cells through which it transmits and receives signals between brain and peripheral nervous system (PNS) (Tinga et al. 2020). Hormones play a crucial role in altering the neurophysiology of pregnancy modifying the functions in various regions of the brain, leading to fundamental change in behaviour. In pregnancy, hormones such as estrogen, progesterone, glucocorticoids like cortisol, oxytocin, and prolactin are secreted in significant amounts to support novel physiological functions of maternal body. These hormones cross the blood-brain barrier complemented by pregnancy-specific placental hormones and tend to change the neurophysiology of the mother. Most of these hormones affect the autonomic and homeostatic systems, hence all the systems require adjustments and adaptations throughout the period (Grattan and Ladyman 2020). Placenta modifies the hormonal profile of the pregnant woman and drives changes in maternal body to optimize the conditions for fetal development even at the cost of maternal physiological systems. An example of placental secretion overriding maternal endocrine systems is the corticotrophin-releasing hormone (CRH) secretion by the placenta, which is stimulated by cortisol, providing a positive feedback mechanism for cortisol release, as opposed to hypothalamic CRH, which is inhibited by cortisol, which provides a negative feedback mechanism (Sasaki et al. 1989). This means that maternal health is compromised, and for adapting these changes and to support the fetal development and at the same time to support mother's physical, physiological, and psychological health, extra efforts and care are required in addition to usual care. Understanding the exact mechanism of how endocrine and neurological systems are interacting and affecting the maternal physiology is a complex as well as a major challenge of current neuroscience.

Neurophysiological changes during pregnancy play a crucial role in adaptation. Because all the organs are innervated, they are sensitive to changes in nervous system and vice versa. Physical and anatomical modifications and mechanical compression of nerves (due to enlargement of uterus and spine alignment) elicit the pain, and nervous system becomes more vulnerable for neuropathies, leading to the frequent occurrence of pain during pregnancy. Besides that, altered levels of sex steroids (estrogen and progesterone) affect the vascular tone, neuronal excitability, cerebral blood flow, and cardiovascular expansion (Sarrel 1999; Lee 2007).

Although these modifications are to compensate maternal changes rather to distress women, they can also adversely make pregnancy more difficult if these changes fail to synchronize with each other and with usual physiology. Traditional approaches such as *Yoga*, meditation, and mind-body practices may aid in better adaptation of these changes by exerting effect on neurophysiology aspects to make this challenging phase better for expectant mother. Therefore, in this chapter we will discuss the health benefits of *Yoga* during pregnancy and potential mechanism to alleviate the physiological, psychological, and physical stress.

6.5 Effect of Stress During Pregnancy

Stress in general is a response to physical and psychological distress which triggers autonomic nervous system (ANS) and immune system as a part of adaptive response and physiological support against fight and flight behavior. During stress, immune system functions bidirectionally with nervous and endocrine systems via interaction of various mediators such as neuropeptides, neurotransmitters, cytokines, and hormones which mediate other responses (mainly fight and flight) in the body to maintain the homeostasis. Hormones play an important role in regulation of the body's response to circumstances and, as a consequence, are a key determinant of how the body reacts to various types of stress. Hormone secretion is regulated by the pituitary gland, which is under the control of the hypothalamus, which is also responsible for controlling hunger, sleep, emotions, and temperature. In response to stress, hypothalamic-pituitary-adrenal (HPA) axis and sympathetic nervous system (SNS) are activated, affecting the physiology. During pregnancy when there is continuous stress, dominance of these systems is continuous. Organ systems that are affected by stress are

- Endocrine system
- Cardiovascular system
- Musculoskeletal system
- Reproductive system
- Integumentary system
- Nervous system
- Immune system
- Respiratory system

SNS dominance activates fight and flight mechanism which activates cascade of hormones to activate physiological responses of organ systems involved in stress response. HPA works on neuroendocrine adaptation of stress by releasing corticotropin-releasing hormones (CRH) from hypothalamus, which stimulate the pituitary to release adrenocorticotropic hormone (ACTH) which in turn stimulates adrenal to release cortisol. Maternal HPA axis undergoes dramatic changes during pregnancy due to continuous stressor and cortisol level rise threefold by the end of third trimester. Adrenal gland gradually becomes hypertrophic, therefore releasing cortisol in excess (Mastorakos and Ilias 2003). In addition to hypothalamus, placenta also secretes CRH, which further results in higher cortisol level in maternal blood stream. Cortisol production from the adrenal glands is normally regulated by central negative feedback, but during pregnancy, placental CRH stimulates both the pituitary and the adrenal glands, which result in an increase in cortisol production. Usually, cortisol does not cross the placental barrier because of placental enzyme HSD11B2 which converts active cortisol into inactive cortisol, thus protecting fetus from excessive exposure of cortisol. Considerably 80–90% proportion of cortisol is optimized by placenta, but if there is excess release of cortisol, it can cross the placenta, and overexposure of developing fetus to excess cortisol can cause lower birth weight, preterm birth, impaired HPA axis at later life stages, and more susceptibility to metabolic, cardiac, and neurodevelopmental diseases (Duthie and Reynolds 2013). Increased cortisol along with epinephrine contributes to the development of other maternal complications, as they affect heart and respiratory rate, trigger gluconeogenesis (hence increase glucose level in blood stream and increased supply to brain), reverse glycolysis (glucose breakdown to produce energy), lipolysis in adipose tissue, and protein degradation in muscle cells during pregnancy, hence increasing the maternal discomfort (Thau et al. 2021). High level of cortisol was also seen associated with reduced cerebral blood flow and disrupted interaction between neural activity and cerebral blood vessels and hence affects neurophysiology and increases cognition disturbances in human brain (Zhang et al. 2021). Animal studies have shown that increased maternal care enhances the negative feedback sensitivity of the HPA axis, leading to a reduction in cortisol secretion in maternal blood and a reduction in offspring fear behaviors (Caldji et al. 1998). *Yoga*, which includes mind-body practice, is known to have immediate effect in stress reduction during pregnancy (Kusaka et al. 2016). Although exact mechanism of how *Yoga* practice during pregnancy can alleviate the stress and anxiety is not known, it is speculated that it regulates the HPA axis and hence balances hormonal secretion which manages the physiologic and psychological health of expectant mother (Fig. 6.2).

Symptomatically, stress produces physical and mental tension and *Yoga* is known to release physical and mental tension and encourage relaxation and reduce the instances of stress by stabilizing autonomic nervous system with greater tendencies toward PNS dominance. *Yoga* practice is known to make practitioner more resilient to stressful conditions rather than reversing the effects. Some of the yogic practices such as SKY are known to increase stress tolerance rather than decrease the stress (Chandra et al. 2017). A number of studies have investigated the effect of *Yoga* on stress, anxiety, depression, and other complications like hypertension and gestational diabetes, and results have confirmed the effectiveness of *Yoga* practice during pregnancy in reducing these adverse behaviour and complications. Feasibility of *Yoga* practice during pregnancy has been reported in complicated and uncomplicated pregnancies. Among high-risk pregnancy population, reduced frequency of hypertension, gestational diabetes, intrauterine growth restriction, and improved fetal outcomes were reported after prenatal *Yoga* practice (Rakhshani et al. 2012). American College of Obstetrics and Gynaecology (ACOG) recommended 150 min of moderate intensity of any aerobic activity per week to pregnant women who

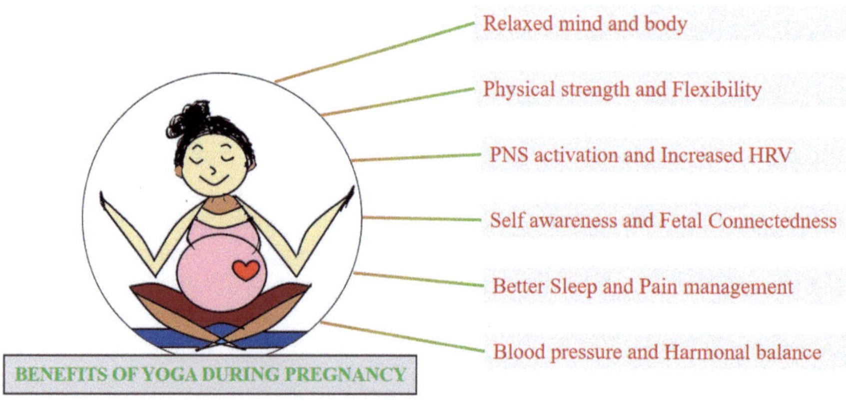

Fig. 6.2 Neurophysiological effect of *Yoga* during pregnancy

doesn't have any comorbidities. Walking, dancing, stationary cycling, stretching, and aerobic exercises are some of the common activities that can be practiced by pregnant women as per the guidelines. All these physical activities are based on physical aspect of health which also promotes mental well-being to some extent. On the other side, *Yoga* being a multidimensional practice tends to improve all aspects of human life such as physical fitness, mental health, breath regulation, through *Asana*, *Pranayama*, meditation, and mindfulness, and is definitely a holistic health approach which should be practiced during pregnancy to improve overall maternal and fetal health. Stress and anxiety management which are prominent during pregnancy are managed efficiently through *Yoga* practices (Bonura 2014). Usually, *Yoga* is being considered like all other physical activities, but it is completely different from usual exercise. A complete *Yoga* protocol starts with warm up or loosening exercise, followed by postural alignment synchronized with breath controlled, breathing regulation, relaxation, and meditation. These techniques make *Yoga* a unique practice as opposed to conventional exercises and perhaps superior if practice is performed under skilled guidance.

6.6 Benefits of *Yoga* over Other Physical Activities

The literal meaning of *Yoga* as expressed in Patanjali's aphorism is balancing mental fluctuations and joining individual consciousness with universal consciousness. It signifies that *Yoga* is beyond just physical well-being but it is more about mental and spiritual well-being. Although there are various forms of modern *Yoga* developed by different experts for modern world, the basic concept of *Yoga* which is awareness is an integral part of any form which is achieved by breath control and meditation (Govindaraj et al. 2016).

As mentioned above, breath awareness while postural alignment, body movements and holding in specific postures, relaxation, and meditation make *Yoga* practice a superior form of physical activity as compared to other exercises.

Precisely talking about the special conditions like pregnancy where along with physical strength, mental and emotional strength is required, *Yoga* can be an ideal lifestyle one can adopt. Based on evidences available, visualization during meditation is reported to improve intrauterine fetal growth and the utero-fetal placental circulation and women felt more connected to fetus during the process (Rakhshani et al. 2015; Gong et al. 2015). Besides, pregnant women also found *Yoga* as better and more effective tool as compared to usual perinatal exercise and also helped in maintaining good psychological health and help in breath regulation (Del Negro et al. 2018). Breathlessness among pregnant women increases with progression of pregnancy. *Pranayama* including deep and slow breathing might help in combating the breathlessness which is usually not included in conventional exercise. Ventral medulla contains as rhythmogenesis center a network of neurons known as pre-Bötzinger complex (PreBötC), responsible for creating breathing rhythm in mammals and is influenced by many neuromodulators and external stimuli. Rhythm generation can be disrupted by removal of this particular region (Yackle et al. 2017). Excitatory and inhibitory interneuron network in PreBötC coordinated all the phases of breathing cycle and also generates essential periodic inspiration drive. Although breathing is a simple and effortless process, it requires sophisticated set of motor neurons for its proper regulation and function (Solomon et al. 2000). Slow and regulated breathing has been used to induce mental relaxation since many years, and it is currently being utilized therapeutically to lower excessive excitement, such as panic attacks. The pre-Bötzinger complex currently appears to be important in the effects of breathing on arousal and mood resulting from physical exercises, *Yoga*, and meditation. The pre-Bötzinger complex is a neuronal sub-population at brainstem responsible for regulating the balance between aroused and calm behaviors. It controls nor-adrenergic neurons in brain area which are involved in attention, arousal, and panic (Von Rosenberg et al. 2017).

Rapid, shallow breathing is the immediate response to stress, which lowers oxygen supply to various organs, including the brain. There is a possibility that this response will disrupt the normal rhythmic breathing pattern and the PreBötC. *Pranayama* practice regulates the breath pattern in different forms for varying duration of time, therefore *Pranayama* may be effective in alleviating stress by normalizing breathing rhythm and oxygen supply to the body. PreBötC is chemosensitive to hypoxia, and neuronal excitability of this particular region get hyped during hypoxia (Jain et al. 1993). Although there is no research available which markedly describes the role of PreBötC in physiological adaptations during pregnancy, yet there is respiratory compromise (Breathlessness) during pregnancy due to hormonal and physical changes, hence breathlessness and gasping may bidirectionally affect PreBötC and ultimately normal breathing rhythm can be disturbed during pregnancy. Deep breathing practice and *Pranayama* during pregnancy may help in relaxation and also increase lung capacity which aids in the normal rhythmic breathing pattern among pregnant women. Various physiological effects of prenatal Yoga are illustrated below (Fig. 6.3).

6.7 *Yoga* as a Primer to Prevent Development of Lifestyle Morbidities Among Non-pregnant Population

Yoga practice is known to promote psychological and physical health conditions, increase self-regulation, and prevent development of lifestyle morbidities. Understanding the science of *Yoga* can contribute to the elucidation of the underlying mechanism-specific modules which can be tailored for different populations. Studies show the role of *Yoga* in mediating medical effects in controlling diseases like diabetes (Patel 1975), hypertension (Jain et al. 1991), asthma (Williams et al. 2005), low backpain (Joseph et al. 1981), etc., and involve in physiological changes by stabilizing nervous system (shift to parasympathetic nervous system) (Nayar et al. 1975), activation of alpha waves, improved respiratory functions (Benson

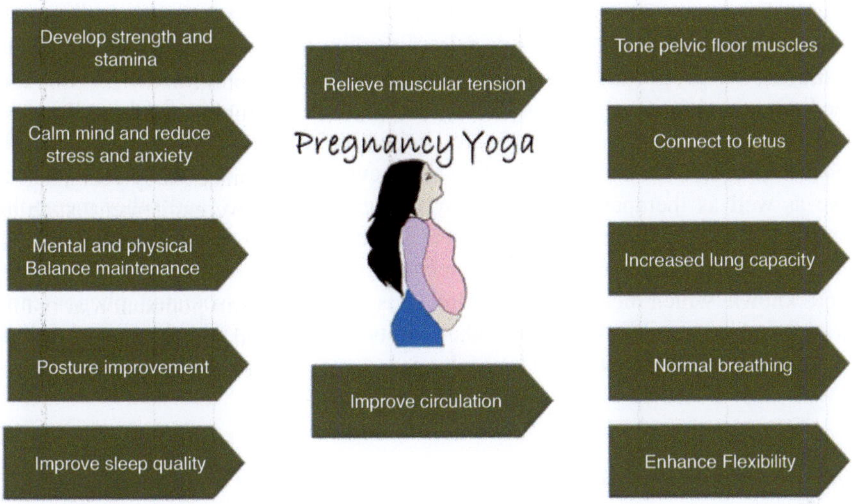

Fig. 6.3 Physiological effects of *Yoga* during pregnancy

et al. 1982), balanced thermoregulatory (Anand et al. 2017), immune, and endocrine (Anand et al. 1961) systems. Electroencephalography readings also showed persistent alpha brain waves (Barnes et al. 2001) in Yogis with increased amplitude of alpha waves during *samadhi* which shows the potential of *Yoga* in altering neurophysiology aspects (Schmidt et al. 1997).

Yoga therapy or regular *Yoga* practice can effectively prevent and reduce lifestyle diseases such as cardiovascular disease, hypertension, and high cholesterol as suggested by the evidence-based studies on *Yoga* and health promotion and disease prevention (Kabat-Zinn 1982; Kristal et al. 2005). A 10-week mindfulness meditation was effective in self-regulating the chronic pain and reducing the mood disturbance among 51 chronic pain patients who were non-responsive to medical treatment (Bijlani et al. 2005). *Yoga*-based lifestyle modification has been investigated by numerous studies which reveal that long-term (1 year) practice of *Yoga* and lifestyle modification by *Yoga* are associated with attenuated weight gain among normal weight and overweight individuals, which implies that *Yoga*-based lifestyle modification can prevent unnecessary weight gain and associated comorbidities (Gard et al. 2014). All these abovementioned studies are early evidences of science of *Yoga* and its potential to significantly alter metabolism, physiology, and physical and mental aspects of practitioners.

Not only the physical health parameters are influenced by the practice of *Yoga* but mental health, cognition, attention, alertness, and mindfulness are also affected which are being further investigated by scientists. Meditation and mindfulness practice is known to exert effects on cognition and various dimensions of the brain. MRI scanning of long-term hatha *Yoga* meditation practitioners showed volumetric difference in the brain structures and demonstrated increased gray matter volume (GMV) as compared to meditation naïve healthy individuals (Young et al. 2019). As it is well known that cognition, intelligence, alertness and memory decline with increased age, investigation of age-related decline of fluid intelligence (ability to learn new things, problem-solving strategies) by Gard T and Lazar SW reveals that meditators and *Yoga* practitioners had lower rate of decline in fluid intelligence, greater network integration, and more resilient networks as compared to controls (Speranza et al. 1998). *Yoga*, according to the evidence provided above, is a preventive as well as therapeutic tool for psychological, lifestyle, and other metabolic imbalance among different populations. Among different populations of people, effects of *Yoga* on pregnancy and associated mechanism are understudied and not well known which need to be explored as pregnancy-associated phenomenon doesn't only affect the expecting mother but its effect is also transferred to developing fetus and its impact on fetus is lifelong.

6.8 Neurophysiological Effect of *Yoga* During Pregnancy

6.8.1 Heart Rate Variability

Pregnancy significantly alters the heart rate and increases cardiac output by 12th gestational week. Effect of autonomic nervous system on cardiac output can be measured by two major peaks (high-frequency (HF) peak and low frequency (LF) peak) of power spectrum analysis of heart rate. Sympathetic nervous system (SNS) influences the LF and parasympathetic nervous system (PNS) influences the HF peaks. Increased LF power indicates SNS dominance, while increased HF power indicates PNS dominance and LF/HF ratio is index for sympathovagal balance b/w two nervous systems (Satyapriya et al. 2009).

Progressive increase in heart rate was reported in normotensive singleton pregnancies as compared to normal controls, while overall HRV and LF/HF ratio decreased significantly which shows vagal dominant shift of autonomic tone during pregnancy (Chu et al. 2017). Autonomic tone is immensely unstable and continuously influenced by emotional state, physical activities, and posture. *Yoga* influences both physical and mental aspects by *Asana* (increase flexibility and viability), *Pranayama* (improves lung capacity), meditation and relaxation (increase physiological rest) and hence improves emotional state as well as posture which might help in stabilizing autonomic tone. A 12-week prenatal *Yoga* practice among 45 healthy pregnant women decreases perceived stress by 31.6% in *Yoga* group and sympathetic tone (LF/HF ratio and LF peak) decreased and parasympathetic tone (HF peak) increased during deep relaxation technique (DRT) (Vaynman et al. 2004). Pregnant women with depressive symptoms also showed increasing HF-HRV and decreasing LF-HRV and LF/HF trend along with decrease in depression and stress symptoms. I-Hua chu et al. suggest that this decrease in HRV may be due to change in breathing pattern in *Yoga* group, therefore regular *Yoga* practice can be suggested to pregnant women with depressive symptoms to cope stress and depression and increase HRV and cardiovascular health (Hayase and Shimada 2018). Salivary alpha amylase significantly lowered and night sleep time increased in pregnant women after *Yoga* practice suggesting relaxing effect of *Yoga*. PNS dominance is helpful in preventing the GDM and hypertension like disorders during pregnancy and PNS activity becomes immediately dominant after practicing *Yoga* (Fujimura et al. 2002). Hence, *Yoga* is helpful in physiological adaptations during pregnancy through neurophysiological modifications.

6.8.2 BDNF-Brain-Derived Neurotrophic Factor

BDNF, a neurotrophic factor, plays a main role in growth, maturation, and survival of nerve cells and it is associated with cognitive improvement. BDNF is involved in the antidepressant pathway and important for neuronal plasticity and learning memory. When sedentary mice and exercised mice were compared for learning memory, it was found that cognitive function (learning and recalling activity) of exercising mice were enhanced, but when receptors for BDNF were blocked, the cognitive function of exercising mice reduced to the level of sedentary mice (Lommatzsch et al. 2006). In peripheral blood, BDNF is stored in platelets and released to repair damaged peripheral nerves and prevent pathological clot formation in any pathological condition (Christian et al. 2016).

During pregnancy, along with other physiological changes, serum BDNF level diminishes in mothers with depressive symptoms (or vice versa) which may play a role in mood disorders in prenatal and perinatal period (Gilmore et al. 2003). Stress is one of the components that lower serum BDNF levels during pregnancy, which may impede the repair of damaged peripheral nerve.

In the second trimester, high cortisol level was found to be associated with low BDNF level which was further found to be associated with the low-birth-weight infants (Gomes da Silva et al. 2016). Brain development process starts in utero, and this process is affected positively and negatively by maternal factors. Maternal nutrient, growth factor, and psychological balance destitutions in early development phase may also be transferred to fetus, impact the development process, and reflect after birth through the lifespan of offspring. Evidence suggests that maternal infection can regulate the neurotrophic factor's expression in fetal brain and hence affect the development of fetal brain and memory (Kiselev and Volik 2019). Maternal stress, depression, and infection negatively regulate the expression of neurotrophic factors in the fetal brain and may impair the normal process of fetal brain development, likewise maternal exercise positively regulates the brain development process and increases the cognition in offspring. Children (7–8 years) of mothers who did mindfulness training during pregnancy had improved visuospatial memory when compared to the age-matched children of the untrained mothers (Consales et al. 2007). Therefore, BDNF is one of the main neurotrophic factors which plays a central role in fetal brain development and its expression is influenced by maternal state which is modified by maternal exercise and *Yoga* practice during pregnancy.

Based on the above evidence it can be concluded that increased BDNF naturally in maternal circulation not only benefits the mother but also the fetus by aiding in brain developmental process in the later life.

6.8.3 GDNF- Glial Cell Line-Derived Neurotrophic Factor

GDNF is a neurotrophic factor for sensory neurons and it promotes the survival of dopaminergic and motor neurons and hence prevents the chronic neurodegeneration. GDNF stimulation modulates the transcriptional expression of genes which are responsible for neuronal plasticity and survival (Miller 2011). In a similar fashion to BDNF, serum level of GDNF also declines during major depression and bipolar depression, and it can be used as a predictor of coping ability (Sabaghi et al. 2019). Maternal aerobic exercise during pregnancy was found to increase cognition, motor coordination, and pain tolerability in offspring, which were positively corelated with the increased serum level of GDNF in offspring (Vichier-Guerre et al. 2017).

6.9 The Impact of *Yoga* on Neurotransmitters During Pregnancy

Neurotransmitters are chemical messengers that play a crucial role in regulating various aspects of brain and body function, including mood, sleep, and pain perception. Studies have shown that practicing *Yoga* during pregnancy can positively impact neurotransmitter levels and regulation in several ways, including:

6.9.1 Serotonin

A neuromodulator plays a role in cognition, autonomic regulation of emotions, neurodevelopment, and neural crest stem-cell migration and proliferation (Shah et al. 2018). It plays an important role in prenatal CNS development, morphogenesis of embryo and activation of raphe neurons (brain stem neuronal nuclei) for descending serotonergic pathways. Abnormal level of serotonin during developmental stage is associated with behavioral deficit in adults and may interfere with normal brain development (Israelyan et al. 2019). Serotonin is basically a chemical which sends signals between nerve cells and other body to regulate various body functions such as sleep, mood, digestion, sexual function, blood clotting, and bone health. Deficiency of serotonin with serotonin deprivation of neurons in CNS may cause depression and deficiency for enteric neurons affect gastrointestinal mobility which is linked to the constipation as studies in the mice (Guarino et al. 2011). Abnormal levels of progesterone and serotonin in epithelial cells of colon were associated with the slow transit and constipation in mice (Crestani et al. 1999). Studies have shown that practicing *Yoga* during pregnancy can increase levels of serotonin in the brain, leading to improved mood and reduced symptoms of depression and anxiety.

6.9.2 Gamma-Aminobutyric Acid (GABA)

GABA is a neurotransmitter that helps regulate anxiety, sleep, and muscle tone. Lower GABA levels are associated with negative, harmful, and fearful behaviour as confirmed in animal studies (Lydiard 2003). However, increased GABA level in brain is known to induce calming effects by counter-balancing the action of excitatory neurotransmitters and modulating neuronal excitability (Gilfarb and Leuner 2022). GABA levels in different parts of the brain fluctuate throughout the pregnancy due to hormonal changes, confirmed by human and animal studies (Streeter et al. 2007). One hour of *Asana* practice by long-term *Yoga* practitioner was seen to significantly increase brain GABA level by 27%, which indicates *Yoga* can significantly impact the brain GABA level and can alleviate stress and anxiety-like behavior (Pal et al. 2014). Although we couldn't find any study reporting the effect of *Yoga* on GABA levels in pregnant women, but *Yoga* might help in alleviating anxiety and mood disorders during pregnancy by increasing brain GABA levels; however, investigation is required in this field.

6.9.3 Dopamine

Dopamine is a neurotransmitter that plays a role in regulating mood, attention, and motivation. Although we could not find any evidence of *Yoga* regulating the dopamine levels in pregnant women, but there is evidence of increased plasma level of dopamine after *Yoga* practice in healthy individuals (Kjaer et al. 2002). Studies have also shown that meditation can increase dopamine levels in the brain, leading to improved mood and cognitive function (Amalia et al. 2020). These findings indicate that *Yoga* practice may impact dopamine levels during pregnancy also, although these speculations need to be confirmed.

6.9.4 Norepinephrine

Norepinephrine is a neurotransmitter that is involved in regulating heart rate, blood pressure, and stress levels. Studies indicate that practicing *Yoga* during pregnancy can reduce norepinephrine levels and help regulate the stress response (Gong et al. 2015).

These are some of the ways in which *Yoga* has been shown to impact neurotransmitter levels and regulation during pregnancy, based on published research in the field. However, more research is required to fully understand the effects of *Yoga* on neurotransmitters during pregnancy. As with any exercise regimen, it's important to consult with a healthcare provider before starting *Yoga* during pregnancy.

6.10 Clinical Implications

Yoga during pregnancy was found to be a safe, feasible, and cost-effective tool for increasing well-being, quality of life, and pregnancy outcomes. *Yoga* not only impacts the mother but also improves the fetal parameters which indicate the *Yoga* is an effective intervention which can be practiced during pregnancy. Physicians should consider prescribing *Yoga* as a complementary intervention during pregnancy.

6.11 Future Implications for Research

Research in the field of *Yoga* and pregnancy has been increasing in recent years. The future directions for this research can be grouped into several areas:

6.11.1 Mechanisms of Action

Researchers are interested in understanding the mechanisms by which *Yoga* practices can improve pregnancy outcomes. This may include exploring the effects of *Yoga* at cellular and molecular markers. Also umbilical cord blood, which is direct connection between mother and fetus, can be examined for various parameters to investigate if fetus is benefited at cellular level or not.

6.11.2 Integrating Yoga *into Clinical Practice*

Research is required to determine the best way to integrate *Yoga* into clinical practice. This may include evaluating the effectiveness of different types of *Yoga* interventions, such as group classes, individualized programs, or home-based practices.

6.11.3 Developing Yoga *Programs*

Research is required to develop *Yoga* protocols which are feasible for pregnant women. There is a need for designing different protocols for normal and high-risk pregnancy with conditions like hypertension, diabetes, short cervical length, etc. It is also necessary to validate frequency, duration, and intensity of particular Yoga practices, through a variety of trials.

6.11.4 Examining the Long-term Effects of Yoga

Long-term studies are required to determine the sustainability of the benefits of *Yoga* during prepartum and postpartum period. This will help to determine whether regular practice is necessary to maintain the benefits over time or not.

6.12 Conclusion

Yoga and mindfulness during pregnancy or otherwise are effective nonpharmacological preventive tools to balance the neurophysiological function of body. Instead of using drugs which can potentially be transferred to fetus by placenta, *Yoga* can be prescribed during pregnancy to the expecting mother under expert consultation. Neurophysiological changes exerted by *Yoga* may be helpful in synchronizing the physiological changes that occur during pregnancy and challenges that women face to adapt these challenges.

References

Amalia R, Syamsuddin S, Idris I, Limoa E, Pelupessy N, Syam A. The effectiveness of prenatal gentle *Yoga* on the recovery of anxiety level and norepinephrine level in pregnant women aged 35 years old. Int J Health Med Sci 2020 3 (1):79–85

Anand BK, Chhina GS, Singh B (1961) Some aspects of electroencephalographic studies in yogis. Electroencephalogr Clin Neurophysiol 13(3):452–456

Anand BK, Chhina GS, Singh B (2017) Some aspects of electroencephalographic studies in yogis. In: Meditation. Routledge, pp 475–479

Barnes VA, Treiber FA, Davis H (2001) Impact of transcendental meditation® on cardiovascular function at rest and during acute stress in adolescents with high normal blood pressure. J Psychosom Res 51(4):597–605

Benson H, Lehmann JW, Malhotra MS, Goldman RF, Hopkins J, Epstein MD (1982) Body temperature changes during the practice of g Tum-mo *Yoga*. Nature 295(5846):234–236

Bijlani RL, Vempati RP, Yadav RK, Ray RB, Gupta V, Sharma R, Mehta N, Mahapatra SC (2005) A brief but comprehensive lifestyle education program based on *Yoga* reduces risk factors for cardiovascular disease and diabetes mellitus. J Altern Complement Med 11(2):267–274

Bonura KB (2014) *Yoga* mind while expecting: the psychological benefits of prenatal *Yoga* practice. Int J Childbirth Educ 29(4):49–54

Caldji C, Tannenbaum B, Sharma S, Francis D, Plotsky PM, Meaney MJ (1998) Maternal care during infancy regulates the development of neural systems mediating the expression of fearfulness in the rat. Proc Natl Acad Sci U S A 95(9):5335–5340

Chandra S, Jaiswal AK, Singh R, Jha D, Mittal AP (2017) Mental stress: neurophysiology and its regulation by Sudarshan Kriya *Yoga*. Int J Yoga 10(2):67

Christian LM, Mitchell AM, Gillespie SL, Palettas M (2016) Serum brain-derived neurotrophic factor (BDNF) across pregnancy and postpartum: associations with race, depressive symptoms, and low birth weight. Psychoneuroendocrinology 74:69–76

Chu IH, Wu WL, Lin IM, Chang YK, Lin YJ, Yang PC (2017) Effects of *Yoga* on heart rate variability and depressive symptoms in women: a randomized controlled trial. J Altern Complement Med 23(4):310–316

Consales C, Volpicelli F, Greco D, Leone L, Colucci-D'Amato L, Perrone-Capano C, di Porzio U (2007) GDNF signaling in embryonic midbrain neurons in vitro. Brain Res 1159:28–39

Crestani F, Lorez M, Baer K, Essrich C, Benke D, Laurent JP, Belzung C, Fritschy JM, Lüscher B, Mohler H (1999) Decreased GABAA-receptor clustering results in enhanced anxiety and a bias for threat cues. Nat Neurosci 2(9):833–839

Del Negro CA, Funk GD, Feldman JL (2018) Breathing matters. Nat Rev Neurosci 19(6):351–367

Duthie L, Reynolds RM (2013) Changes in the maternal hypothalamic-pituitary-adrenal axis in pregnancy and postpartum: influences on maternal and fetal outcomes. Neuroendocrinology 98(2):106–115

Fujimura H, Altar CA, Chen R, Nakamura T, Nakahashi T, Kambayashi JI, Sun B, Tandon NN (2002) Brain-derived neurotrophic factor is stored in human platelets and released by agonist stimulation. Thromb Haemost 87(4):728–734

Gard T, Taquet M, Dixit R, Hölzel BK, de Montjoye YA, Brach N, Salat DH, Dickerson BC, Gray JR, Lazar SW (2014) Fluid intelligence and brain functional organization in aging *Yoga* and meditation practitioners. Front Aging Neurosci 6:76

Gilfarb RA, Leuner B (2022) GABA system modifications during periods of hormonal flux across the female lifespan. Front Behav Neurosci 16:802530

Gilmore JH, Jarskog LF, Vadlamudi S (2003) Maternal infection regulates BDNF and NGF expression in fetal and neonatal brain and maternal–fetal unit of the rat. J Neuroimmunol 138(1–2):49–55

Gomes da Silva S, De Almeida AA, Fernandes J, Lopim GM, Cabral FR, Scerni DA, de Oliveira-Pinto AV, Lent R, Arida RM (2016) Maternal exercise during pregnancy increases BDNF levels and cell numbers in the hippocampal formation but not in the cerebral cortex of adult rat offspring. PLoS One 11(1):e0147200

Gong H, Ni C, Shen X, Wu T, Jiang C (2015) *Yoga* for prenatal depression: a systematic review and meta-analysis. BMC Psychiatry 15:1–8

Govindaraj R, Karmani S, Varambally S, Gangadhar BN (2016) *Yoga* and physical exercise—a review and comparison. Int Rev Psychiatry 28(3):242–253

Grattan DR, Ladyman SR (2020) Neurophysiological and cognitive changes in pregnancy. Handb Clin Neurol 171:25–55

Guarino M, Cheng L, Cicala M, Ripetti V, Biancani P, Behar J (2011) Progesterone receptors and serotonin levels in colon epithelial cells from females with slow transit constipation. Neurogastroenterol Motil 23(6):575–e210

Hayase M, Shimada M (2018) Effects of maternity *Yoga* on the autonomic nervous system during pregnancy. J Obstet Gynaecol Res 44(10):1887–1895

Hux VJ, Catov JM, Roberts JM (2014) Allostatic load in women with a history of low birth weight infants: the national health and nutrition examination survey. J Womens Health (Larchmt) 23(12):1039–1045

Hux VJ, Roberts JM, Okun ML (2017) Allostatic load in early pregnancy is associated with poor sleep quality. Sleep Med 33:85–90

Israelyan N, Del Colle A, Li Z, Park Y, Xing A, Jacobsen JP, Luna RA, Jensen DD, Madra M, Saurman V, Rahim R (2019) Effects of serotonin and slow-release 5-hydroxytryptophan on gastrointestinal motility in a mouse model of depression. Gastroenterology 157(2):507–521

Jain SC, Rai L, Valecha A, Jha UK, Bhatnagar SO, Ram K (1991) Effect of *Yoga* training on exercise tolerance in adolescents with childhood asthma. J Asthma 28(6):437–442

Jain SC, Uppal A, Bhatnagar SO, Talukdar B (1993) A study of response pattern of non-insulin dependent diabetics to *Yoga* therapy. Diabetes Res Clin Pract 19(1):69–74

Joseph S, Sridharan K, Patil SK, Kumaria ML, Selvamurthy W, Joseph NT, Nayar HS (1981) Study of some physiological and biochemical parameters in subjects undergoing yogic training. Indian J Med Res 74:120–124

Kabat-Zinn J (1982) An outpatient program in behavioral medicine for chronic pain patients based on the practice of mindfulness meditation: theoretical considerations and preliminary results. Gen Hosp Psychiatry 4(1):33–47

Kiselev S, Volik I (2019) Mindfulness training during pregnancy has positive effect on development of visuospatial abilities in children. Abstracts/Psychoneuroendocrinology 100S. 1:S64

Kjaer TW, Bertelsen C, Piccini P, Brooks D, Alving J, Lou HC (2002) Increased dopamine tone during meditation-induced change of consciousness. Cogn Brain Res 13(2):255–259

Kristal AR, Littman AJ, Benitez D, White E (2005) *Yoga* practice is associated with attenuated weight gain in healthy, middle-aged men and women. Altern Ther Health Med 11(4):28–33

Kusaka M, Matsuzaki M, Shiraishi M, Haruna M (2016) Immediate stress reduction effects of *Yoga* during pregnancy: one group pre–post test. Women Birth 29(5):e82–e88

Lee LK (2007) Physiological adaptations of pregnancy affecting the nervous system. Semin Neurol 27(5):405–410. © Thieme Medical Publishers

Lommatzsch M, Hornych K, Zingler C, Schuff-Werner P, Höppner J, Virchow JC (2006) Maternal serum concentrations of BDNF and depression in the perinatal period. Psychoneuroendocrinology 31(3):388–394

Lydiard RB (2003) The role of GABA in anxiety disorders. J Clin Psychiatry 64:21–27

Mastorakos G, Ilias I (2003) Maternal and fetal hypothalamic-pituitary-adrenal axes during pregnancy and postpartum. Ann N Y Acad Sci 997(1):136–149

McEwen BS (1998) Protective and damaging effects of stress mediators. N Engl J Med 338(3):171–179

McEwen BSJM (2003) Interacting mediators of allostasis and allostatic load: towards an understanding of resilience in aging. Metabolism 52:10–16

Miller CA (2011) Stressed and depressed? Check your GDNF for epigenetic repression. Neuron 69(2):188–190

Nayar HS, Mathur RM, Kumar RS (1975) Effects of yogic exercises on human physical efficiency. Indian J Med Res 63(10):1369–1376

Pal R, Singh SN, Chatterjee A, Saha M (2014) Age-related changes in cardiovascular system, autonomic functions, and levels of BDNF of healthy active males: role of yogic practice. Age 36:1–7

Patel C (1975) 12-month follow-up of *Yoga* and bio-feedback in the management of hypertension. Lancet 305(7898):62–64

Rakhshani A, Nagarathna R, Mhaskar R, Mhaskar A, Thomas A, Gunasheela S (2012) The effects of *Yoga* in prevention of pregnancy complications in high-risk pregnancies: a randomized controlled trial. Prev Med 55(4):333–340

Rakhshani A, Nagarathna R, Mhaskar R, Mhaskar A, Thomas A, Gunasheela S (2015) Effects of *Yoga* on utero-fetal-placental circulation in high-risk pregnancy: a randomized controlled trial. Adv Prev Med 2015:373041

Sabaghi A, Heirani A, Kiani A, Yosofvand N, Sabaghi S (2019) Effects of aerobic exercise during pregnancy on neurobehavioral performances and serum levels of GDNF in adult male mice offspring. Neurophysiology 51(6):430–437

Sarrel PM (1999) The differential effects of oestrogens and progestins on vascular tone. Hum Reprod Update 5(3):205–209

Sasaki A, Shinkawa O, Yoshinaga K (1989) Placental corticotropin-releasing hormone may be a stimulator of maternal pituitary adrenocorticotropic hormone secretion in humans. J Clin Invest 84(6):1997–2001

Satyapriya M, Nagendra HR, Nagarathna R, Padmalatha V (2009) Effect of integrated *Yoga* on stress and heart rate variability in pregnant women. Int J Gynecol Obstet 104(3):218–222

Schmidt T, Wijga A, Von Zur Mühlen A, Brabant G, Wagner TO (1997) Changes in cardiovascular risk factors and hormones during a comprehensive residential three month kriya *Yoga* training and vegetarian nutrition. Acta Physiol Scand Suppl 640:158–162

Shah R, Courtiol E, Castellanos FX, Teixeira CM (2018) Abnormal serotonin levels during perinatal development lead to behavioral deficits in adulthood. Front Behav Neurosci 12:114

Solomon IC, Edelman NH, Neubauer JA (2000) Pre-Botzinger complex functions as a central hypoxia chemosensor for respiration in vivo. J Neurophysiol 83(5):2854–2868

Soma-Pillay P, Nelson-Piercy C, Tolppanen H, Mebazaa A (2016) Physiological changes in pregnancy: review articles. Cardiovasc J Afr 27(2):89–94

Speranza G, Verlato G, Albiero A (1998) Autonomic changes during pregnancy: assessment by spectral heart rate variability analysis. J Electrocardiol 31(2):101–109

Streeter CC, Jensen JE, Perlmutter RM, Cabral HJ, Tian H, Terhune DB, Ciraulo DA, Renshaw PF (2007) Yoga asana sessions increase brain GABA levels: a pilot study. J Altern Complement Med 13(4):419–426

Talbot L, Maclennan K (2016) Physiology of pregnancy. Anaesth Intensive Care Med 17(7):341–345

Thau L, Gandhi J, Sharma S (2021) Physiology, cortisol. StatPearls (Internet)

Tinga AM, de Back TT, Louwerse MM (2020) Non-invasive neurophysiology in learning and training: mechanisms and a SWOT analysis. Front Neurosci 14:589

Vaynman S, Ying Z, Gomez-Pinilla F (2004) Hippocampal BDNF mediates the efficacy of exercise on synaptic plasticity and cognition. Eur J Neurosci 20(10):2580–2590

Vichier-Guerre C, Parker M, Pomerantz Y, Finnell RH, Cabrera RM (2017) Impact of selective serotonin reuptake inhibitors on neural crest stem cell formation. Toxicol Lett 281:20–25

Von Rosenberg W, Chanwimalueang T, Adjei T, Jaffer U, Goverdovsky V, Mandic DP (2017) Resolving ambiguities in the LF/HF ratio: LF-HF scatter plots for the categorization of mental and physical stress from HRV. Front Physiol 8:360

Williams KA, Petronis J, Smith D, Goodrich D, Wu J, Ravi N, Doyle EJ Jr, Juckett RG, Kolar MM, Gross R, Steinberg L (2005) Effect of Iyengar *Yoga* therapy for chronic low back pain. Pain 115(1–2):107–117

Yackle K, Schwarz LA, Kam K et al (2017) Breathing control center neurons that promote arousal in mice. Science 355(6332):1411–1415

Young AI, Benonisdottir S, Przeworski M, Kong A (2019) Deconstructing the sources of genotype-phenotype associations in humans. Science 365(6460):1396–1400

Zhang Y, Zhou T, Feng S, Wang W, Liu H, Wang P, Sha Z, Yu X (2021) The chronic effect of cortisol on orchestrating cerebral blood flow and brain functional connectivity: evidence from Cushing's disease. Metabolism 115:154432

Chapter 7
Psychosocial Implications of *Yoga* for Healthcare Workers During Covid-19

Rudra B. Bhandari, Neha P. Sangodkar, and Acharya Balkrishna

Contents

Abstract The Covid-19 pandemic has adversely impacted healthcare workers' physical, psychological, and social well-being and exposed them to a diverse set of psychosocial problems (PSPs)—depression, anxiety, post-traumatic stress disorder (PTSD), burnout, work-life conflict (WLC), social isolation, stigma, and poor quality of life (QoL). This review aims to overview the post-Covid PSPs of healthcare workers (HCWs), mind-body practices—poses, conscious breath regulations, seals, locks, mindfulness, and meditations and the scientific rationale behind their effectiveness for prevention and moderation of PSPs. Finally, an integral *Yoga* protocol (IYP) for prevention/moderation of PSPs of HCWs is derived as a clinical implication of the study. However, the proposed IYP for managing PSPs needs validation via further experimental studies.

Keywords Depression · Anxiety · Post-traumatic stress disorder · Burnout · *Yoga* · Mindfulness

---

R. B. Bhandari (✉) N. P. Sangodkar · A. Balkrishna
Department of Yoga Science, University of Patanjali, Patanjali Yogpeeth, Haridwar, Uttarakhand, India
e-mail: acharyaji@divyayoga.com

A. Anand (ed.), *Neuroscience of Yoga*,
https://doi.org/10.1007/978-981-97-2851-0_7

7.1 Introduction

The healthcare professionals (HCPs) engaged in the nursing, diagnosis, and treatment of the Covid-19 patients have met numerous psychosocial health issues (Schneider et al. 2021). Medical healthcare workers (HCWs) during the Covid-19 pandemic had high prevalence rates of severe insomnia, anxiety, depression, somatization, and obsessive-compulsive symptoms (Santarone et al. 2020). Among the significant psychosocial problems (PSPs), the prevalence of depression, anxiety, post-traumatic stress disorder (PTSD), and burnout has been reported as 24.3%, 25.8%, 45%, and 18%, respectively, along with impaired quality of life (QoL), work-life balance (WLB), social company, and social perception. Studies regarding the 2003 SARS outbreak in Hongkong, Taiwan, and Canada have discussed how the battle against SARS led to enormous psychological morbidity amongst frontline HCWs (Chong et al. 2004; McAlonan et al. 2007; Sim and Chua 2004). During the 2003 SARS outbreak in Taiwan, nurses working in the severe acute respiratory syndrome (SARS) unit and directly exposed to infected patients' care suffered more depressive symptoms and insomnia (Su et al. 2007). World health organization (WHO)'s survey reported that more than 93% of SARS-Cov-2 infections worldwide demand psychosocial mental health services (Correa-Palacio et al. 2020). Increased workload and working hours, inadequate personal protective equipment (PPE), and over-exposure to pandemic findings in the media are possible reasons for the adverse psychosocial outcomes seen in HCWs (Schneider et al. 2021). The results of the pandemic's consequences on HCWs show how the Covid-19 pandemic offers a significant threat to their physical, psychological, and social well-being in various ways. HCWs frequently had complex and conflicting thoughts, such as feeling professional responsibility along with the fear of this new Covid-19 disease and guilt about potentially exposing their families to infection during the Covid-19 emergency leading to the high prevalence of stress, anxiety, depression, post-traumatic stress disorder, and insomnia among them (Pappa et al. 2020). The significant attempts to prevent and manage the psychosocial issues faced by HCWs include conventional medications, social support, counselling, cognitive-behavioural therapy (CBT), paramedical approaches, and other diverse healing modalities from complementary and alternative medicine (CAM). The CAM modalities used to prevent and moderate HCWs' psychosocial issues include mind-body practices, energy healing, body manipulative therapies, herbal formulations, and dietary and nutritional supplements. The majority of the research made to assess the impact of mind-body practices—mindfulness, transcendental meditation (TM), *Vipassana*, *Zen*, *Isha*, *Sudarshan* Kriya Yoga (SKY), *Sahaj Yoga*, *Soham* Meditation, *Yoga Nidra*, Iyengar, *Kundalini Yoga*, on the psychosocial issues have concluded their significant effects in the prevention and moderation of the psychosocial problems (PSPs) (Antonova et al. 2021; Shaku et al. 2014; Rain et al. 2021; Elder et al. 2014; Michalsen et al. 2005; Sesnich et al. 2017; Bhandari 2022). However, most of the interventional studies conducted to assess the effect of mind-body practices on PSPs of the HCWs cover selected *Yoga* technique/s rather than the integral protocol of the mind-body practices addressing the social, mental, physical, and spiritual dimensions of human existence. Indeed, the modern coinage of the mind-body intervention is a dimension of CAM and includes traditional *Yoga* wisdom and diverse practices. It thereby warrants the need to develop an integral and scientifically sound mind-body protocol to address the psychosocial issues of the Covid HCWs. With this background, this study aims at discussing the prevalence, aetiology, concepts, and possible yogic management of the PSPs of HCWs. Examining the factors influencing the psychological well-being of HCWs can aid in developing measures to prevent and mitigate psychosocial problems and improve psychosocial environments.

7.2 Psychosocial Problems

7.2.1 Depression

The Diagnostic and Statistical Manual of Mental Disorders, Fifth Edition, has defined depression as five or more of the following symptoms present for two or more weeks that cause significant emotional distress and impairment in functioning (American Psychiatric Association 2013). Symptoms are depressed or sad mood, short-temperament, disinterest or enjoyment in hobbies or activities that were previously enjoyed, feeling of worthlessness or guilt, thoughts of death or suicide, poor concentration and decisive power, tiredness or fatigue, restlessness, over/under eating, weight loss/gain, and impaired sleep pattern (Tolentino and Schmidt 2018). In a recent study, depressive psychopathology was linked to elevated baseline scores on immune activation and inflammation after 3-month hospitalization for Covid-19 pneumonia (Mazza et al. 2021). The biomarkers of post-Covid-19 depression include increased levels of interleukin 6 (IL-6), soluble interleukin six receptor (SIL-6R), interleukin one β (IL-1β), tumour necrosis factor α (TNF-α), interferon-gamma (IFN-γ), interleukin 10 (IL-10), interleukin 2 (IL-2), soluble interleukin 2 (SIL-2R), C-reactive protein (CRP), monocyte chemoattractant protein-1 (MCP-1), serum amyloid a (SAA1), and metabolites of the kynurenine pathway and decreased brain-derived neurotrophic factor (BDNF) and tryptophan (TRP) (Lorkiewicz and Waszkiewicz 2021).

7.2.2 Anxiety

A large body of literature argues that those experiencing pandemic-related psychological distress have elevated levels of anxiety which can last well beyond the pandemic (Wheaton et al. 2012). Pandemic anxiety was coined to describe the distress people are experiencing due to adjusting to the new normal (Cheng, Wang, &

Ebrahimi, 2021). Covid-19-induced anxiety symptoms are feeling nervous, restlessness, or tense, sensing impending danger, panic, increased heart rate, hyperventilation, sweating, trembling, feeling weak or tired, poor concentration and deviation from present awareness, worry, sleeping difficulty, disinterest, fearfulness, and experiencing upset gut. The biomarkers of post-Covid-19 anxiety disorders include increased levels of D-dimer, interleukin-6 (IL-6), CRP, and lymphocyte count (Spoorthy et al. 2020). Regular somatic practices reduce IL-6, IL-1, cortisol, TNF-α, and CRP and increase NK cells, T cells, B cells, and CD4 (da Silveira et al. 2021).

7.2.3 Post-traumatic Stress Disorder (PTSD)

Previous studies have evidenced increased PTSD after epidemic or natural calamities—SARS, earthquake, or tornado, including Covid-19. PTSD is a condition that develops after experiencing psychological trauma when these responses persist for over a month and cause substantial distress and disruption in one's life. Furthermore, the risk increases in the following weeks as these individuals lack immediate social support due to the need to self-quarantine (Joly et al. 2021; Restauri and Sheridan 2020). Covid-19-induced PTSD symptoms are experiencing recurrent and intrusive memories of the trauma, flashbacks or nightmares of the trauma, disinterest in previously enjoyable activities, feelings of guilt, anger, or shame, and sleeping difficulty (d'Ettorre et al. 2021). PTSD is also associated with increased activity in brain networks involved in threat detection and negative emotional responses; reduced expression in networks involved in executive control, problem-solving, emotional regulation, and reward-related brain circuits (Hayes et al. 2012). Different biomarkers of PTSD include increased corticotropin-releasing factor (CFR) and adrenocorticotropic hormone (ACTH), IL-6, TNF-α, dysfunctional HPA axis, glucocorticoid receptor (GCR/GR) sensitivity, and allele of apolipoprotein E gene (ApoE4); decreased gamma-aminobutyric acid (GABA), dopamine transporter gene (DAT), trimethylated histone H3 lysine four amino acid (H3K27), glial fibrillary acidic protein (GFAP), NK cells, and T-helper cells (Th), and serotonergic dysfunction (Baker et al. 2012).

7.2.4 Work-Life Conflict

Work-life balance is maintaining and creating friendly and stress-free working environments, balancing paid work and personal engagements, and encouraging employee loyalty and productivity (Devadhasan 2018). A working individual's day includes work, family, and personal time. When any of these activities is prioritized over others, in terms of time and attention, with little attention to the others, work-life conflict (WLC) arises (Gragnano et al. 2020). HCWs experienced increased workload demand for medical professionals, lack of life saving equipment, higher demand than their supply, and reduced intake of medical professionals by hospitals due to their need for higher remunerations and welfare packages and services to many patients (Giorgi et al. 2019). Another study by Wuhan researchers in China surveyed 103 health workers infected with Covid-19 and reported that most (84.5%) got infected due to their work environment (Shreffler et al. 2020). The work environment is a crucial factor influencing the psychological health of healthcare personnel. Providing sufficient personal protection equipment, instituting prevention policies, and a balanced workload may determine changes in work pressure among healthcare personnel (Shreffler et al. 2020).

7.2.5 Burnout

Burnout is a psychological syndrome of cynicism, exhaustion, and inefficacy experienced in response to chronic job stressors (Maslach and Leiter 2007). When HCWs experience stress levels beyond their capacities, it causes mental fatigue due to a disturbed neuro-endocrinal system, which we call burnout (Chen et al. 2021). Burnout, in general, is the result of prolonged interpersonal stressors at work. HCW develops symptoms such as irritability, mood swings, anxiety, insomnia, and a sense of failure as symptoms of burnout due to increased work volume and intensity, limited resources, longer shifts, disturbed sleep, and occupational risks associated with exposure to Covid-19, which upsurged burnout in them (Tian et al. 2022). Psychological distress induced due to Covid-19-related deaths, intense despair, and extra-long working hours contribute towards HCWs' burnout exponentially. The long-term hyperactivity of the HPA axis during burnout may change to hypoactivity after long-term exposure to stressful situations. Different biomarkers of burnout related to HPA, immune system, autonomic nervous system (ANS), metabolic processes, antioxidant defence, and hormones include increased ACTH, cortisol, CRP, cholesterol, prolactin, and fibrinogen (Deneva and Ianakiev 2019). These are reflected in peripheral blood by releasing catecholamines via the ANS and cortisol via the HPA axis, increasing blood pressure and heart rate. The immune system gets temporarily suppressed along with hyper catabolism. People with burnout report increased incidences of flu-like illness, back pain, physical fatigue, irritability, and gastrointestinal problems. A survey of 7500 physicians from eight countries during the Covid-19 pandemic reveals that the burnout has an exponentially high rate (Kane 2020). In addition to the stress caused by uncertainty and fear of the unknown, numerous studies have demonstrated that organizational support for healthcare personnel was negatively correlated with their work satisfaction, emotional response, and conflict (Siu 2002; Niu et al. 2019; Marjanovic et al. 2007). Possible factors causing a high risk of job burnout include high mortality rate, intensive patient care, and inappropriate job conditions in terms of high workload coupled with a lack of time to address the patient's needs adequately. This posed a significant challenge to healthcare professionals juggling duty and safety by creating burnout among them (Tian et al. 2022). Stress and burnout impact the physical and mental health of the individual and compromise their quality of care, satisfaction, and affirmation (Cocchiara et al. 2019).

7.2.6 Social Isolation

Social isolation can be defined structurally as the absence of social interactions, contacts, and relationships with family and friends, with neighbours on an individual level, and with society at large on a broader level (Coplan et al. 2021). The upsurge in Covid-19 cases has heavily burdened, impaired, and overwhelmed the healthcare systems (Yao et al. 2020). To control the rapid human-to-human transmission of SARS-CoV-2, there had been lockdown enforcement, isolation, and closure of public places. Since HCWs were in the vicinity of infected patients and experienced social isolation and stigma, 88.3% of the HCWs experienced stress or emotional tension during isolation. Confinement and lack of social and physical contact with friends and family frequently caused boredom, frustration, and isolation from the rest of the world, distressing many HCWs. The study showed that people experiencing distress were apprehensive about their health and transmitting it to their loved ones. The impact of the mitigation measures, such as quarantine, self-isolation, and physical distancing, has contributed to a global mental health pandemic, including depression, anxiety, panic attacks, and PTSD (Vindegaard and Benros 2020; Fiorillo and Gorwood 2020). HCWs responsible for the care of Covid-19 patients are more likely to get secluded from the rest of the community due to the fear of infection spreading. It is also critically important for them to maintain active social connections. Disseminating accurate information may help counteract Covid-19, social isolation and loneliness in people and root out the social discrimination faced by frontline healthcare providers. This will protect their mental well-being and effectively control this public health crisis. A study was conducted to reduce social isolation and loneliness compatible with Covid-19 shielding and social distancing measures (Christopher et al. 2021).

7.2.7 Stigma

Goffman first coined the term stigma in 1963, which is now redefined as a social process that sets to exclude those who are perceived to be a potential source of disease and may pose a threat to adequate social living in a society in the ongoing context of the Covid-19 pandemic (Bhanot et al. 2020). HCWs working in a higher risk of infected environment had a greater fear of passing SARS-Cov-2 to their family, friends, colleagues, and others and hence were often stigmatized and rejected by others. At the same time, misinformation overload, rumours, continuous rise in infected patients, and negligence by patients to the Covid-19 guidelines further provoked the turmoil (Pandey et al. 2021). HCWs were utter victims of social disapproval during the Covid-19 pandemic. Numerous frontline healthcare providers working in hospitals and laboratories face discrimination from hotel staff and struggle to find food and shelter. Neighbours and people in the community have shown displeasure in accommodating frontline healthcare providers in their homes despite HCWs working with all necessary precautions. HCWs felt greater stigmatization than the general public. They reported more anger, frustration, guilt, annoyance, fear, helplessness, isolation, loneliness, nervousness, sadness, and worry.

Additionally, anyone who provides healthcare facilities in a setting with many Covid-19 cases is untouchable. Unfortunately, the healthcare providers were labelled, set apart, and faced the loss of status and discrimination because of the stigma attached to Covid-19 (Tandon 2020). To thwart the stigma attached to Covid-19, it is imperative to disseminate information related to Covid-19. Moreover, accurate data plays a crucial role in enabling the environment to make appropriate decision-making towards fighting this public and social crisis with perseverance.

7.2.8 *Quality of Life*

In a problematic setting, HCWs face many stressful events that can compromise QoL (Kisely et al. 2020). The high demand to maximize work performance and productivity ignored HCWs' need to release emotional and physical tension, causing their health compromise (Buselli et al. 2020). Compared with the public, HCWs are highly susceptible to negative psychological responses exacerbated by the risk of contact with infected patients. This infection has an unusual tendency to arouse fear and subsequent ineffective psychological and social response adaptation, threatening the quality of life (QoL). Experiencing continuous stress can adversely impact HCWs' QoL. Several reviews have identified various protective and preventive risk factors associated with social distress in HCWs. The most commonly reported factors are lack of clear communication, social support, practical support (e.g. providing appropriate work attire and access to adequate PPE), and sufficient rest (Arora and Grey 2020). Covid-19 rapidly became a global health emergency, causing physical and psychological concerns as people face unexpected deaths or death threats. HCWs having close contact with Covid-19 patients were regularly exposed to the virus, witnessed increased illnesses, frustration, insomnia, WLC, aggression, poor psychosocial environment, impaired interpersonal relationship, poor compassion and resilience, fear of infection and death, odd and unexpected work hours and supply shortages impacting QoL adversely. Practical approaches for improving WLC include CBT, interpersonal psychotherapy, dialectic behavioural therapy, support therapy, and self-help manuals. Moreover, there are an increasing number of preventive strategies that address individual, family, and social risk factors to develop a positive self-image.

7.3 Clinical Implications: *Yoga* for Psychosocial Problems

7.3.1 Yoga *for Depression*

Mindfulness is the process of witnessing present-moment sensations, thoughts, emotions, and experiences in a non-judgmental manner (Maynard et al. 2017). It has been reported as beneficial for managing depression both in clinical (Sephton et al. 2007) and non-clinical samples (Brown and Ryan 2003). Consistent *Yoga* alleviates depression, increases serotonin, and decreases monoamine oxidase, an enzyme that degrades neurotransmitters and cortisol. High-frequency *Yoga* breathing (*Kapalbhati*) increases the low-frequency power of HRV by implying an increased sympathetic tone (Wingert and Kantrowitz 2002). Substantial studies demonstrated the potential benefits of *Yoga* interventions on depression symptoms (Duan-Porter et al. 2015). Increased GABA production during meditation and *Yoga Asana* sessions is associated with the high pre-frontal cortex (PFC) (Elias et al. 2000). Transcendental meditation (TM) practice significantly ameliorates stress-induced anxiety disorders by increasing GABA-genic tone (Nidich et al. 2009). The SKY method reduced depression, stress, and anxiety among healthcare workers engaged in diagnosing, treating, and nursing care of Covid-19 patients (Divya et al. 2021).

7.3.2 Yoga *for Anxiety*

Neurotransmitters play an essential role in modulating and regulating behaviour and anxiety. Several studies have found a link between GABA activity and stress (Krishnakumar et al. 2015). In patients with anxiety disorders, GABA activity is reduced in the hippocampus, lingual gyrus, middle temporal gyrus, and visual cortex (Cahn and Polich 2006). *Yoga* causes inhibition of the hypothalamus's posterior or sympathetic area, optimizes the sympathetic responses to stressful stimuli, and restores autonomic reflex mechanisms with stress (Kiecolt-Glaser et al. 2010). Yogic breath regulations improve neurocognitive, metabolic, respiratory, and autonomic functions by inducing a significant parasympathetic tone (Telles and Gupta 2020). Selected yogic practices such as conscious breath regulations—abdominal breathing, *Pranayama–Paranakarshan*, *Ujjyai*, and *Bhastrika*; neuromuscular locks—*Moola Bandha* (sustained perineal contraction), *ShaktichalaniMudra* (effortful and sustained external anal sphincter contraction); and meditations—*Omkar*, TM, and solar meditation with *Gayatri Mantra* recitation inhibit the areas responsible for rage, fear, and aggression while stimulating the rewarding pleasure centres in the median forebrain and other places, resulting in pleasure and bliss. This inhibitory activity results in low anxiety, heart rate (HR), breath rate (BR), blood pressure (BP), and cardiac output (CO) in *Yoga* and meditation practitioners (McCall 2007).

The practice of both components of mindfulness involving an orientation towards the present moment characterized by curiosity, openness, acceptance, and self-regulation of attention was noted as beneficial to moderate anxiety (Lee et al. 2017). Conversely, excessive orientation towards the past or future when dealing with stressors has been linked to feelings of depression and anxiety.

7.3.3 **Yoga** *for Post-traumatic Stress Disorder*

There is evolving evidence supporting the biological and neural mechanisms underlying mind-body practices for managing stress-related illnesses (Brown and Gerbarg 2009). Long-term meditators show positive changes in parts of their brain that deal with stress and anxiety (Afonso et al. 2020). Regular physical activity/yogic postures can excite the cerebral cortex and increase the supply of oxygen and nutrients to preserve or restore normal nervous system function, thereby increasing GABA levels (Chen et al. 2020). Meditators displayed higher alpha activity than their controls, associated with a more relaxed and alert state of mind, showing its usefulness to moderate clinically significant anxiety (Krishnakumar et al. 2015). The increased alpha activity enhanced the number of oscillations in the alpha band, decreasing signs and symptoms of stress (Ulett et al. 1953). Another research study showed higher theta wave activity in mildly anxious persons with long-term meditation practice than non-meditators (Inanaga 1998). A pilot study of Mindfulness-Based Exposure Therapy (MBET) brought positive changes in the PTSD symptoms by increasing amygdala and fusiform gyrus responses (King et al. 2016). Substantial research on *Yoga* and PTSD has shown that mind-body interventions positively impact health outcomes, QoL, and PTSD by normalizing the imbalanced autonomic activity and stimulating parasympathetic tone (Bormann et al. 2012).

7.3.4 **Yoga** *for Burnout*

Yoga improves physical and mental health, promote self-care, and prevent burnout among HCWs. Eight weeks of *Yoga* intervention helped manage stress and enhance emotional resilience, empathy, perceived stress, self-regulation, and self-compassion in medical students (Bond et al. 2013). Medical nurses participating in a mindfulness training programme reported less burnout and anxiety than their controls after a 10-week intervention (Horner et al. 2014). *Yoga* and meditation helped reduce stress in employees complaining of burnout (de Bruin et al. 2017). Intensive care nurses have busy and stressful jobs, putting them at risk of burnout (Antonova et al. 2015). A study has shown that *Yoga* and meditation can effectively strengthen well-being in intensive care nurses (Jarden et al. 2019).

7.3.5 Yoga *for WLC*

Yoga teaches us to live rhythmic and unconditional dutiful life by perceiving entire universal entities as a part of an undivided whole. The spirit behind *Karma Yoga* makes one aware of their diverse roles in family, workplace, community, religious places, and other life settings. Moreover, regular practice of *Yoga* techniques addressing bioenergy, cognitive, emotive, and behavioural domains of human existence enhances and channelizes bioenergy, deepens and harmonizes breath currents, modulates cardiopulmonary pace and synchronicity, and induces heightened coping with adversities, mental steadiness, emotional stability, and paranormal perceptions (Sullivan et al. 2018). The practice of SKY, TM, mindfulness, *Yoga Nidra*, *Soham*, and *Savita Dhyan* induces alpha brain waves, deepens synchronicity between cardio and pulmonary pace, enhances HRV and secretions of GABA, serotonin, endorphins, and anandamide resulting in empowered immune functions, mental peace, and intellectual geniuses. Enhanced bioenergy, mental steadiness, emotional stability, and elevated coping skills reduce stress and ease timely decision-making and roles' discharging to better WLB (Bhandari 2017). Bharathi and Mala have reported *Yoga* and meditation as personal enhancers that help women in information technology (IT) achieve WLB (Bharathi and Mala 2016). Additionally, WLB impairment due to Covid-19 induced physical or psychosocial illnesses—pneumonia, hypertension, depression, anxiety, PTSD, stigma, and WLC can be reduced by implementing necessary online *Yoga* intervention among HCWs (Sharma et al. 2020).

7.3.6 Yoga *for Social Isolation*

As mentioned in Hatha Yogic Texts, the six unsupportive behaviours for the initiation of *Yoga* practice cover overeating, overexertion, overindulgence, and inflexibility from set norms and social relationships. Essentially, *Yoga* is an inner journey to be made secretly and constantly in social isolation. The regular practice of integral *Yoga* induces introspection, inner happiness, and self-satisfaction by limiting social relations. Traditional *Yoga* practitioners across the globe did not face mental suffocation and emotional tension during their quarantined stay. Still, they enjoyed this timing to advance their mind-body practices. They happily accepted the imposed social isolation as a preventive behaviour to minimize the spread of SARS-Cov-2 infection. Regular practice of poses, breath regulation, concentration, meditation, and devotional chants induce secretion of painkilling and healing neuro-hormones—GABA, endorphins, encephalin, serotonin, oxytocin, and anandamide that inhibit psychic distractions, extroversion, and urge of social interaction, especially for the catharsis of psychic deposits and morbidities. The induction of pain killing and healing neuro-hormones and parasympathetic tone as the effect of integral *Yoga* practice maximize coping with stress, depression, burnout, anxiety, PTSD, fear, fatigue, grief, work-life conflict, and impaired QoL. The imposed social isolation during the Covid-19 pandemic has been perceived as an opportunity by regular practitioners to maintain celibacy, self-study, and verbal silence needed for advanced *Yoga* practice and experiences.

Interestingly, social isolation also limited the frequency of fast-food intakes from hotels, restaurants, food courts, and other food supply shops/chains. The social isolation compelled the quarantined personnel to prepare their natural food in less variety. This dietary pattern and habit are suitable for health promotion and healing as prioritized in the *Yoga* system.

Conversely, the persons with extroverted personalities relied on external food supplies and the social company faced intense mental suffocation and emotional stress during imposed social isolation as the global regulatory provision to limit the spread of SARS-Cov-2 infection. *Yama* (non-violence, truthfulness, non-stealing, non-possessiveness, and celibacy) and *Niyama* (cleanliness, contentment, austerity, self-study, and divine surrender) are psychosocial behaviours to sustain and advance in the *Yoga* journey. The regular practice of these yogic psychosocial behaviours—self-study, coping with adversities, psychosomatic cleanliness, non-possessiveness, celibacy, and faith in supreme power can make the pandemic-imposed social isolation a golden opportunity for advancing *Yoga* living, resulting in boosted psychosomatic immunity.

The practice of poses induce psychosomatic steadiness and power to cope with adversities and pain. The method of breath regulation deepens and delays breath currents, enhances energy levels, depletes psychic morbidities, and heightens concentration. The practice of yogic seals, gestures, and sensory withdrawal cut bioenergy outflow and induce focused cognitive and conative workflow. Guided or auto-regulated meditation such as Mindfulness, *Soham*, *Kundalini*, TM, *Yoga Nidra*, SKY, and *Savita Dhyan* develop intellectual sharpness, emotional stability, mental calmness, inner peace, and overwhelming fulfilment by limiting the urge for social isolation. Therefore, the practice of integral *Yoga* protocol guided by the master with sound experimental and experiential knowledge and skills is highly recommended to prevent possible psychosocial adverse effects of imposed social isolation and quarantine during the Covid-19 pandemic.

7.3.7 **Yoga** *for Stigma*

The yogic principle of realizing the self as a part of an undifferentiated whole intensifies the spirit behind the unconditional roles for collective welfare, harmony, and peace. The *Yoga* practice with this spirit awakes the cognitive geniuses to discriminate the truth behind stigma and be aligned with selfless duties rather than caring for time-bound stigma. The empowered cardiopulmonary efficiency (Vempati and Telles 2002) due to integral *Yoga* practice and abundant cellular oxygen supply eases stress coping (Kirkwood et al. 2005). Additionally, the emotive, cognitive, and energetic *Yoga* practices significantly enhance empathy, emotional conscientiousness, holism, and feeling of privilege transfer as the improved markers of psycho-socio-spiritual well-being (Moadel et al. 2007). Mindfulness-based cognitive therapy (MBCT) includes many specific mindfulness practices such as focused attention to breathing, body scanning, prosocial meditation (such as affection and compassion), and gentle *Hatha Yoga* eases cognitive remodelling and moderates the recurrence of depression (Williams et al. 2014). The practice of *Pranakarshan Pranayama* and *Savita Dhyan* synchronized with tongue, perineal, and anal locks and preceded by joint loosening and spine aligning postural series may be a novel, effective, harmless, cost-friendly, and scientifically valid option to overcome the impacts of Covid-19-induced stigma among HCWs.

7.3.8 **Yoga** ***for Quality of Life***

Anxiety, depression, stress, PTSD, and poor QoL are the most common reasons behind the popularity and increasing demand for CAM—mind-body intervention, energy healing, natural dietary supplements, bodywork therapies, and traditional whole medical systems (Pilkington et al. 2005). Among the CAM dimensions, mind-body practice is a subset of yogic living to induce well-being, relaxation, self-confidence, mental tranquillity, attentiveness, work efficiency, and an optimistic outlook towards life and subsequent worldly happenings (Arora and Bhattacharjee 2020). *Yoga* decreases oxidative stress markers and upregulates telomere expression leading to psycho-physiological homeostasis, robust immunity, and optimal well-being (Duraimani et al. 2015). *Anulom Vilom* stimulates parasympathetic tone resulting in decreased BR, cortisol, BP, and HR and increased relaxation, present awareness, and blood flow to the gut and other visceral organs (Si et al. 2021). The practice of slow breathing exercises—abdominal and victorious (*Ujjayi*)—synchronized with meditation induces relaxation and improves the QoL (Goldstein et al. 2018). Mindfulness training has also been shown to reduce stress and negative mood states and enhance emotional well-being and QoL in physically ill patients (Chiesa and Serretti 2011). A well-designed *Yoga* therapy protocol has been shown to reduce stress, anxiety, depression, and PTSD and improve QoL and sleep patterns (Woodyard 2011). Besides psychosomatic illness and social maladjustment, the QoL also gets impaired by the constant greed and lust induced by worldly attachments and delusion. The teachings of *Yoga* philosophy (*Jnyan Yoga*) on the natural phenomena, evolution, involution, law of action, and natural qualities of body, mind, and soul, as explained in *Shrimadbhagvad Geeta,* may restructure cognition and correct ill perceptions behind worldly happenings as the triggers of poor QoL. The operational modalities of *Jnyan Yoga—Atamabodha* and *Tatwabodha* are exact potential practices for cognitive restructuring, correcting false perceptions, and optimal time management (Acharya 2000).

Based on the discussion above on the positive effects of diverse *Yoga* practices, the integral *Yoga* protocol as depicted in Table 7.1 is proposed to prevent/overcome the PSPs faced by HCWs during the Covid-19 pandemic as the potential clinical implication of this study (Table 7.1).

Table 7.1 Integral *Yoga* protocol

Dimensions	Yogic practices	Time
Somatic	***Asana*** **Loosening and Energizing Series (LES):** Toes stretch, ankle stretch, ankle rotation, knee stretches, knee rotation, half-butterfly, hip joint stretch and rotation, semi-spinal twists with extended legs, fingers stretch from middle and distal phalangeal joints, palm stretches, wrist stretches and rotations, shoulder stretches and rotations, neck movements: up-down, sidewise—tilts and twists, course, chest opening, spinal stretch, hands rotation from shoulders, facial movements—protraction, retraction, clinching, grinding teeth, Spinal series/alligator twisting (Stiles 2005; Brahmachari 1965) *Parvatasana* (Mountain pose) *Tadasana* (Palm Tree Pose) *Padhastasana* (Head to Toe Pose) *Ustrasana* (Camal Pose) *Utkatasana* (Chair pose) *Yogmudrasana* (Rabbit pose with hands extended back with interlocked fingers) *Ardhatadasana* (Semi Palm Tree Pose) *Bhujangasana* (Cobra pose) *Shalabhasana* (Locust Pose) *Vakrasana* (Seated Semi Spinal Twist) *Vajrasana* (Thunderbolt Pose) *Makarasana* (Crocodile Pose) (Burley 2000; Saraswati 2013) *Note*: Practice of the poses mentioned above be synchronized with breathing patterns and mental awareness by caring for patients' comfort limits	30 min
Pranic	**Breathing practices** Uni-nostril Breathing (UNB) Alternate Nostril Breathing (ANB) (Edwards 2005) *Kapalbhati* (High-Frequency *Yoga* Breathing) *Bhastrika* (Forceful Deep Breathing) *Kapol Shakti Vikasak* (KSV) (Crown Breathing) *Ujjayi* (Coherent Breathing) *Bhramari* (Resonance Breathing) ***Mudras*:** Ashwini/*Shaktichalani* (Anal Lock), *Jnyan*, *Shanmukhi*, *Prana*, and *Apana*. ***Bandhas*:** *Moola* (Perineal Lock), *Jalandhar* (Throat Lock), and *Uddiyan* (Abdominal Lock) **Note:** Breathing practices be synchronized with specified *Mudras* and *Bandhas* (Saraswati 2013)	15 min
Mental	*Savita Dhyan* (Meditation of rising sun on navel, heart, and crown centres) Mindfulness Meditation (Kostanski and Hassed 2008) *Attmabodha* and *Tattwabodha* (Contemplation on self-existence, duty, and supreme power) (Acharya 2000)	15 min
Emotional	Nadyog (Acharya 2000)	5 min

7.4 Future Direction for the Research

The efficacy of the proposed integral *Yoga* protocol to prevent/mitigate Covid-19-induced psychosocial problems of healthcare workers can be tested in further experimental studies.

7.5 Conclusion

The workload, odd working hours, increased patients flow and their deaths, uncertain pathophysiology and treatments of SARS-Cov-2 infection, logistic and technological suffocations, fear of infection and death, social stigma, rampant media rumours, physical distancing, and social isolation jointly contributed to the onset and progression of psychosocial problems among HCWs. Depression, anxiety, burnout, PTSD, WLC, social isolation, stigma, and poor quality of life are frequently reported psychosocial health problems of the Covid-19 HCWs.

Covid-19-induced depression symptoms are hopelessness, emptiness, fearfulness, or sadness; excessive or too little sleep, disinterest or pleasure in most activities—sex, sports, or hobbies; frustration, irritability, or angry outbursts; tiredness and low energy; agitation, restlessness, unexplained pain, feeling of guilt or unworthy, poor concentration, memory, and decision-making; slowed speaking, impaired body movements, over/under eating, frequent suicidal thoughts and attempts (Gloster et al. 2020). Covid-19-induced anxiety symptoms are feeling nervous, restless, or tense, sensing impending danger, panic, increased heart rate, hyperventilation, sweating, trembling, feeling weak or tired, trouble concentrating or thinking about anything other than present worry, sleeping difficulty, loss of interest, fearfulness, and experiencing upset gut. PTSD is a condition that develops after experiencing psychological trauma when these responses persist for over a month and cause substantial distress and disruption in one's life. Covid-19-induced PTSD symptoms include recurrent and intrusive memories of the trauma, flashbacks or nightmares of the trauma, disinterest in previously enjoyable activities, guilt, anger, shame, and difficulty sleeping. WLC is the inability to maintain and create friendly and stress-free working environments, balancing paid work and personal engagements, and encouraging employee loyalty and productivity. HCWs experienced increased workload, demand for medical professionals, lack of equipment, higher demand than their supply, and reduced intake of medical professionals by hospitals due to their need for higher remunerations and welfare packages and services to many patients. Burnout is a psychological syndrome of cynicism, exhaustion, and inefficacy experienced in response to chronic job stressors. HCWs experienced persistent chronic stress beyond their capacities resulting in mental fatigue associated with the disturbed neuro-endocrine system.

Social isolation is the absence of social interactions, contacts, and relationships with family and friends, neighbours on an individual level, and society at large.

Lockdown enforcement, isolation, and closure of public places to control the rapid human-to-human transmission of SARS-CoV-2 had heavily burdened, impaired, and overwhelmed the HCWs. Stigma is a social process that excludes those perceived as potential disease sources. It may threaten good social living in the ongoing context of the Covid-19 pandemic. HCWs working in a higher risk of infected environment had a greater fear of passing SARS-Cov-2 to their family, friends, colleagues, and others and hence were often stigmatized and rejected by others. The ongoing stressful events faced by HCWs during the Covid-19 pandemic adversely affected their physical, mental, and social health and well-being, leading to compromised QoL.

Yoga is an ancient oriental practice that aims at purifying the body, breath, bioenergy, cognition, and emotion to promote holistic wellness as an indispensable foundation for spiritual elevation. Interestingly, *Yoga* is widely practised in modern times for recreation, health promotion, disease prevention, immunomodulation, healing, and age reversal. *Yoga* includes an array of practices—cleansing techniques (*Kriyas*), pose (*Asana*), conscious breath regulation (*Pranayama*), seal (*Bandha*), gesture (*Mudra*), sensory withdrawal (*Pratyahara*), concentration (*dharana*), and meditation (*Dhyan*). Regular practice of the proposed integral *Yoga* protocol may prevent and moderate PSPs faced by HCWs by optimizing psycho-neuroendocrine interactions, musculoskeletal health, immune system, cardiopulmonary functions, metabolic processes, bioenergy, and worldly outlook.

References

Acharya SS (2000) Super Science of Gayatri. Yugantar Chetna Press, Haridwar, India

Afonso RF, Kraft I, Aratanha MA, Kozasa EH (2020) Neural correlates of meditation: a review of structural and functional MRI studies. Front Biosci (Schol Ed) 12(1):92–115. https://doi.org/10.2741/S542

American Psychiatric Association (2013) Diagnostic and statistical manual of mental disorders-5TM, 5th edn. American Psychiatric Association, Washington, DC

Antonova E, Schlosser K, Pandey R, Kumari V, Armocida B, Formenti B, Ussai S, Palestra F, Missoni E, Afonso RF, Kraft I, Aratanha MA, Kozasa EH, Ahmadpanah M, Paghale SJ, Bakhtyari A, Kaikhavani S, Aghaei E, Nazaribadie M, Holsboer-Trachsler E, Brand S, Alexander GK, Rollins K, Walker D, Wong L, Pennings J (2015) Yoga for Self-Care and Burnout Prevention Among Nurses. Work Heal Saf 63(10):462–471. https://doi.org/10.1177/2165079915596102

Antonova E, Schlosser K, Pandey R, Kumari V (2021) Coping With COVID-19: Mindfulness-Based Approaches for Mitigating Mental Health Crisis. Front psychiatry 12:563417. https://doi.org/10.3389/fpsyt.2021.563417

Arora T, Grey I (2020) Health behaviour changes during COVID-19 and the potential consequences: a mini-review. J Health Psychol 25(9):1155–1163

Baker DG, Nievergelt CM, O'Connor DT (2012) Biomarkers of PTSD: neuropeptides and immune signaling. Neuropharmacology 62(2):663–673

Bhandari R (2017) Yogic intervention for coping with distress. J Clin Diagnostic Res 11(12):OC44–OC49. https://doi.org/10.7860/JCDR/2017/29332.10944

Bhandari R (2022) Online Yoga and Ayurveda Intervention as Tertiary Prevention of Psychological Comorbidities in COVID-19 Survivors: A Randomized Controlled Trial. Ann Neurosci 29(4):233–244. https://doi.org/10.1177/09727531221117623

Bhanot D, Singh T, Verma SK, Sharad S (2020) Stigma and discrimination during COVID-19 pandemic. Front Public Health 8:577018. https://doi.org/10.3389/fpubh.2020.577018

Bharathi SV, Mala EP (2016) A study on the determinants of work–life balance of women employees in information technology companies in India. Glob Bus Rev 17(3):665–683, https://doi.org/10.1177/0972150916630847

Bond AR, Mason HF, Lemaster CM, Shaw SE, Mullin CS, Holick EA et al (2013) Embodied health: the effects of a mind Á body course for medical students. Med Educ Online 18:1–8. https://doi.org/10.3402/meo.v18i0.20699

Bormann JE, Liu L, Thorp SR, Lang AJ (2012) Spiritual wellbeing mediates PTSD change in veterans with military-related PTSD. Int J Behav Med 19(4):496–502. https://doi.org/10.1007/s12529-011-9186-1

Brahmachari D (1965) Yogic Suksma Vyayama: the complete sequence of 48 yogc exercises Hindi Version. Dhirendra Yog Prakashan, Delhi, India

Brown R, Gerbarg P (2009) *Yoga* breathing, meditation, and longevity. Ann N Y Acad Sci 1172:54–62. https://doi.org/10.1111/j.1749-6632.2009.04394.x

Brown KW, Ryan RM (2003) The benefits of being present: mindfulness and its role in psychological well-being. J Pers Soc Psychol 84(4):822–848. https://doi.org/10.1037/0022-3514.84.4.822

Burley M (2000) Haṭha-Yoga: Its Context, Theory, and Practice, 1st edn. Motilal Banarsidass, Delhi

Buselli R, Corsi M, Baldanzi S, Chiumiento M, Del Lupo E, Dell'Oste V et al (2020) Professional quality of life and mental health outcomes among health care workers exposed to Sars-Cov-2 (Covid-19). Int J Environ Res Public Health 17(17):6180. https://doi.org/10.3390/ijerph17176180

Cahn BR, Polich J (2006) Meditation states and traits: EEG, ERP, and neuroimaging studies. Psychol Bull 132(2):180–211. https://doi.org/10.1037/0033-2909.132.2.180

Chen T, Wu D, Chen H, Yan W, Yang D, Chen G et al (2020) Clinical characteristics of 113 deceased patients with coronavirus disease 2019: retrospective study. BMJ 368:m1295. https://doi.org/10.1136/bmj.m1295

Chen R, Sun C, Chen JJ, Jen HJ, Kang XL (2021) A large-scale survey on trauma, burnout, and posttraumatic growth among nurses during the COVID-19 pandemic. Int J Ment Health Nurs 30(1):102–116. https://doi.org/10.1111/inm.12796

Cheng C, Wang H-Y, Ebrahimi OV (2021) Adjustment to a "New Normal:" Coping Flexibility and Mental Health Issues During the COVID-19 Pandemic. Front psychiatry 12:626197. https://doi.org/10.3389/fpsyt.2021.626197

Chiesa A, Serretti A (2011) Mindfulness-based interventions for chronic pain: a systematic review of the evidence. J Altern Complement Med 17(1):83–93. https://doi.org/10.1089/acm.2009.0546

Chong M-Y, Wang W-C, Hsieh W-C, Lee C-Y, Chiu N-M, Yeh W-C, Huang T-L, Wen J-K, Chen C-L (2004) Psychological impact of severe acute respiratory syndrome on health workers in a tertiary hospital. Br J Psychiatry 185(2):127–133. https://doi.org/10.1192/bjp.185.2.127

Christopher YKW, Adam TT, Kapur M, Alice FF, Nunn R, Galante J, Phillips V, Gentry S, Usher- J (2021) Interventions to reduce social isolation and loneliness during COVID-19 physical distancing measures : A rapid systematic review. PLoS One 16(2):1–28. https://doi.org/10.1371/journal.pone.0247139

Cocchiara RA, Peruzzo M, Mannocci A, Ottolenghi L, Villari P, Polimeni A, Guerra F, La Torre G (2019) The Use of Yoga to Manage Stress and Burnout in Healthcare Workers: A Systematic Review. J Clin Med 8(3):284. https://doi.org/10.3390/jcm8030284

Coplan RJ, Hipson WE, Bowker JC (2021) Social Withdrawal and Aloneliness in Adolescence: Examining the Implications of Too Much and Not Enough Solitude. J Youth Adolesc 50(6):1219–1233. https://doi.org/10.1007/s10964-020-01365-0

Correa-Palacio AF, Hernandez-Huerta D, Gómez-Arnau J, Loeck C, Caballero I, Chandra PS, Shiva L, Nagendrappa S, Ganjekar S, Thippeswamy H (2020) Affective psychosis after

COVID-19 infection in a previously healthy patient: a case report. Psychiatry Res 290:113136. https://doi.org/10.1016/j.psychres.2020.113136

d'Ettorre G, Ceccarelli G, Santinelli L, Vassalini P, Innocenti G Pietro, Alessandri F, Koukopoulos AE, Russo A, d'Ettorre G, Tarsitani L (2021) Post-Traumatic Stress Symptoms in Healthcare Workers Dealing with the COVID-19 Pandemic: A Systematic Review. Int J Environ Res Public Health 18(2):601. https://doi.org/10.3390/ijerph18020601

da Silveira MP, da Silva Fagundes KK, Bizuti MR, Starck É, Rossi RC, de Resende e Silva DT (2021) Physical exercise as a tool to help the immune system against COVID-19: an integrative review of the current literature. Clin Exp Med 21(1):15–28. https://doi.org/10.1007/s10238-020-00650-3

de Bruin EI, Formsma AR, Frijstein G, Bögels SM (2017) Mindful2Work: effects of combined physical exercise, *Yoga*, and mindfulness meditations for stress relieve in employees. A proof of concept study. Mindfulness 8(1):204–217. https://doi.org/10.1007/s12671-016-0593-x

Deneva T, Ianakiev YKD (2019) Burnout syndrome in physicians—psychological. Medicina (Kaunas) 55(5):209. https://doi.org/10.3390/medicina55050209

Devadhasan BD (2018) A report on the importance of work-life balance. Int J Appl Eng Res 10(1):21659–21665

Divya K, Bharathi S, Somya R, Darshan MH (2021) Impact of a yogic breathing technique on the well-being of healthcare professionals during the COVID-19 pandemic. Glob Adv Health Med 10:2164956120982956. https://doi.org/10.1177/2164956120982956

Duan-Porter W, Coeytaux RR, McDuffie JR, Goode AP, Sharma P, Mennella H, Nagi A, Williams JW (2016) Evidence Map of Yoga for Depression, Anxiety, and Posttraumatic Stress Disorder. J Phys Act Health 13(3):281–288. https://doi.org/10.1123/JPAH.2015-0027

Duraimani S, Schneider RH, Randall OS, Nidich SI, Xu S, Ketete M, Rainforth MA, Gaylord-King C, Salerno JW, Fagan J (2015) Effects of lifestyle modification on telomerase gene expression in hypertensive patients: A pilot trial of stress reduction and health education programs in African Americans. PLoS One 10(11):1–18. https://doi.org/10.1371/journal.pone.0142689

Edwards S (2005) A Psychology of Breathing Methods. Int J Ment Health Promot 7(4):30–36. https://doi.org/10.1080/14623730.2005.9721958

Elder C, Nidich S, Moriarty F, Nidich R (2014) Effect of Transcendental Meditation on Employee Stress, Depression, and Burnout : A Randomized Controlled Study. Perm J 18(1):19–23. https://doi.org/10.7812/TPP/13-102

Elias AN, Guich S, Wilson AF (2000) Ketosis with enhanced GABAergic tone promotes physiological changes in transcendental meditation. Med Hypotheses 54(4):660–662. https://doi.org/10.1054/mehy.1999.0921

Fiorillo A, Gorwood P (2020) The consequences of the COVID-19 pandemic on mental health and implications for clinical practice. Eur. Psychiatry 63:e32

Giorgi G, Lecca LI, Alessio F, Finstad GL, Bondanini G, Lulli LG, Arcangeli G, Mucci N (2019) COVID-19-Related Mental Health Effects in the Workplace : A Narrative Review. Int J Environ Res Public Health 17(21):1–22. https://doi.org/10.3390/ijerph17217857

Gloster AT, Lamnisos D, Lubenko J, Presti G, Squatrito V, Constantinou M, Nicolaou C, Papacostas S, Aydın G, Chong YY, Chien WT, Cheng HY, Ruiz FJ, Garcia-Martin MB, Obando-Posada DP, Segura-Vargas MA, Vasiliou VS, McHugh L, Höfer S, Baban A, Neto DD, da Silva AN, Monestès JL, Alvarez-Galvez J, Paez-Blarrina M, Montesinos F, Valdivia-Salas S, Ori D, Kleszcz B, Lappalainen R, Ivanović I, Gosar D, Dionne F, Merwin RM, Kassianos AP, Karekla M (2020) Impact of COVID-19 pandemic on mental health: An international study. PLoS One 15(12):e0244809. https://doi.org/10.1371/journal.pone.0244809

Goldstein E, Topitzes J, Brown RL, Barrett B (2018) Mediational pathways of meditation and exercise on mental health and perceived stress: A randomized controlled trial. J Health Psychol 25(12):1816–1830. https://doi.org/10.1177/1359105318772608

Gragnano A, Simbula S, Miglioretti M (2020) Work–life balance: weighing the importance of work–family and work–health balance. Int J Environ Res Public Health 17(3):907

Hayes JP, Vanelzakker MB, Shin LM (2012) Emotion and cognition interactions in PTSD: a review of neurocognitive and neuroimaging studies. Front Integr Neurosci 6(10):89. https://doi.org/10.3389/fnint.2012.00089

Horner JK, Piercy BS, Eure L, Woodard EK (2014) A pilot study to evaluate mindfulness as a strategy to improve inpatient nurse and patient experiences. Appl Nurs Res 27(3):198–201. https://doi.org/10.1016/j.apnr.2014.01.003

Inanaga K (1998) Frontal midline theta rhythm and mental activity. Psychiatry Clin Neurosci 52(6):555–566. https://doi.org/10.1046/j.1440-1819.1998.00452.x

Jarden R, M S, RJ S (2019) Strengthening workplace wellbeing : Perceptions of intensive care nurses. Nurs Crit Care 24(1):15–23. https://doi.org/10.1111/nicc.12386

Joly F, Rigal O, Guittet L, Lefèvre-Arbogast S, Grellard JM, Binarelli G, Lange M, Rieux C, Fernette M, Tron L, Gernier F, Travers R, Morel A, Richard D, Griffon B, Leconte A, Bastien E, Quilan F, Pépin LF, Jardin F, Leheurteur M, Clarisse B, Lequesne J, Faveyrial A (2021) Post-traumatic stress symptomatology and adjustment of medical oncology practice during the COVID-19 pandemic among adult patients with cancer in a day care hospital. Cancer 127(24):4636–4645. https://doi.org/10.1002/cncr.33856

Kane L (2020) US and International Physicians' COVID-19 Experience Report: Risk, Burnout, Loneliness. In: Medscape. https://www.medscape.com/viewarticle/936876?form=fpf. Accessed 14 Feb 2021

King AP, Block SR, Sripada RK, Rauch SAM, Porter KE, Favorite TK, Giardino N, Liberzon I (2016) A pilot study of mindfulness-based exposure therapy in OEF/OIF combat veterans with PTSD: Altered medial frontal cortex and amygdala responses in social-emotional processing. Front Psychiatry 7(9):1–13. https://doi.org/10.3389/fpsyt.2016.00154

Kirkwood G, Rampes H, Tuffrey V, Richardson J, Pilkington K (2005) Yoga for anxiety: A systematic review of the research evidence. Br J Sports Med 39(12):884–891. https://doi.org/10.1136/bjsm.2005.018069

Kisely S, Warren N, McMahon L, Dalais C, Henry I, Siskind D (2020) Occurrence, prevention, and management of the psychological effects of emerging virus outbreaks on healthcare workers: rapid review and meta-analysis. BMJ 369:1642. https://doi.org/10.1136/bmj.m1642

Kostanski M, Hassed C (2008) Mindfulness as a concept and a process. Aust Psychol 43(1):15–21. https://doi.org/10.1080/00050060701593942

Krishnakumar D, Hamblin MR, Lakshmanan S (2015) Meditation and Yoga can Modulate Brain Mechanisms that affect Behavior and Anxiety- A Modern Scientific Perspective. Anc Sci 2(1):13. https://doi.org/10.14259/as.v2i1.171

Lee AKW, Gansler DA, Zhang N, Jerram MW, King JA, Fulwiler C (2017) Relationship of mindful awareness to neural processing of angry faces and impact of mindfulness training: A pilot investigation. Psychiatry Res - Neuroimaging 264(16):22–28. https://doi.org/10.1016/j.pscychresns.2017.03.016

Lorkiewicz P, Waszkiewicz N (2021) Biomarkers of Post-COVID Depression. J Clin Med 10(18):4142. https://doi.org/10.3390/jcm10184142

Marjanovic Z, Greenglass ER, Coffey S (2007) The relevance of psychosocial variables and working conditions in predicting nurses' coping strategies during the SARS crisis: an online questionnaire survey. Int J Nurs Stud 44(6):991–998. https://doi.org/10.1016/j.ijnurstu.2006.02.012

Maslach C, Leiter M (2007) Burnout. In: Stress: Concepts, Cognition, Emotion, and Behavior: Handbook of Stress. pp 358–362

Maynard BR, Solis MR, Miller VL, Brendel KE (2017) Mindfulness-based interventions for improving cognition, academic achievement, behavior, and socioemotional functioning of primary and secondary school students. Campbell Syst Rev 13(1):1–144. https://doi.org/10.4073/CSR.2017.5

Mazza MG, Palladini M, De Lorenzo R, Magnaghi C, Poletti S, Furlan R, Ciceri F, Rovere-Querini P, Benedetti F (2021) Persistent psychopathology and neurocognitive impairment in COVID-19 survivors: Effect of inflammatory biomarkers at three-month follow-up. Brain Behav Immun 94:138–147. https://doi.org/10.1016/j.bbi.2021.02.021

McAlonan GM, Lee AM, Cheung V, Cheung C, Tsang KWT, Sham PC, Chua SE, Wong JGWS (2007) Immediate and sustained psychological impact of an emerging infectious disease outbreak on health care workers. Can J Psychiatry 52(4):241–247. https://doi.org/10.1177/070674370705200406

McCall T (2007) *Yoga* as medicine: the yogic prescription for health and healing. Bantam Dell, New York

Michalsen A, Grossman P, Acil A, Langhorst J, Lüdtke R, Esch T, Stefano GB, Dobos GJ (2005) Rapid stress reduction and anxiolysis among distressed women as a consequence of a three-month intensive yoga program. Med Sci Monit Int Med J Exp Clin Res 11(12):CR555–561

Moadel AB, Shah C, Wylie-Rosett J, Harris MS, Patel SR, Hall CB, Sparano JA (2007) Randomized controlled trial of yoga among a multiethnic sample of breast cancer patients: effects on quality of life. J Clin Oncol 25(28):4387–4395. https://doi.org/10.1200/JCO.2006.06.6027

Nidich SI, Rainforth M V, Haaga DAF, Hagelin J, Salerno JW, Travis F, Tanner M, Gaylord-King C, Grosswald S, Schneider RH (2009) A randomized controlled trial on effects of the Transcendental Meditation program on blood pressure, psychological distress, and coping in young adults. Am J Hypertens 22(12):1326–1331. https://doi.org/10.1038/ajh.2009.184

Niu S-F, Kuo S-F, Tsai H-T, Kao C-C, Traynor V, Chou K-R (2019) Prevalence of workplace violent episodes experienced by nurses in acute psychiatric settings. PLoS One 14(1):e0211183. https://doi.org/10.1371/journal.pone.0211183

Pandey K, Thurman M, Johnson SD, Acharya A, Johnston M, Klug EA, Olwenyi OA, Rajaiah R, Byrareddy SN (2021) Mental Health Issues During and After COVID-19 Vaccine Era. Brain Res Bull 176(5):161–173. https://doi.org/10.1016/j.brainresbull.2021.08.012

Pappa S, Ntella V, Giannakas T, Giannakoulis VG (2020) Prevalence of depression, anxiety, and insomnia among healthcare workers during the COVID-19 pandemic: a systematic review and meta-analysis. Brain, Behav Immun 88(1):901–907. https://doi.org/10.1016/j.bbi.2020.05.026

Pilkington K, Kirkwood G, Rampes H, Richardson J (2005) Yoga for depression: the research evidence. J Affect Disord 89(1–3):13–24. https://doi.org/10.1016/j.jad.2005.08.013

Rain M, Subramaniam B, Avti P, Mahajan P, Anand A (2021) Can Yogic Breathing Techniques Like Simha Kriya and Isha Kriya Regulate COVID-19-Related Stress? Front Psychol 12(4):635816. https://doi.org/10.3389/fpsyg.2021.635816

Restauri N, Sheridan AD (2020) Burnout and Posttraumatic Stress Disorder in the Coronavirus Disease 2019 (COVID-19) Pandemic: Intersection, Impact, and Interventions. J Am Coll Radiol 17(7):921–926. https://doi.org/10.1016/j.jacr.2020.05.021

Santarone K, McKenney M, Elkbuli A (2020) Preserving mental health and resilience in frontline healthcare workers during COVID-19. Am J Emerg Med 38(7):1530–1531. https://doi.org/10.1016/j.ajem.2020.04.030

Saraswati S (2013) Asana Pranayama Mudra Bandha. Yoga Publications Trust, Bihar, India

Schneider J, Talamonti D, Gibson B, Forshaw M (2021) Factors mediating the psychological well-being of healthcare workers responding to global pandemics: A systematic review. J Health Psychol 27(8):1875–1896. https://doi.org/10.1177/13591053211012759

Sephton SE, Salmon P, Weissbecker I, Ulmer C, Floyd A, Hoover K, Studts JL (2007) Mindfulness meditation alleviates depressive symptoms in women with fibromyalgia: Results of a randomized clinical trial. Arthritis Care Res 57(1):77–85. https://doi.org/10.1002/art.22478

Sesnich G, Garrido M, Hernández M, Gamonal J (2017) Longitudinal and Immediate Effect of Kundalini Yoga on Salivary Levels of Cortisol and Activity of Alpha - Amylase and Its Effect on Perceived Stress. Int J Yoga 10(2):73–80. https://doi.org/10.4103/ijoy.IJOY

Shaku F, Tsutsumi M, Goto H, Saint Arnoult D (2014) Measuring the effects of Zen training on quality of life and mental health among Japanese monk trainees: a cross-sectional study. J Altern Complement Med 20(5):406–410. https://doi.org/10.1089/acm.2013.0209

Sharma K, Anand A, Kumar R (2020) The role of Yoga in working from home during the COVID-19 global lockdown. Work 66(4):731–737. https://doi.org/10.3233/WOR-203219

Shreffler J, Petrey J, Huecker M (2020) The Impact of COVID-19 on Healthcare Worker Wellness : A Scoping Review. West J Emerg Med 21(5):1059–1066. https://doi.org/10.5811/westjem.2020.7.48684

Si MY, Xiao WJ, Pan C, Wang H, Huang YM, Lian J, Mak WWS, Leng ZW, Su XY, Tang QP, Jiang Y, Feng LZ, Yang WZ, Wang C (2021) Mindfulness-based online intervention on mental health and quality of life among COVID-19 patients in China: an intervention design. Infect Dis Poverty 10(1):69. https://doi.org/10.1186/s40249-021-00836-1

Sim K, Chua HC (2004) The psychological impact of SARS: a matter of heart and mind. C Can Med Assoc J = J l'Association medicale Can 170(5):811–812. https://doi.org/10.1503/cmaj.1032003

Siu O-L (2002) Predictors of job satisfaction and absenteeism in two samples of Hong Kong nurses. J Adv Nurs 40(2):218–229. https://doi.org/10.1046/j.1365-2648.2002.02364.x

Spoorthy MS (2020) Mental health problems faced by healthcare workers due to the COVID-19 pandemic–A review. Asian J Psychiatr 51(1):102119. https://doi.org/10.1016/j.ajp.2020.102119

Stiles M (2005) Structural Yoga Therapy: Adopting to the Individuals. Weiser Books, San Francisco, CA

Su T-P, Lien T-C, Yang C-Y, Su YL, Wang J-H, Tsai S-L, Yin J-C (2007) Prevalence of psychiatric morbidity and psychological adaptation of the nurses in a structured SARS caring unit during outbreak: a prospective and periodic assessment study in Taiwan. J Psychiatr Res 41(1–2):119–130. https://doi.org/10.1016/j.jpsychires.2005.12.006

Sullivan MB, Erb M, Schmalzl L, Moonaz S, Noggle Taylor J, Porges SW (2018) Yoga Therapy and Polyvagal Theory: The Convergence of Traditional Wisdom and Contemporary Neuroscience for Self-Regulation and Resilience. Front Hum Neurosci 12:67. https://doi.org/10.3389/fnhum.2018.00067

Tandon R (2020) The COVID-19 pandemic, personal reflections on editorial responsibility. Asian J Psychiatry 50:102100. https://doi.org/10.1016/j.ajp.2020.102100

Telles S, Gupta RK (2020) Psychophysiological effects and the applications of *Yoga* breathing practices. In: Handbook of research on evidence-based perspectives on the psychophysiology of *Yoga* and its applications. IGI Global, Hershey PA

Tian T, Meng F, Pan W, Zhang S, Cheung T, Ng CH et al (2022) Mental health burden of frontline health professionals treating imported patients with COVID-19 in China during the pandemic. Psychol Med 52(2):398–399. https://doi.org/10.1017/S0033291720002093

Tolentino JC, Schmidt SL (2018) DSM-5 criteria and depression severity: Implications for clinical practice. Front Psychiatry 9:450. https://doi.org/10.3389/fpsyt.2018.00450

Ulett GA, Gleser G, Winokur G, Lawler A (1953) The EEG and reaction to photic stimulation as an index of anxiety-proneness. Electroencephalogr Clin Neurophysiol 5(1):23–32. https://doi.org/10.1016/0013-4694(53)90049-2

Vempati RP, Telles S (2002) Yoga-based guided relaxation reduces sympathetic activity judged from baseline levels. Psychol Rep 90(2):487–494. https://doi.org/10.2466/pr0.2002.90.2.487

Vindegaard N, Benros ME (2020) COVID-19 pandemic and mental health consequences: Systematic review of the current evidence. Brain Behav Immun 89:531–542. https://doi.org/10.1016/j.bbi.2020.05.048

Wheaton MG, Abramowitz JS, Berman NC, Fabricant LE, Olatunji BO (2012) Psychological Predictors of Anxiety in Response to the H1N1 (Swine Flu) Pandemic. Cognit Ther Res 36(3):210–218. https://doi.org/10.1007/s10608-011-9353-3

Williams JMG, Crane C, Barnhofer T, Brennan K, Duggan DS, Fennell MJ V, Hackmann A, Krusche A, Muse K, Rohr IR Von, Shah D, Crane RS, Eames C, Jones M, Radford S, Silverton S, Sun Y, Weatherley-jones E, Whitaker CJ, Russell D, Russell IT, Williams JMG, Crane C, Barnhofer T, Bren- K, Duggan DS, Fennell MJ V, Hackmann A, Muse K, Rohr IR Von, Shah D, Russell IT, Investigator P, Barnhofer T, Crane C, Crane RS, Duggan DS, Melanie J V, Russell D, Silverton S (2014) Mindfulness-Based Cognitive Therapy for Preventing Relapse in Recurrent Depression : A Randomized Dismantling Trial. J Consult Clin Psychol 82(2):275–286. https://doi.org/10.1037/a0035036

Wingert P, Kantrowitz B (2002) Young and depressed. Newsweek 140(52-58):60

Woodyard C (2011) Exploring the therapeutic effects of yoga and its ability to increase quality of life. Int J Yoga 4(2):49–54. https://doi.org/10.4103/0973-6131.85485

Yao H, Chen JH, Xu YF (2020) Patients with mental health disorders in the COVID-19 epidemic. The Lancet Psychiatry 7(4):e21. https://doi.org/10.1016/S2215-0366(20)30090-0

Chapter 8
Psychometric Analysis in Mind-Body Interventions (MBIs) and Role of Ayurgenomics

Shifa Narula, Shalini Kumari Kashyap, Mahendra Kumar, Deepali Mathur, Divya Singh, and Saras Jyoti Sharma

Contents

S. Narula
Amity Institute of Biotechnology, Amity University, Noida, Uttar Pradesh, India

S. K. Kashyap
Department of Kayachikitsa, Vaidya Yagya Dutt Sharma Ayurvedic Mahavidyalaya, Khurja, Uttar Pradesh, India

M. Kumar (✉)
Department of Immunopathology, Postgraduate Institute of Medical Education and Research, Chandigarh, India
e-mail: kumar.mahendra@pgimer.edu.in

D. Mathur
Neurocare Centre, Bhubaneswar, Odisha, India

D. Singh
Centre for Integrative Medicine and Research, AIIMS, Delhi, India

S. J. Sharma (✉)
AYUSH Department, Haryana Yog Aayog, Panchkula, Haryana, India

A. Anand (ed.), *Neuroscience of Yoga*,
https://doi.org/10.1007/978-981-97-2851-0_8

Abstract The psychometric analysis assesses changes in human thoughts and behavior, including changes in cognitive, emotional, attentional, and behavioral abilities through a battery of tests having a defined standard. The important determinants for the implementation of psychometric analysis in mind-body interventions (MBIs) are time, duration, kind of practice, and uniform methodology for assessment of expected outcomes of the intervention. The set parameters can reveal the effect of MBIs in terms of changes in various psychometric parameters and their limitations. The uniform set of instructions and parameters will also help to reduce the discrepancies observed in outcomes of interventions both within and between periods in cognitive, attentional, physical, social, and emotional variables. The major component of this chapter describes the impact of MBIs on psychometric parameters. Moreover, the chapter also highlights the role of the combination of genomics and *Ayurveda*, i.e., "Ayurgenomics," in the understanding of the scientific basis for MBIs-induced changes in psychometric behaviors.

Keywords Psychometric · Mindfulness · Yoga · Cognitive · Ayurgenomics

8.1 Introduction

Psychometric testing is a scientific method designed for the assessment of psychometric attributes and abilities of an individual, through using a set of standardized psychometric tests and questionnaires. The data generated by these tests are analyzed to measure an individual's cognitive, behavioral, personality, and other psycho-physiological traits. The psychometric analysis provides more accurate and objective data and thus is an integral part of recruitment and decision-making processes in the various sectors including education, work, research, and healthcare (British Psychological Society 2019; American Psychological Association 2019). In a clinical setting, psychometric analysis is used to assess the relevance of therapies and interventions for disease management. Recently, many clinical studies attempted to generate psychometric scores or assessment systems to categorize patients in terms of morbidity in various diseases like systemic lupus erythematosus and chronic obstructive pulmonary disease (Alsubheen et al. 2021; Cleanthous et al. 2021). Even psychometric scoring has been used as an important tool for the diagnosis of minor hepatic encephalopathy (PessidjoDjomatcho et al. 2021).

Mind and body intervention (MBI) has been an ancient form of therapy all over the world and is well described in history. The MBIs, such as relaxation, mindfulness, *Yoga*, meditation, imagery, hypnosis, aromatherapy, psychotherapy, music, art, dance, and exercise, utilize the mind's capabilities to communicate with the body, thus influencing the overall psycho-physiological health. Mind and body interventions when conjugated with conventional medicines have been shown to improve the outcomes of disease management (Noordali et al. 2017; Montgomery et al. 2017; Lolla 2018). The psychometric analysis tools are of great significance in evaluating the efficacy of MBIs in epidemiological and clinical studies. It may also be used to examine subjects' capacity, ability or disability, and resilience to a particular clinical condition (Lichtenberger and Kaufman 2013; Lolla 2018; Gallego et al. 2016). In order to establish the accuracy of a psychometric test, it should include a few basic criteria such as "validity" to measure what it intended to measure, "reliability" that it is stable over time, and "sensitivity" to detect changes in the outcome measures. These psychometric properties determine the quality, accuracy, and efficiency of a testing tool (Rust and Golombok 2009).

8.2 Importance of Psychometric Testing and Validation in MBIs

Psychometric testing plays a pivotal role in the assessment studies of mind-body interventions. Several research groups have developed psychometric assessment tools as defined sets of questionnaires for the measurement of various traits influenced by MBIs. Birdee and group in 2016 developed a psychometric tool Yoga Self-Efficacy Scale (YSES) for the assessment of *Yoga* self-efficacy. They recruited 309 *Yoga* practitioners as subjects to study the psychometric properties of 12-item YSES by using a secure online survey, and the psychometric properties such as internal consistency and construct validity of the tool were maintained for all the *Yoga* styles (Birdee et al. 2015). Similarly, Thind and group in 2017 conducted a clinical trial with adult *Yoga* participants ($n = 185$) in a baseline survey. They developed and tested a scale for the assessment of the physical, mental, and spiritual health benefits of *Yoga* practices (Thind et al. 2017). George and Ilavarasu in 2021 developed a psychometric testing instrument, named Music Receptivity Scale (MRS), to study music receptivity and its psychometric validation. The tool was designed to evaluate the extent of music internalization in subjects, which could help to differentiate between musicians and non-musicians based on various aspects of music psychology such as interest, emotions, attention, and lyrical and music appraisal. They did enrolment in three sets of experiments and recruited 63 subjects in the pilot, 313 in the general population, and 44 in the experts set. The psychometric analysis of data generated by all three sets was separately performed (George and Ilavarasu 2021).

A series of clinical studies have been conducted using already available, well-established, and validated psychometric instruments to evaluate the effect of MBIs in various diseases including cancer. For instance, Rao and group in 2008 studied the influence of *Yoga* interventions on the stress levels of breast cancer patients undergoing surgical treatment. They recruited 98 female patients in a single center, randomized controlled trial (RCT); psychometric tests and immune assays were carried out pre-operation and after 4 weeks of surgery. The psychometric assessment for stress, anxiety, depression, and quality of life was conducted using the following tools: State-Trait Anxiety Inventory (STAI), Beck's Depression Inventory (BDI), and Functional Living Index of Cancer (FLIC). The psychometric analysis suggested a significant decrease in the overall stress levels including anxiety and depression, thereby showing improvement in the quality of life in the *Yoga* intervention group post-surgery (Rao et al. 2008). In another study, the authors established the modulatory effect of *Yoga* practice on the quality of sleep and neuroendocrine-immune response in patients with metastatic breast cancer. They enrolled 91 patients with metastatic breast cancer in the clinical trial and randomized them into two treatment groups "integrated *Yoga*-based stress reduction program" (n = 45) and "education and supportive therapy sessions" (n = 46) for the duration of 3 months. They performed psychometric assessment coupled with pathological examinations before and after interventions and the data showed a positive impact on neuroendocrine responses and sleep quality in the integrated *Yoga* program group (Rao et al. 2017). In 2019, another group evaluated the impact of *Yoga* on the anxiety, depression, fatigue, and sleep qualities of cancer patients. They used the Brief Fatigue Inventory, Hospital Anxiety and Depression Scale, and Pittsburgh Sleep Quality Index Scale for psychometric assessment. The data collection and analysis were done in two phases, i.e., before and after 90 min per week *Yoga* protocol for the duration of 8 weeks. The data analysis showed a significant change in anxiety and depression levels; however, the data also suggested that the levels of fatigue and sleep quality remained unchanged in patients undergoing cancer treatment (Mirandola et al. 2019).

Moreover, a lot of research data is available to establish the effect of mind and body interventions in various other pathologies and psycho-physiological settings. A recent psychometric study established the positive influence of mindfulness on temperamental characteristics and the reduction of anxiety and depression in primary school students. In this multi-arm RCT, 41 subjects were recruited and experimental groups were subjected to 8 weeks of mindfulness training comprising 60 min sessions per week. For the measurement of temperamental traits Italian Questionnaires of Temperament (QUIT) was used; Test for Anxiety and Depression in Childhood and Adolescence (TAD) was used to measure anxiety and depression levels. The analysis of psychometric data revealed that mindfulness interventions considerably reduced anxiety levels and influenced the temperamental dimensions of students (Poli et al. 2022). Another group investigated the effect of short mindfulness retreats on the levels of stress and inflammatory bio-mediators. They recruited 95 healthy subjects in this RCT, and the intervention arm was given a three-day mindfulness program. The psychometric assessment was done using the Mindful Attention Awareness Scale (MAAS), State-Trait Anxiety Inventory Form Y (STAI-Y), and Perceived Stress Scale (PSS). The biochemical tests coupled with psychometric analysis revealed a significant reduction in the pro-cytokines IL-6 and IL-8 levels along with reduced stress and anxiety levels (Gardi et al. 2022). Bhanushali and the group conducted a single-arm exploratory study in 2020 to determine the effect of *Yoga* and mindfulness meditation programs on people with high altitude-related sickness. They recruited 48 participants in an advanced meditation program for the duration of 4 days, and their biochemical, anthropometric, and psychometric tests were assessed before and after the advanced meditation protocol. For psychometric analysis, the following inventories were used: State-Trait

Anxiety Inventory (STAI) for measuring state anxiety, the Warwick-Edinburgh Mental Well-being Scale (WEMWBS) to check the mental health status, and Oxford Happiness Questionnaire (OHQ) to evaluate the current level of happiness of the subjects. Findings of the study highlighted the beneficiary impact of advanced meditation programs on biochemical and neuro-cognitive parameters of subjects at high altitudes (Bhanushali et al. 2020).

The cost-effectiveness of mind-body interventions has also been compared with usual conventional care. Hartfiel and group in 2017 conducted an RCT to analyze the self-efficacy and cost-effectiveness of *Yoga* for the management of back pain and musculoskeletal conditions in 151 National Health Service (NHS) employees. They used repeated-measures generalized linear modeling for the Roland-Morris Disability Questionnaire (RDQ) and the Keele STarT Back Screening Tool. The analysis from healthcare perspective, showed Yoga a significant cost effective care (yielding a ratio of £2103 per quality adjusted life year(QALY) compared to ususal care cost of £20 000). From societal perspective Yoga was preferred treatment over usual care (Hartfiel et al. 2017). Different MBI's program and various scales for psychometric analysis are summarized in Table 8.1.

Even though an extensive amount of research has been carried out using these psychometric analysis tools for the measurement of various dimensional traits influenced by MBIs, there are limitations to the accuracy and validity of the same. One of the critical issues with psychometric tests is the extent of variations that could arise due to false inputs or diversions among the subjects undergoing assessment (Silvia et al. 2014, 2016; Perez-Achiaga et al. 2009; Gendolla 2000). Therefore, for the validation of generated outcomes, psychometric tests are often used in combination with other physiological analytical tests.

8.3 Designing of Psychometric Tests for MBIs

The most used questionnaire in mind-body-based interventions includes Mindfulness-Based Stress Reduction (Brown and Ryan 2003; Baer et al. 2004), Dialectical Behavioral Therapy (DBT) (Linehan 1993), Acceptance and Commitment Therapy (ACT) (Hayes et al. 1996), Mindfulness-Based Cognitive Therapy (MBCT) (Segal et al. 2002), and Mindfulness Acceptance Commitment (MAC) (Gardner and Moore 2007, 2012). The concept of mindfulness or *Yoga* can be one-dimensional and multidimensional (Burg et al. 2017), although many perceive it as an attribute or health condition (Chiesa and Serretti 2014; Smith 2019, 2021). Some scales assess only the mind level; however, others include physical sensations and experiences (Grossman 2008). The scales have been designed based on the self or personal experience who have good experience and practice MBIs. This might influence the psychometric outcomes of MBIs and might reduce the

Table 8.1 MBI's program using various scales for psychometric analysis

S. no.	First author and year	Title	Language	Population	Scale	Dimensions (no. of items)	Conclusion
1.	Birdee et al. (2015)	Development and psychometric properties of the *Yoga* Self-Efficacy Scale (YSES)	English	$N = 309$	*Yoga* Self-Efficacy Scale (YSES)	12 items	YSES is developed to measure self-efficacy for *Yoga* practice which may give an understanding of steps for implementing and maintaining MBIs for well-being
2.	Thind et al. (2017)	Development and Validation of the Outcome Expectations for *Yoga* Scale	English	$N = 185$	Outcome Expectations for *Yoga* (OEY) Scale	20 items	This test is used to evaluate the physical, mental, and spiritual well-being after MBIs
3.	Rao et al. (2017)	Effect of *Yoga* on sleep quality and neuroendocrine immune response in metastatic breast cancer patients	English	$N = 257$	Pittsburgh Insomnia Rating Scale	65 items	The outcome revealed the modulation of neuroendocrine responses and quality of sleep in patients with advanced breast cancer after intervention
4.	Hartfiel et al. (2017)	Cost-effectiveness of *Yoga* for managing musculoskeletal conditions in the workplace	English	$N = 163$	Roland-Morris Disability Questionnaire (RDQ) and the KeeleSTarT Back Screening Tool		Enhance Health-Related Quality of Life (HQRL) and may improve the health condition associated with back pain, lower sickness absence due to musculoskeletal disorders, and is likely to be cost-effective

5.	Bhanushali et al. (2020)		English	$N = 48$	*Prakriti* (*TriDosha*) analysis		Advance meditation at high altitude showed biochemical and neuro-cognitive benefits
6.	George and Ilavarasu (2021)	Development and Psychometric Validation of the Music Receptivity Scale	English	$N = 322$	Music Receptivity Scale (MRS)	12 items	Music Receptivity Scale (MRS) is developed to study music receptivity and its psychometric validation. They discussed the effect of MRS in the field of music psychology

consistency of the intervention. When designing psychometric tests for MBIs, it is important to consider the following implications:

1. Validity: The test or questionnaire must be accurate in measuring what it is intended to measure, such as changes in emotional well-being or stress and anxiety levels.
2. Reliability: The test should be consistent in results over time so that any minute or deep change can be observed attributed to the MBIs rather than measurement error.
3. Sensitivity: The test should be able to detect slight changes in the outcome measure, such as changes in emotional well-being, that occur as an effect of the MBI.
4. Feasibility: The test should be practical to administer in the real-world setting where the MBI will be delivered.
5. Cultural sensitivity: The test should be culturally sensitive and consider different cultures and their language.
6. Privacy and ethical concerns: The test should ensure the privacy and confidentiality of the participants and comply with ethical guidelines for research.
7. Consider the specific MBI: The design of the psychometric test should be tailored to the specific MBI being studied and consider the theoretical framework and mechanisms of action underlying the MBI.

The psychometric parameters and strong reliability of the test will be a potential tool in assessing the multidimensional properties of the MBIs and may provide a framework for implementing various public health-related policies to clinicians and researchers working on health-promoting programs.

8.4 Ayurgenomics: A New Era of Molecular *Ayurveda*

Ayurveda is a natural medical system that was established in India during the Vedic era in primeval times (Pandey et al. 2013; Sharma and Keith 2020). The word "Ayurveda" is derived from the Sanskrit language and is split into two words *Ayus* and *Veda*. *Ayus* refers to life and *Veda* refers to knowledge or science. Therefore, it is perceived as the knowledge or science of life. Unfortunately, many of the theories related to *Ayurveda* were elusive for a long time which led to inconsistent explanation in the current literature (Sumantran and Tillu 2013).

According to "The *Tridosha* Theory of *Ayurveda*," there are three primary forces that are responsible for overseeing the physiology of everyone which are known as *Doshas*. These *Doshas* are named *Vata, Pitta*, and *Kapha*. Each of them exhibits specific attributes and plays crucial roles in regulating the physiology of an individual. *Vata Dosha* plays a vital role in transferring different biomolecules present in the body to the nervous system (Prasher et al. 2016, 2017). Furthermore, it is involved in several physiological functions such as cell division, cell signaling, excretion, and movement. *Pitta*, on the other hand, is in charge of controlling several body processes, including digestion, metabolism, energy balance, thermoregulation, vision, pigmentation, and host protection (Wallace 2020). Lastly, *Kapha* is indulged in the storage of substances and the development and maintenance of cells in the individual (Sumantran and Tillu 2011).

The *Tridosha* itself is composed of the five *Mahabhutas,* which are air (*Vayu*), water (*Jal*), fire (*Agni*), earth (*Bhumi*), and space (*Aakash*). One or more of these five *Mahabhutas* has a dominant influence on the others, with the other four having a lesser effect (Rotti et al. 2014a). There can never be a situation when one of the *Mahabhutas* is entirely absent. Life cannot exist without all five *Mahabhutas*. Moreover, these *Tridosha* must be present in balanced amounts in an individual to maintain good health and well-being of the individual (Gupta 2015). Disbalance in either *Vata*, *Pitta*, or *Kapha* levels can lead to the development of health hazards, the onset of *doshic* illnesses and diseases in the individual (Shilpa and Venkatesha Murthy 2011). Space and *Vayu* combine to form the *Vata Dosha* (air), and *Tejas* or *Agni* (fire) and *Ap* (water) constitute the *Pitta Dosha*, whereas *Ap* (water) and *Prithvi* (earth) make up the *Kapha Dosha* (Sharma and Prajapati 2020).

Prakriti refers to a person's constitution before birth with all three *Doshas* (*Vata*, *Pitta*, and *Kapha*) present in balanced and appropriate amounts. It remains the same throughout the lifetime of a person and doesn't change. Several factors including the genetic makeup of an organism, the environment in which the fetus develops and grows, maternal diet, and the lifestyle of the mother are responsible for attributing a unique percentage to these three *Doshas* in an individual. However, differences in culture, ethnicity, and ancestral characteristics, have a strong impact on *Prakriti*. *Vikruti*, on the other hand, refers to the stage after a person is created or born. A healthy environment promotes a healthy embryo. If this environment is maintained idyllic after birth, the child grows up into a healthy individual. The three *Doshas*, however, disrupt normal physiology in an unfavorable environment, which promotes the emergence of several diseases. All three *Doshas* are present in different amounts in *Prakriti* and *Vikruti* (Sharma and Prajapati 2020).

Ayurgenomics is a combination of "Ayurveda" and "Genomics" and a field of contemporary research where some scientific proof is being produced for the basis of Ayurveda (Patwardhan and Bodeker 2008). The fundamental idea of Ayurgenomics is that since the *Tridosha* system is present in all creatures, there must be a mechanism for it to be inherited. *Prakriti*, to put it simply, must be a phenotypical experience resulting from a certain genomic expression. Ayurgenomics has been introduced to integrate concepts in *Ayurveda* with modern genetics (gene profiling). It correlates the combination of three *Doshas* (*Vata*, *Pitta*, and *Kapha*) with the expression profile of specific genes and physiological characteristics (Prasher et al. 2016; Wallace 2020; Govindaraj et al. 2015). The *Doshas* reflect many aspects of an individual's physiology and metabolic pathways involving several enzymes. Protein enzymes are inherited, and although understanding the expression of enzymes involved in complex metabolic pathways is extremely challenging, significant progress has been made in recognizing the metabolic processes of certain illnesses and the specific genes associated with them.

8.5 Physiological Parameters as a Determinant of Psychometric Testing and Ayurgenomics

Ayurgenomics is a new and recently introduced domain of research. So, there is a lack of studies highlighting the direct correlation of genomic data with various changes in psychometric parameters. However, it is a well-known fact that all psychometric parameters are directly or indirectly reflections of our physiological alteration in the body. Moreover, a handful of literature has also documented a definite association of genomic data with physiological changes in the body. So, these pieces of evidence provide indirect evidence for the association of Ayurgenomics with psychometric parameters. In this section of the chapter, we have tried to assemble data about the association of *Prakriti* with metabolism, chronic illness, and currently recognized genome types.

Accumulating literature reveals that there is an association of *Prakriti* with the genetic makeup of an organism (Sharma and Prajapati 2020; Hankey 2005; Vianu 2015). Bhushan and coauthors assessed various alleles of human leucocyte antigen (HLA) DRB1 in people with differing *Prakriti* phenotypes and realized that the allele frequency is altered in different *Prakriti* types. Consistent with previous studies, they too observed a strong relationship between HLA type and *Prakriti* phenotype. There was no expression of HLA DRB1*02 allele in *Vata* type and HLA DRB1*13 allele in *Kapha* type. Similarly, the rate of occurrence of HLA DRB1*10 allele in *Kapha* was higher as compared to phenotypes *Pitta* and *Vata* (Bhushan et al. 2005).

Prasher et al. studied the relationship between different *Prakriti* groups—*Vata*, *Pitta*, and *Kapha*—and expression of genome. The authors investigated the expression of the CYP2C19 gene in healthy participants differing in *Prakriti* types. CYP2C19 gene plays a key role in metabolizing and processing nearly all sorts of drugs. Their findings demonstrated a strong relationship between CYP2C19 and the *Prakriti* phenotype. Furthermore, they showed upregulated expression of genes in *Pitta* dominants which are involved in the inflammatory pathways. Metabolic disease and inflammatory biomarkers were predominantly expressed in *Kapha* males compared to *Vata* males. However, Prothrombin, a biomarker of blood coagulation, in *Kapha* males was found to be decreased. Similarly, *Kapha* males expressed lower levels of fibrinolytic genes and elevated levels of genes involved in Adenosine triphosphate (ATP) production than *Pitta* and *Vata*. A lipid profile was conducted in these *Prakriti* participants and the results showed upregulated values of total cholesterol, triglycerides, high-density lipoprotein (HDL), and low-density lipoprotein (LDL) in *Kapha* phenotype as compared to *Pitta* and *Vata phenotypes*. Participants with *Pitta* phenotype showed upregulated expression of hemoglobin gene as compared to *Vata* and *Kapha* phenotypes which suggested that there are differences in gene expression between differing *Prakriti* subtypes and a strong link exists between them (Prasher et al. 2008). Ghodke et al. in 2011 investigated the same CYP2C19 gene and correlated its expression in different *Prakriti* phenotypes. Their findings revealed that the expression of this gene is significantly reduced in *Kapha* types and elevated in *Pitta* types (Ghodke et al. 2011).

Agarwal et al. studied the connection between EGLN1 gene (variants rs479200 (C/T) and rs480902 (T/C)) in different *Prakriti* phenotypes at high altitude. EGLN1 gene, mainly involved in the cellular response to hypoxia, was greatly expressed in *Kapha* phenotype than *Vata* or *Pitta*. They also demonstrated that TT genotype of rs479200 allele of EGLN1 gene was highly expressed in *Kapha*, whereas in *Pitta*, it was minimally expressed at high altitudes (Aggarwal et al. 2010). Findings by Juyal and colleagues revealed an increased expression of inflammatory genes in *Vata* than in *Pitta* or *Kapha*. Similarly, in *Pitta* and *Kapha*, the protein sequences indulged in the oxidative stress regulation path were upregulated (Juyal et al. 2012).

A lot of *Prakriti* diagnostic methods have been used by various studies, but unfortunately, none of them, except a few, have been assimilated (Rastogi and Chiappelli 2012; Rotti et al. 2014b). Studies conducted to identify connections between genes associated with immune-mediated pathways and with *Kapha*, *Pitta*, or *Vata Prakriti* types have demonstrated similar findings. CD25 expressing B cells, which are regulated and activated by natural killer cells expressing CD56, are shown to be upregulated in higher amounts in *Kapha* types than in *Vata* or *Pitta*. B cells play a crucial role in adaptive immune response and are responsible for providing humoral immunity (Gururaj et al. 2004).

Govindaraj et al. discovered variation in 52 single nucleotide polymorphisms (SNPs) in these three phenotypes of *prakriti*. SNPs are the most frequent type of genomic change among people. Every SNP portrays a variation in a particular nucleotide. They found that the SNP in the PGM1 gene (rs11208257) is similar in *Pitta* than in *Kapha* and *Vata* (Govindaraj et al. 2015). PGM1 (Phosphoglucomutase 1) is an enzyme that regulates glycolytic pathway and expedites the two-way bi-connection of glucose 1-phosphate (G-1-P) and glucose 6-phosphate (G-6-P) (60).

The ancient writings of *Ayurveda* also contain suggestions for sustaining lives with harmony of one's *Prakriti* for continuing well-being of an individual in a customized way, as *Prakriti* is tied with specific physical and mental inclinations that influence vulnerability to illnesses (Gururaj et al. 2004). Although there are many literature available on each of the aforementioned topics, there is still a need for summarizing the present body of knowledge on *Ayurveda-Prakriti* and its connection with metabolism, chronic illness, and genotypes. All these concepts of association between *Prakriti* and genomic study will possibly be the focus of future interventions and research programs.

We also extrapolate from this information to foresee a day when it will be feasible to genotypically test infants to identify their *prakriti*, which may then assist us choose the kind of healthy lifestyle that will allow these newborns to live long, disease-free lives. This will be a ground-breaking development for individualized preventative medicine in people.

8.6 Psychological Personality Inventory Versus Ayurgenomic Inventory

- A psychological personality inventory is a test that measures an individual's personality traits, such as extroversion, neuroticism, and agreeableness. An Ayurgenomic inventory is a test that measures an individual's Ayurvedic constitution, or *Dosha*, which is believed to reflect the individual's physical and mental characteristics and tendencies.
- The correlation between psychological personality inventories and Ayurgenomic inventories is not well studied. However, it is possible that there may be some overlap between the two types of inventories. For example, a person with a *Vata* Ayurgenomic constitution may be more prone to anxiety and nervousness, which could be related to high scores on neuroticism in a personality inventory.
- The main difference between psychological personality inventories and Ayurgenomic inventories is the theoretical framework and cultural context in which they are developed and used. Psychological personality inventories are developed within the Western psychological tradition and are used to understand and predict an individual's behavior and personality traits. Ayurgenomic inventories, on the other hand, are developed within the *Ayurvedic* tradition of medicine and are used to understand an individual's physical and mental characteristics and tendencies to provide personalized health recommendations.

8.7 Clinical Implication

There is widespread use of MBIs treatment for many psychiatric and noncommunicable disorders. The popularity of MBIs in the world has grown exponentially, especially after the COVID-19 pandemic. Psychometric testing is an important tool for assessing the impact of MBIs and it should have all the qualities of an ideal testing, including validity, reliability, sensitivity, feasibility, and cultural sensitivity. There are still some bridges missing or need to connect the MBIs as a clinical intervention.

8.8 Future Direction and Conclusion

Integrating Ayurgenomics and a set of standard psychometric tests will help develop a reliable roadmap for understanding human biology in relation to health and disease. The application of Ayurgenomics would facilitate the development of psychometric testing and analysis after the screening of the participant based on the *Prakriti* assessment. Based on the Ayurgenomics data, an individual's prevalent *Prakriti* can be refined using specific genomic factors associated with expression of genes, genetics, epigenetic, and biochemical factors.

A set of standard guidelines should be followed when designing, evaluating, validating, and applying psychometric analysis for MBI programs, in conjunction with Ayurgenomics for their application as clinical interventions.

References

Aggarwal S, Negi S, Jha P, Singh PK, Stobdan T, Pasha MA et al (2010) EGLN1 involvement in high-altitude adaptation revealed through genetic analysis of extreme constitution types defined in Ayurveda. Proc Natl Acad Sci U S A 107(4):18961–18966. https://doi.org/10.1073/pnas.1006108107

Alsubheen SA, Wshah A, Goldstein R, Brooks D (2021) Psychometric properties of patient-reported outcome measures assessing self-efficacy in patients with chronic obstructive pulmonary disease (COPD): a systematic review. COPD 18(2):254–263

American Psychological Association (2019) Understanding psychological testing and assessment. https://www.apa.org/helpcenter/assessment

Baer RA, Smith GT, Allen KB (2004) Assessment of mindfulness by self-report. Assessment 11(3):191–206

Bhanushali D, Tyagi R, Limaye Rishi Nityapragya N, Anand A (2020) Effect of mindfulness meditation protocol in subjects with various psychometric characteristics at high altitude. Brain Behav 10(5):e01604

Bhushan P, Kalpana J, Arvind C (2005) Classification of human population based on HLA gene polymorphism and the concept of *Prakriti* in Ayurveda. J Altern Complement Med 11:349–353. https://doi.org/10.1089/acm.2005.11.349

Birdee GS, Sohl SJ, Wallston K (2015) Development and psychometric properties of the *Yoga* Self-Efficacy Scale (YSES). BMC Complement Altern Med 16:3

British Psychological Society (2019) About psychological testing. https://ptc.bps.org.uk/tests-and-testing/psychological-tests/about-psychological-testing

Brown KW, Ryan RM (2003) The benefits of being present: mindfulness and its role in psychological well-being. J Pers Soc Psychol 84(4):822–848

Burg JM, Probst T, Heidenreich T, Michalak J (2017) Development and psychometric evaluation of the body mindfulness questionnaire. Mindfulness 8(3):807–818

Chiesa A, Serretti A (2014) Are mindfulness-based interventions effective for substance use disorders? A systematic review of the evidence. Subst Use Misuse 49(5):492–512

Cleanthous S, Bongardt S, Marquis P, Stach C, Cano S, Morel T (2021) Psychometric analysis from EMBODY1 and 2 clinical trials to help select suitable fatigue PRO scales for future systemic lupus erythematosus studies. Rheumatol Ther 8(3):1287–1301

Gallego J, Aguilar-Parra JM, Cangas AJ, Rosado A, Langer ÁI (2016) Efecto de intervenciones-mente/cuerposobrelosniveles de ansiedad, estrés y depresiónenfuturosdocentes de edu. Revista de Psicodidactica 21(1):87–101

Gardi C, Fazia T, Stringa B, Giommi F (2022) A short mindfulness retreat can improve biological markers of stress and inflammation. Psychoneuroendocrinology 135:105579

Gardner FL, Moore ZE (2007) The psychology of enhancing human performance: the mindfulness-acceptance commitment (MAC) approach. Springer, New York

Gardner FL, Moore ZE (2012) Mindfulness and acceptance models in sport psychology: a decade of basic and applied scientific advancements. Can Psychol 53(4):309–318

Gendolla GHE (2000) On the impact of mood on behavior: an integrative theory and a review. Rev Gen Psychol 4(4):378–408

George M, Ilavarasu J (2021) Development and sychometric validation of the Music Receptivity Scale. Front Psychol 11:585891

Ghodke Y, Joshi K, Patwardhan B (2011) Traditional medicine to modern pharmacogenomics: Ayurveda Prakriti type and CYP2C19 gene polymorphism associated with the metabolic variability. Evid Based Complement Alternat Med 2011:249528. https://doi.org/10.1093/ecam/nep206

Govindaraj P, Nizamuddin S, Sharath A, Jyothi V, Rotti H, Raval R et al (2015) Genome-wide analysis correlates Ayurveda Prakriti. Sci Rep 5:15786. https://doi.org/10.1038/srep15786

Grossman P (2008) On measuring mindfulness in psychosomatic and psychological research. J Psychosom Res 64(4):405–408

Gupta PD (2015) Pharmacogenetics, pharmacogenomics and Ayurgenomics for personalized medicine: a paradigm shift. Indian J Pharm Sci 77:135–141. https://doi.org/10.4103/0250-474x.156543

Gururaj A, Barnes CJ, Vadlamudi RK, Kumar R (2004) Regulation of phosphoglucomutase 1 phosphorylation and activity by a signaling kinase. Oncogene 23:8118–8127. https://doi.org/10.1038/sj.onc.1207969

Hankey A (2005) The scientific value of Ayurveda. J Altern Complement Med 11(2):221–225. https://doi.org/10.1089/acm.2005.11.221

Hartfiel N, Clarke G, Havenhand J, Phillips C, Edwards RT (2017) Cost-effectiveness of *Yoga* for managing musculoskeletal conditions in the workplace. Occup Med (Lond) 67(9):687–695

Hayes SC, Wilson KW, Gifford EV, Follette VM, Strosahl K (1996) Experiential avoidance and behavioral disorders: a functional dimensional approach to diagnosis and treatment. J Consult Clin Psychol 64(6):1152–1168

Juyal RC, Negi S, Wakhode P, Bhat S, Bhat B, Thelma BK (2012) Potential of Ayurgenomics approach in complex trait research: leads from a pilot study on rheumatoid arthritis. PLoS One 7:e45752. https://doi.org/10.1371/journal.pone.0045752

Lichtenberger EO, Kaufman AS (2013) Essentials of WAIS-IV assessment, 2nd edn. Wiley, Hoboken

Linehan MM (1993) Cognitive-behavioral treatment of borderline personality disorder. Guilford Press, New York

Lolla A (2018) Mantras help the general psychological well-being of college students: a pilot study. J Relig Health 57(1):110–119

Mirandola M, Sabogal Rueda MD, Andreis F, Meriggi F, Codignola C, Gadaldi E, Prochilo T, Libertini M, Di Biasi B, Abeni C, Noventa S, Rota L, Ogliosi C, Zaniboni A (2019) *Yoga* protocol for cancer patients: A systematic exploration of psychophysiological benefits. Rev Recent Clin Trials 14(4):261–268

Montgomery GH, Sucala M, Dillon MJ, Schnur JB (2017) Cognitive-behavioral therapy plus hypnosis for distress during breast radiotherapy: a randomized trial. Am J Clin Hypn 60(2):109–122

Noordali F, Cumming J, Thompson JL (2017) Effectiveness of mindfulness-based interventions on physiological and psychological complications in adults with diabetes: a systematic review. J Health Psychol 22(8):965–983

Pandey MM, Rastogi S, Rawat AKS (2013) Indian traditional Ayurvedic system of medicine and nutritional supplementation. Evid Based Complement Altern Med 2013:376327. https://doi.org/10.1155/2013/376327

Patwardhan B, Bodeker G (2008) Ayurvedic genomics: establishing a genetic basis for mind-body typologies. J Altern Complement Med 14(5):571–576. https://doi.org/10.1089/acm.2007.0515

Perez-Achiaga N, Nelson S, Hassiotis A (2009) Instruments for the detection of depressive symptoms in people with intellectual disabilities: a systematic review. J Intellect Disabil 13:55–76

PessidjoDjomatcho L, Kowo MP, Ndam AN et al (2021) Normalisation of the psychometric encephalopathy score within the Cameroonian population. BMC Gastroenterol 21:287

Poli A, Maremmani AGI, Gemignani A, Miccoli M (2022) Randomized trial on the effects of a mindfulness intervention on temperament, anxiety, and depression: a multi-arm psychometric study. Behav Sci (Basel) 12(3):74

Prasher B, Negi S, Aggarwal S, Mandal AK, Sethi TP, Deshmukh SR et al (2008) Whole genome expression and biochemical correlates of extreme constitutional types defined in Ayurveda. J Transl Med 6:48. https://doi.org/10.1186/1479-5876-6-48

Prasher B, Gibson G, Mukerji M (2016) Genomic insights into Ayurvedic and Western approaches to personalized medicine. J Genet 95:209–228. https://doi.org/10.1007/s12041-015-0607-9

Prasher B, Varma B, Kumar A, Khuntia BK, Pandey R, Narang A et al (2017) Ayurgenomics for stratified medicine: TRISUTRA consortium initiative across ethnically and geographically diverse Indian populations. J Ethnopharmacol 197:274–293. https://doi.org/10.1016/j.jep.2016.07.063

Rao RM, Nagendra HR, Raghuram N, Vinay C, Chandrashekara S, Gopinath KS, Srinath BS (2008) Influence of *Yoga* on mood states, distress, quality of life and immune outcomes in early stage breast cancer patients undergoing surgery. Int J Yoga 1(1):11–20

Rao RM, Vadiraja HS, Nagaratna R, Gopinath KS, Patil S, Diwakar RB, Shahsidhara HP, Ajaikumar BS, Nagendra HR (2017) Effect of *Yoga* on sleep quality and neuroendocrine immune response in metastatic breast cancer patients. Indian J Palliat Care 23(3):253–260

Rastogi S, Chiappelli F (2012) Development and validation of a prototype *Prakriti* analysis tool (PPAT): inferences from a pilot study. Ayu 33:209–218. https://doi.org/10.4103/0974-8520.105240

Rotti H, Guruprasad KP, Nayak J, Kabekkodu SP, Kukreja H, Mallya S et al (2014a) Immunophenotyping of normal individuals classified on the basis of human *Dosha* Prakriti. J Ayurveda Integr Med 5:43–49. https://doi.org/10.4103/0975-9476.128857

Rotti H, Raval R, Anchan S, Bellampalli R, Bhale S, Bharadwaj R et al (2014b) Determinants of Prakriti, the human constitution types of Indian traditional medicine and its correlation with contemporary science. J Ayurveda Integr Med 5:167–175. https://doi.org/10.4103/0975-9476.140478

Rust J, Golombok S (2009) Modern psychometrics: the science of psychological assessment. Routledge, East Sussex

Segal ZV, Williams JMG, Teasdale JD (2002) Mindfulness based cognitive therapy for depression: a new approach to preventing relapse. Guilford Press, New York

Sharma H, Keith WR (2020) *Ayurveda* and epigenetics. Medicina (Kaunas) 56(12):687. https://doi.org/10.3390/medicina56120687

Sharma R, Prajapati PK (2020) Predictive, preventive and personalized medicine: leads from Ayurvedic concept of *Prakriti* (human constitution). Curr Pharmacol Rep 6:441–450. https://doi.org/10.1007/s40495-020-00244-3

Shilpa S, Venkatesha Murthy CG (2011) Understanding personality from Ayurvedic perspective for psychological assessment: a case. Ayu 32:12–19. https://doi.org/10.4103/0974-8520.85716

Silvia PJ, Nusbaum EC, Eddington KM, Beaty RE, Kwapil TR (2014) Effort deficits and depression: the influence of anhedonic depressive symptoms on cardiac autonomic activity during a mental challenge. Motiv Emot 38(6):779–789

Silvia PJ, Mironovová Z, McHone AN, Sperry SH, Harper KL, Kwapil TR, Eddington KM (2016) Do depressive symptoms "blunt" effort? An analysis of cardiac engagement and withdrawal for an increasingly difficult task. Biol Psychol 118:52–60

Smith JC (2019) Third-generation mindfulness & the universe of relaxation: professional version. Kendall Hunt Publishing Company, Dubuque

Smith JC (2021) The psychology of relaxation, meditation, and mindfulness. In: Principles and practice of stress management, vol 39

Sumantran V, Tillu GS (2011) Ayurvedic pharmaceutics and insights on personalized medicine. researchgate.net. undefined

Sumantran VN, Tillu G (2013) Insights on personalized medicine from Ayurveda. J Altern Complement Med 19:370–375. https://doi.org/10.1089/acm.2011.0698

Thind H, Sillice MA, Fava JL, Lantini R, Horowitz S, Jennings E, Rosen RK, Carmody J, Becker BM, Marcus BH, Bock BC (2017) Development and validation of the outcome expectations for *Yoga* scale. Am J Health Behav 41(6):796–802

Vianu V (2015) Invited article foreword. J ACM 62:297–303. https://doi.org/10.1145/2831493

Wallace RK (2020) Ayurgenomics and modern medicine. Medicina 56:661. https://doi.org/10.3390/medicina56120661

Chapter 9
Neuro-molecular Effects of *Yoga* and Endurance Exercises: Similarities and Dissimilarities

Kanupriya Sharma, Khushboo Arya, Saurabh Kumar, and Sheetal Sharma

Contents

Kanupriya Sharma and Khushboo Arya contributed equally with all other contributors.

K. Sharma
Department of Neurology, Postgraduate Institute of Medical Education and Research, Chandigarh, India

K. Arya
Department of Experimental Medicine and Biotechnology, Postgraduate Institute of Medical Education and Research, Chandigarh, India

Department of Yoga and Life Sciences, SVYASA, Bengaluru, India

S. Kumar
CCRYN—Collaborative Center for Mind-Body Intervention through Yoga, Post Graduate Institute of Medical Education and Research, Chandigarh, India

S. Sharma (✉)
Department of Experimental Medicine and Biotechnology, Postgraduate Institute of Medical Education and Research, Chandigarh, India
e-mail: sharma.sheetal@pgimer.edu.in

A. Anand (ed.), *Neuroscience of Yoga*,
https://doi.org/10.1007/978-981-97-2851-0_9

Abstract Ageing can be described as decline at molecular, cellular and physiological levels in a progressive fashion and is a risk factor for diseases like cancer and cardiovascular and neurological disorders. Brain ageing parallels the ageing of other organs with a gradual decline in memory, learning, motor coordination and attention impairment. Evidence suggests that oxidative stress may considerably contribute to the ageing process. Such oxidative stress may have some residual epigenetic effect and impact neural function at the molecular level. Considerable evidence has shown better health through *Yoga* and endurance exercises at physiological and psychological levels. Yet, not much is known about the exact molecular mechanism or pathway that operates in the induction of beneficial effects of either exercise or *Yoga* on human health. This chapter traverses the impact of exercise and *Yoga* at the neuro-molecular level and discusses the similarities and dissimilarities between the two regimes.

Keywords Endurance exercises · Epigenetics · Lifestyle · Neuro-molecular · Physical activity

9.1 Introduction

Physical activity is considered an essential part of our daily routine in order to maintain good health of brain and body. Plenty of evidence supports the effects of exercise on overall health and brain functioning. Literature suggests the importance of physical activity, as it has become an economically viable therapeutic strategy and is progressively common for overall physical and mental health development (Nay et al. 2021). Regular practice of the physical activity helps to achieve better physical and mental health. Physical inactivity may be a contributor to the development and progression of many diseases like cardiovascular diseases (CVDs), neurodegenerative diseases, cancer, diabetes and other lifestyle disorders. WHO has recommended that 150 min of moderate intensity weekly physical activity would greatly reduce the global disease burden. Sedentariness can be described as any behaviour which implies energy expenditure below 1.5 metabolic equivalents (METs) while sitting, laying or reclining (Bull et al. 2020).

Yoga lifestyle and environmental factors interfere with gene activity most likely through epigenetic mechanisms. Alteration in chromatin which leads to the modification of gene transcription, thus influencing the gene function and health of the individual, is referred to as epigenetic mechanisms and is saved as "epigenetic memory." DNA methylation, histone modifications and small non-coding RNAs play an important role in epigenetics. There has been enormous progress in understanding the molecular mechanisms modulated by exercise or *Yoga* and its influence on the gene programme and its functional consequences. Studies have shown that physical activity alters gene pathways associated with immune response, inflammation and cardiovascular diseases (Ertek and Cicero 2012; Gleeson et al. 2011; Fernández-Ruiz 2019; Scartoni et al. 2020). Physical activity can be categorized into various forms, each with variable energy requirements and its effects, some are more focused on endurance building like endurance exercise or aerobics while other forms of physical activity like *Yoga* and *tai chi* focus on the breathing and mental aspects alongside the physical parameters. Both kinds of physical activities improve the overall health of an individual, but the mechanism of practice varies widely. In this chapter, we will discuss similarities or dissimilarities between the two forms of physical activity, at both molecular and epigenetic levels.

9.2 Effects of Ageing on Neural Function

Ageing can be defined as a progressive decline in the organisms' function which is a risk factor for major diseases including cancer, cardiovascular and neurological diseases (Delgado-Morales et al. 2017). Brain ageing parallels the ageing of other organs with shrinkage of frontal and temporal lobes leading to a gradual decline in memory, learning, motor coordination and attention (Peters 2006). Impacts of brain ageing at the molecular and cellular level involves: flawed mitochondrial function, accumulation of oxidatively damaged proteins intracellularly, damage to lipids and nucleic acids, energy metabolism dysregulation, impairment of cellular "waste disposal," compromised signalling of stress response, weakened mechanisms of DNA repair, -impaired neural network activity, exhaustion of stem cells and inflammation. A decline in the levels of neurotransmitters like dopamine and serotonin along with neurotrophic factors, i.e. BDNF, may be implicated with a decrease in neurogenesis and synaptic neuroplasticity. These changes in the neurotransmitters and neurotrophic factors can be further implicated with the decrease in cognitive function with increasing age (Peters 2006). A study on DNA methylation showed that genes with downregulated expression in Alzheimer's disease also show changes in DNA cytosine methylation in blood and brain, therefore suggesting epigenetic changes that start appearing with ageing and can influence the gene expression (Horvath et al. 2012). At the molecular level, genes associated with mitochondrial functioning show a downregulation while an upregulation in stress response pathways and apolipoprotein D was reported (Haigis and Yankner 2010; Dassati et al. 2014). Overall, genes responsible for the glial activity, myelin production, and immune and stress response show upregulated gene expression. In contrast, genes involved in mitochondrial function, neural plasticity and ubiquitin-proteasome pathways are downregulated in ageing brains (Tripathi 2012). During leisure time, physically active individuals were found to be biologically younger than those who had a sedentary lifestyle (Sedentary Lifestyles Associated With Accelerated Aging Process 2023). This group studied 2401 white twin volunteers and reported that telomere length of leukocyte had a positive correlation with physical activity in leisure time which remained significant after adjustment with age, sex, BMI, socioeconomic status and physical activity at work (Cherkas et al. 2008). There is growing evidence to suggest that an active lifestyle has a preventive effect on the functioning of the brain in the elderly population (Rolland et al. 2010).

9.3 Impact of Sedentary Lifestyle on Neuro-molecular Function

The brain is the most metabolically active organ which requires a large amount of oxygen for its function and thus a large producer of free oxygen radicals (ROS) (Winterbourn et al. 1985), leading to oxidative stress, and, therefore, the epicentre of neurodegeneration. The brain's high energy demand, rich lipid content and weak ROS scavenging system make it most vulnerable to oxidative damage. Superoxide anion (O_2^-), hydrogen peroxide (H_2O_2) and highly reactive hydroxyl radical (HO•) are the free radicals involved in neurodegeneration. Antioxidant system neutralizes the deleterious effects of ROS under normal conditions; however, the production of free radicals more than the neutralizing capacity of antioxidant system leads to extensive oxidation of protein and lipid peroxidation leading to degeneration of cells, diminishing Long-Term Potentiation (LTP), synaptic signalling and mechanisms involved in neural plasticity (Salim 2017). Studies have shown that brain regions, which include hippocampus, amygdala and granular cells of cerebellum, are prone to damage induced by oxidative stress (Salim et al. 2011).

Physical inactivity may be associated with an increase in oxidative stress whereby sedentariness leads to a number of chronic diseases (Laufs et al. 2005; Sánchez-Rodríguez et al. 2021). As oxidative stress is known to be associated with ageing, cardiovascular diseases and neurodegeneration, it may be that physical activity may exert its beneficial effects on each of these diseases through its effects on the management of redox status. It has been found that people who are extremely inactive exhibit lower levels of endogenous antioxidants (Moreno-Camacho et al. 2019). Evidence suggests that excessive sedentary behaviour not only increases the risk for chronic diseases like T2DM (Hamilton et al. 2014) but also increases the risk of cognitive decline (Raichlen et al. 2022). However, physical activity has shown a positive impact under both the above-mentioned conditions showing improvement in glycaemic control (Shah et al. 2021) and in cognitive parameters in a wide range of individuals, i.e. cognitively declined or individuals with dementia (Kumar et al. 2022; Mandolesi et al. 2018).

9.4 Impact of Epigenetics on Neuro-molecular Function Through Physical Activity

Epigenetics includes dynamic DNA methylation, differential 5-methylcytosine oxidized derivations, post-translational modifications of histone, histone variant incorporation into the chromatin and regulation of microRNA (miRNA). Epigenome changes allow gene readout alteration in the CNS cells which alters their function and physiology (Landgrave-Gómez et al. 2015). Neurodegenerative and cerebrovascular disorders could be a result of epigenetic alterations accumulated over time (Martínez-Iglesias et al. 2022). It has been observed that epigenetic dysregulation is involved in addiction, depression, stress and neurodegeneration (Abel and Poplawski 2014). A study identified that monozygotic twins have different DNA methylation and differentially methylated regions in one with Major Depressive Disorder (MDD) compared to the unaffected sibling (Byrne et al. 2013). Levenson et al. revealed that increased acetylation of H3 histone helps in the formation of memory and further showed increased LTP induction by administration of HDAC inhibitor in mice (Levenson et al. 2004). Similarly, another study showed that maintenance of LTP could be reduced by administration of DNMT inhibitor (Sui et al. 2012). Several key genes involved in neuronal synthesis and maintenance are known to be regulated by CpG methylation, such as neurotropic factor BDNF (Martinowich et al. 2013). A study corroborated this finding by showing that epigenetic regulation can reverse the age-dependent changes in BDNF (Zeng et al. 2011). Both genetics (~25%) and environment (~75%) influence an individual's longevity (Akbarian et al. 2013). Further, the influence of environmental factors, in particular stress on alteration of epigenetic markers, has also been well documented (Szyf 2012; Provençal et al. 2013; Turecki and Meaney 2016). Early-life adverse experiences, exposure to traumatic events leading to post-traumatic stress disorder (PTSD), and mood disorders are related to whole-genome DNA methylation (McGowan and Kato 2008; Roth and Sweatt 2011; Roth et al. 2011). Further, depressive disorders are related to modification in histone acetylation and changes in sncRNA activity (Su et al. 2015; Misztak et al. 2018). Exposure to stress during early life could lead to the methylation of genes like glucocorticoid receptors and serotonin 1A receptors (McGowan et al. 2009; Le François et al. 2015). A study showed that exposure to prenatal maternal stress leads to changes in the pattern of DNA methylation of blood cells, which may influence the immune function of the offspring (Cao-Lei et al. 2014). A study by McGowan et al. (2009) found that victims of suicide who have a history of abuse during childhood had hypermethylation of glucocorticoid receptors. A study by Vriens et al. explored the effect of screen time which was correlated with sedentary behaviour. They found that higher screen time was associated with higher miR-222 and miR146a levels in saliva, which led the authors to propose these miRNAs as target candidates to explore the effects of sedentary behaviour on children's health (Lee et al. 2021). A study by Wu et al. recruited 59 obese and 39 controls and assessed for methylation at 7 CpG islands in FAIM2 (associated with obesity) promoter region, wherein it was found that sedentary behaviour was significantly associated with FAIM2 promoter methylation. Further, a significant difference between the methylation level in obese and lean participants was observed based on high or moderate physical activity level, i.e. <150 min/week (Wu et al. 2015).

There is a dynamic epigenetic remark to external signals which can be modified depending on the availability of metabolites and co-factors during adulthood, and it appears that diet also impacts this process. When post-weaning mice were fed with metabolites like folate, methionine and choline deficient diet, this led to hypermethylation of promoters and lower expression of AMPA (amino-3-hydroxy-5-methyl-4-isoxazolepropionic acid) receptors and also a reduction in hippocampal dependent memory (Tomizawa et al. 2015). Modifying diet like calorie restriction has improved

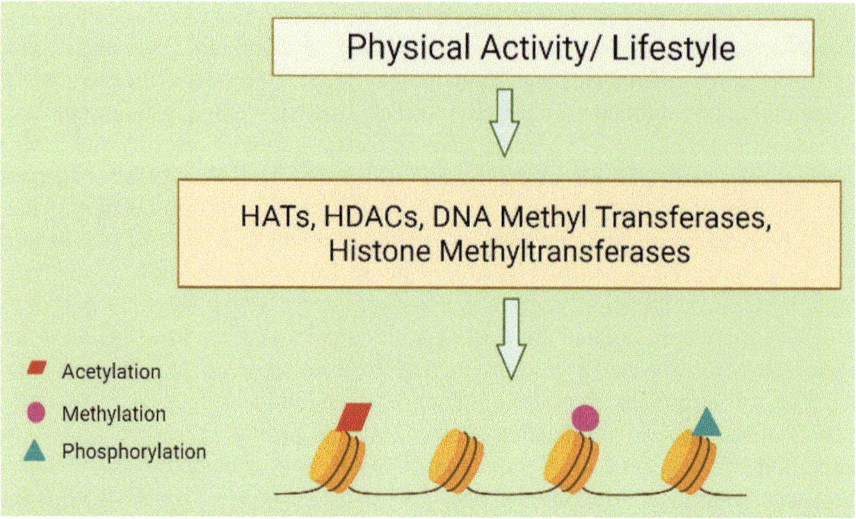

Fig. 9.1 Epigenetic modulation through enzymatic action induced by lifestyle and physical activity (Created with BioRender.com)

cognition and neurogenesis and has induced a de novo pattern in the ageing brain (Hadad et al. 2018). These studies suggest that lifestyle and epigenetic influence on the neuro-molecular function are profoundly intertwined, and therefore modulation of lifestyle can significantly influence the epigenetic mechanisms.

Physical activity is a lifestyle parameter which contributes towards epigenetic changes. Studies show that regular physical activity has a positive impact on the expression of histone proteins and DNA methylation and may also be associated with the prevention of certain diseases like cancer and diabetes. Also, in the elderly, these changes are associated with the antioxidant activity of the cells (Rezapour et al. 2018). In fact, it was observed that the elderly who were physically active had methylation patterns which increased the insulin sensitivity and genes with energy metabolism, myogenesis, contractile properties and oxidative stress resistance in the skeletal muscles of aged individuals (Sailani et al. 2019) (Fig. 9.1).

9.5 Molecular, Cellular and Epigenetic Effects of Endurance Exercise

Generally, there are two types of exercises, resistance and endurance. When an individual performs an exercise against a relatively high load for a shorter time, it is the strength or resistance form of exercise, whereas endurance or aerobic form of exercise is carried out over a long time against a relatively low load (Hughes et al. 2018). The one who performs endurance exercise develops improved oxygen consumption and delivery, whereas training with resistance exercise develops stronger and bigger muscles. Considerable evidence has shown better health, both physiological and psychological, in physically active individuals with some studies demonstrating the epigenetic response associated with endurance exercises. Interestingly, miR-20a and miR-21 were found to be upregulated in athletes after 90 min of endurance exercise, suggesting their role in exercise adaptation (Baggish et al. 2011). Another study on 12 young athletes showed that after a progressive cycling ergometer exercise test for ten rounds of 2 min, a total of 19 miRNAs were altered, which were involved in endocytic and p53 signalling pathway and also JAK-STAT pathway, and this was found to be associated with atherosclerotic vasculature, showing the anti-atherosclerotic effect of miRNA regulation through exercise (Radom-Aizik et al. 2014). One study on healthy subjects who were not involved in heavy physical activity earlier underwent a four-week training programme, and the methylation profile of their DNA samples were compared before and after the intervention; the sites which were associated with cardiovascular physiology, focal adhesion, signalling of calcium and MAPK pathways showed methylation response to exercise which is involved in the maintenance of cardiovascular health (Denham et al. 2015). At the neuro-molecular level, an in vivo study showed that after one week of voluntary wheel running, acetylation of global H3 was increased in the hippocampus (involved in learning and memory) and cerebellum (involved in motor function), and this was found to be correlated with increase in BDNF levels in running mice, concluding that exercise induces changes in genes of synaptic plasticity, epigenetic regulatory pathways and signalling mechanism in both cerebellum and the hippocampal region of adolescent mice brain (Abel and Rissman 2013). Evidences suggest that with increasing age, there is a decline in BDNF (Erickson et al. 2010; Oh et al. 2016), which has a role in adult nervous system in order to support neuronal survival, differentiation and plasticity of existing neurons (Bathina and Das 2015), and decline in BDNF is an important target of regulation by ageing, and a number of studies have revealed exercise-induced variation in BDNF (Walsh and Tschakovsky 2018; Miranda et al. 2019), showing that exercise enhances cognition through the regulation of neurotropic factors epigenetically (De la Rosa et al. 2019).

These studies are suggestive of the role of molecules and the mechanisms which could be affected if exercise or physical activities are involved in the lifestyle. The mechanism could be in the recovery from the disease state or could be associated with the prevention of the disease.

9.6 Molecular, Cellular and Epigenetic Effects of *Yoga*

Yoga is an emerging field which combines physical practice along with breathing practices and meditation. The empirical evidence suggests the role of *Yoga* in regulating the endocrine system in several ways, by the increase in the level of serotonin and melatonin and decrease in the level of cortisol through regular practice. *Pranayama* has respiratory benefits, and *Asana* and *Dhyana* reduce sympathetic activation. Its effect results in increase of GABA levels and it also regulates the HPA axis. The regular practice of *Yoga* indicates its preventative effects on several metabolic syndromes and cardiovascular diseases by lowering the oxygen consumption rate, resting heart rate and decreasing BMR and BMI (Kuppusamy et al. 2020). It is considered a holistic approach towards health. Evidence has suggested that *Yoga* helps in the management and improvement of health at physical and physiological levels along with psychological aspects. A 2016 study explored the effects of *Yoga* on women suffering from psychological distress and reported a reduced methylation in TNF region in *Yoga* group that was given 8 weeks twice weekly one-hour *Yoga* sessions in comparison to waitlisted control group (Harkess et al. 2016a). A single arm exploratory trial in 2017 reported that after *Yoga* and meditation-based lifestyle intervention (YMLI) for 12 weeks among 96 healthy individuals, there was an improvement in cardinal and metabotropic biomarkers of cellular ageing, 8-hydroxy-2' -deoxyguanosine (8-OHdG), ROS, cortisol and IL-6 significantly reduced and β-endorphin, telomerase, BDNF and sirtuin 1 increased significantly after YMLI practice (Tolahunase et al. 2017a). A study by Su Qu et al. in 2013 explored global gene expression changes in peripheral blood mononuclear cells (PBMCs) of healthy individuals who practised either a *Yoga* programme which included *Yoga* postures, breathing exercises and *Sudarshan Kriya* Meditation, and that of a control group who practised nature walks or listened to relaxing music. The results showed that the *Yoga* and meditation practice led to rapid gene expression alteration which may be related to long-term cell biology and higher-level health effects (Qu et al. 2013). An animal study published in 2017 showed that when rhythms were imposed in Anterior Cingulate Cortex of mouse by optogenetics, it imposed similar effects as meditation training, and it further led to decreased anxiety-like behaviours when optogenetic stimulation was provided at different frequencies in comparison to the control group animals (Weible et al. 2017). *Yoga*-like practices such as *Tai chi*, which is also a mind-body-based practice, have shown an improvement in the lifestyle of cancer survivors through the gene regulation by reducing cortisol and also showed a reduction in systolic blood pressure after a week of intervention (Amini et al. 2020).

The studies discussed above highlight the association of *Yoga* practices with an impact on the gene expression at molecular and cellular levels, which could be mediated through the epigenetic mechanism. These studies highlight the influence of including *Yoga* as a part of lifestyle at the molecular and epigenetic levels.

9.7 Exercise Versus *Yoga*

Yoga and exercise are two lifestyle interventions, which have an influence on the overall health of an individual. The effects of both practices are mediated through epigenetic regulation. But besides being physical activity-based lifestyle interventions, the overall impact of these two kinds of practices varies. *Yoga*, which is a breath regulation-based practice, involves physical (*Asana*), breathing practices (*Pranayama*) and meditative practices.

Yoga, commonly associated with spirituality, consciousness and well-being, was originally considered a variety of physical exercise. Gradually, public opinion began to incline towards the unprecedented idea that *Yoga* represents physical, mental, emotional and spiritual well-being. Besides being appreciated as a mind-body medicine for enhancing the mind's capacity (Erwin Wells et al. 2011), *Yoga* is often reviewed as a therapy focusing on brain, mind, body and behaviour and the interrelation between them (Wahbeh et al. 2008). Indeed, *Yoga* today is recognized as an alternative and adjunctive system of therapy. *Yoga* as a healthcare system is so widely accepted that scientists find it imperative to understand the mechanisms that make *Yoga* so effective. So far, evidence-based studies have documented positive effects of *Yoga* on inflammatory response (Kiecolt-Glaser et al. 2010; Gautam et al. 2019; Kaminsky et al. 2017; McCarthy et al. 2017; Harkess et al. 2016b), oxidative stress (Yadav et al. 2019; Tolahunase et al. 2017b) and neuroplasticity (Gautam et al. 2019; Nirwan et al. 2021). There are reports suggesting that *Yoga* offers relief in DNA damage as well (Nirwan et al. 2021; Bisht et al. 2020) and is linked with the increase in relative telomere length (Ornish et al. 2013; Tolahunase et al. 2018). Evidence supports the role of *Yoga* in maintaining the cardio-vagal tone, bringing a balance between the sympatho-adrenal and parasympathetic systems, and in down-regulating the responses of the HPA axis (Gautam and Dada 2023) (Fig. 9.2).

Exercise, on the other hand, focuses on neuro-muscular or cardio-respiratory training. It can be aerobic/anaerobic or a combination of both. It is planned and structured for the overall conditioning of the physical body—still is also correlated with an increase in general well-being as well as quality of life (Pozuelo-Carrascosa et al. 2017; Galper et al. 2006). It is also observed that physical exercise safeguards from stress and stress-induced diseases and disorders. It may enhance neuroendocrine and physiological responses, as well as improve anti-inflammatory factors and neuroplasticity (Silverman and Deuster 2014; Fragala et al. 2011).

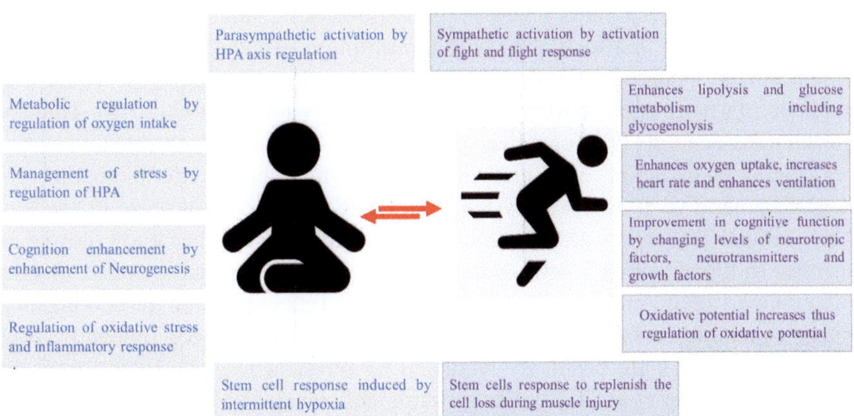

Fig. 9.2 Endurance exercise versus *Yoga*

With all that considered, physical exercise leads to an immediate escalation in heart rate with a rise in sympathetic outflow to skeletal muscle vasculature, followed by withdrawal of parasympathetic activity (Fisher et al. 2015). On the contrary, decreased sympathoadrenal activity and enhanced parasympathetic tone along with a reduction in stress hormone secretion are associated with *Yoga* (Jerath et al. 2014). Inflammatory response may be diminished upon regular exercise and may upregulate endogenous anti-inflammatory response. However, acute exercise may stimulate pro-inflammatory responses and induce oxidative stress (Simpson et al. 2015; Fiuza-Luces et al. 2018). Evidence-based studies have revealed that *Yoga* may play a significant role in reducing oxidative stress (Yadav et al. 2019; Tolahunase et al. 2017b). Nevertheless, both physical exercise and *Yoga* reduce the levels of cytokines (Kiecolt-Glaser et al. 2010; McCarthy et al. 2017; Harkess et al. 2016b; Spielman et al. 2016). Evidence also suggests that regulation of BDNF (neurogenic marker) may enhance the inherent DNA repair capability of neurons (Song et al. 2015). In addition to that, both *Yoga* and physical exercise are known for upregulating BDNF (Nirwan et al. 2021; Nascimento et al. 2015; Cahn et al. 2017; Naveen et al. 2013, 2016). Also, it may be deduced that levels of peripheral BDNF may mirror the levels in the brain, as BDNF traverses the blood-brain barrier (Cattaneo et al. 2016). Despite the viewpoint that both *Yoga* and physical exercise may offset stress-induced neurodegenerative processes through preservation of neuroplasticity in the brain and tapering cellular ageing, a few studies have attempted to explore the mechanisms through which *Yoga* contributes to the maintenance of BDNF levels, whereas several studies have investigated mechanisms through which physical exercise affects the epigenetics of BDNF (Walsh et al. 2020; Neeper et al. 1995; El Hayek et al. 2019). Besides, physical exercise also promotes neurogenesis by moving radial neural stem/precursor cells (latent reservoir for neurons) from dormant to active state (Lugert et al. 2010). It is also known to maintain the homeostasis of the stem cell pool (Lugert et al. 2010; Dong et al. 2019)

We have already discussed the effect of physical exercise on BDNF expression. Further, it can also enhance the levels of activated MeCP2. Gomez-Pinilla et al. (2011) showed upregulated levels of phospho-MeCP2 after 7 days of exercise, which additionally led to BDNF gene transcription. Another study mentions the enhancement of neurogenesis through running, thereby accelerating the recovery of synaptic plasticity and recovery of memory capacity (Alam et al. 2018). The ten-eleven translocation (TET) enzymes (like Tet1 and Tet2) are responsible for catalysing the demethylation of 5-methylcytosine (5-mC; epigenetic biomarker for neurodegenerative diseases). Sølvsten et al. (2018) showed an increase in levels of hippocampal Tet1 in rats after 2 weeks of wheel training. Another study showed an increase in hippocampal 5-hmC content after 4 weeks of physical exercise, thereby offsetting an ageing-induced decrease in Tet1 and Tet2 expression, promoting cognitive functions in mice (Jessop and Toledo-Rodriguez 2018). Evidently, both MeCP2 and TETs are positively regulated by physical exercise (Pareja-Galeano et al. 2014; Zhang et al. 2020). Furthermore, deficiency of MeCP2 leads to depletion of microglia as neuroinflammatory diseases progress (Cronk et al. 2015).

Adding to this, Mela et al. (2020) demonstrated the effect of physical exercise on suppressing neuroinflammation by reducing the level of senescent microglia.

Physical exercise is also known to prevent oxidative damage of DNA, thus reversing related neuronal loss (Lu et al. 2017). DNA methylation is known to mediate the decrease in the expression and function of oxidative-damage-preventing agents like Nrf2 (nuclear erythroid 2-related factor 2) and SOD2 (superoxide dismutase 2) (Griñan-Ferré et al. 2016; Balasubramanian et al. 2021). In 2018, Wu et al. (2018) showed that physical exercise notably enhances the level and activity of Nrf2, which thus promotes antioxidant downstream genes like SOD2.

9.8 Clinical Implications

A physically active lifestyle during adulthood induces changes at the epigenetic levels which may contribute to the prevention of diseases like neurodegenerative disorders, cardiovascular diseases, certain types of cancer, metabolic disorders like atherosclerosis and type 2 diabetes. Physical activities like *Yoga* and endurance exercise may operate through epigenetic regulation, which has been shown through numerous available evidence. However, *Yoga* and endurance exercise have different mechanisms of action through which they regulate the mechanism of maintenance of good health.

9.9 Future Directions

We acknowledge the need for further exploration of both physical exercise' and *Yoga*'s potential for improvement in the symptoms of neurodegenerative diseases and reversal of ageing. Current literature provides us with a scope and direction for further inspection and investigation to build up a comprehensive guide for such regimes when used as adjunctive therapies.

9.10 Conclusion

Yoga and physical exercise share common features and differences in mechanisms and effects. The physiology and molecular pathways through which these two regimens function is understandably different, and a lot remains unexplored in the neuro-molecular field of *Yoga*. Though a lot more work is desired in the area of physical exercise as well, nevertheless, studies so far have provided us with a particular insight into the impact that physical exercise has on molecular mechanisms.

References

Abel T, Poplawski S (2014) Epigenetic advances in clinical neuroscience. Dialogues Clin Neurosci 16(3):273–275

Abel JLB, Rissman EF (2013) Running-induced epigenetic and gene expression changes in the adolescent brain. Int J Dev Neurosci 31(6):382–390

Akbarian S, Beeri MS, Haroutunian V (2013) Epigenetic determinants of healthy and diseased brain aging and cognition. JAMA Neurol 70(6):711–718

Alam MJ, Kitamura T, Saitoh Y, Ohkawa N, Kondo T, Inokuchi K (2018) Adult neurogenesis conserves hippocampal memory capacity. J Neurosci 38(31):6854–6863

Amini MR, Sheikhhossein F, Naghshi S, Djafari F, Askari M, Shahinfar H et al (2020) Effects of berberine and barberry on anthropometric measures: a systematic review and meta-analysis of randomized controlled trials. Complement Ther Med (Internet) (cited 2023 Jan 31);49. https://pubmed.ncbi.nlm.nih.gov/32147051/

Baggish AL, Hale A, Weiner RB, Lewis GD, Systrom D, Wang F et al (2011) Dynamic regulation of circulating microRNA during acute exhaustive exercise and sustained aerobic exercise training. J Physiol 589(16):3983–3994

Balasubramanian N, Sagarkar S, Choudhary AG, Kokare DM, Sakharkar AJ (2021) Epigenetic blockade of hippocampal SOD2 via DNMT3b-mediated DNA methylation: implications in mild traumatic brain injury-induced persistent oxidative damage. Mol Neurobiol 58(3):1162–1184

Bathina S, Das UN (2015) Brain-derived neurotrophic factor and its clinical implications. Arch Med Sci 11(6):1164–1178

Bisht S, Banu S, Srivastava S, Pathak RU, Kumar R, Dada R et al (2020) Sperm methylome alterations following *Yoga*-based lifestyle intervention in patients of primary male infertility: a pilot study. Andrologia 52(4):2–5

Bull FC, Al-Ansari SS, Biddle S, Borodulin K, Buman MP, Cardon G et al (2020) World Health Organization 2020 guidelines on physical activity and sedentary behaviour. Br J Sports Med (Internet). (cited 2023 Jan 31);54(24):1451–62. https://pubmed.ncbi.nlm.nih.gov/33239350/

Byrne EM, Carrillo-Roa T, Henders AK, Bowdler L, McRae AF, Heath AC et al (2013) Monozygotic twins affected with major depressive disorder have greater variance in methylation than their unaffected co-twin. Transl Psychiatry 3(6):e269–e266

Cahn BR, Goodman MS, Peterson CT, Maturi R, Mills PJ (2017) *Yoga*, meditation and mind-body health: increased BDNF, cortisol awakening response, and altered inflammatory marker expression after a 3-month *Yoga* and meditation retreat. Front Hum Neurosci 11:1–13

Cao-Lei L, Massart R, Suderman MJ, Machnes Z, Elgbeili G, Laplante DP et al (2014) DNA methylation signatures triggered by prenatal maternal stress exposure to a natural disaster: project ice storm. PLoS One 9(9):e107653

Cattaneo A, Cattane N, Begni V, Pariante CM, Riva MA (2016) The human BDNF gene: peripheral gene expression and protein levels as biomarkers for psychiatric disorders. Transl Psychiatry 6(11):e958

Cherkas LF, Hunkin JL, Kato BS, Richards JB, Gardner JP, Surdulescu GL et al (2008) The association between physical activity in leisure time and leukocyte telomere length. Arch Intern Med (Internet) (cited 2023 Jan 31);168(2):154–8. https://pubmed.ncbi.nlm.nih.gov/18227361/

Cronk JC, Derecki NC, Ji E, Xu Y, Lampano AE, Smirnov I et al (2015) Methyl-CpG binding protein 2 regulates microglia and macrophage gene expression in response to inflammatory stimuli. Immunity 42(4):679–691

Dassati S, Waldner A, Schweigreiter R (2014) Apolipoprotein D takes center stage in the stress response of the aging and degenerative brain. Neurobiol Aging 35(7):1632–1642

De la Rosa A, Solana E, Corpas R, Bartrés-Faz D, Pallàs M, Vina J et al (2019) Long-term exercise training improves memory in middle-aged men and modulates peripheral levels of BDNF and Cathepsin B. Sci Rep 9(1):1–11

Delgado-Morales R, Agís-Balboa RC, Esteller M, Berdasco M (2017) Epigenetic mechanisms during ageing and neurogenesis as novel therapeutic avenues in human brain disorders. Clin Epigenetics 9(1):1–18

Denham J, O'Brien BJ, Marques FZ, Charchar FJ (2015) Changes in the leukocyte methylome and its effect on cardiovascular-related genes after exercise. J Appl Physiol 118(4):475–488

Dong J, Pan YB, Wu XR, He LN, Liu XD, Feng DF et al (2019) A neuronal molecular switch through cell-cell contact that regulates quiescent neural stem cells. Sci Adv 5(2):eaav4416

El Hayek L, Khalifeh M, Zibara V, Abi Assaad R, Emmanuel N, Karnib N et al (2019) Lactate mediates the effects of exercise on learning and memory through sirt1-dependent activation of hippocampal brain-derived neurotrophic factor (BDNF). J Neurosci 39(13):2369–2382

Erickson KI, Prakash RS, Voss MW, Chaddock L, Heo S, McLaren M et al (2010) Brain-derived neurotrophic factor is associated with age-related decline in hippocampal volume. J Neurosci 30(15):5368–5375

Ertek S, Cicero A (2012) Impact of physical activity on inflammation: effects on cardiovascular disease risk and other inflammatory conditions. Arch Med Sci (Internet). (cited 2023 Jan 31);8(5):794. /pmc/articles/PMC3506236/

Erwin Wells R, Phillips RS, McCarthy EP (2011) Patterns of mind-body therapies in adults with common neurological conditions. Neuroepidemiology 36(1):46–51

Fernández-Ruiz I (2019) Exercise protects against cardiovascular disease by modulating immune cell supply. Nat Rev Cardiol (Internet) (cited 2023 Jan 31);17(1):5. https://www.nature.com/articles/s41569-019-0311-1

Fisher JP, Young CN, Fadel PJ (2015) Autonomic adjustments to exercise in humans. Compr Physiol 5(2):475–512

Fiuza-Luces C, Santos-Lozano A, Joyner M, Carrera-Bastos P, Picazo O, Zugaza JL et al (2018) Exercise benefits in cardiovascular disease: beyond attenuation of traditional risk factors. Nat Rev Cardiol 15(12):731–743

Fragala MS, Kraemer WJ, Denegar CR, Maresh CM, Mastro AM, Volek JS (2011) Neuroendocrine-immune interactions and responses to exercise. Sports Med 41(8):621–639

Galper DI, Trivedi MH, Barlow CE, Dunn AL, Kampert JB (2006) Inverse association between physical inactivity and mental health in men and women. Med Sci Sports Exerc 38(1):173–178

Gautam S, Dada R (2023) Molecular mechanisms underlying the effects of *Yoga*. https://services.igi-global.com/resolvedoi/resolve.aspx?doi=104018/978-1-7998-3254-6.ch007 (Internet) 1AD (cited 2023 Jan 31);103–23. https://www.igi-global.com/chapter/molecular-mechanisms-underlying-the-effects-of-*Yoga*/261146

Gautam S, Tolahunase M, Kumar U, Dada R (2019) Impact of *Yoga* based mind-body intervention on systemic inflammatory markers and co-morbid depression in active rheumatoid arthritis patients: a randomized controlled trial. Restor Neurol Neurosci 37(1):41–59

Gleeson M, Bishop NC, Stensel DJ, Lindley MR, Mastana SS, Nimmo MA (2011) The anti-inflammatory effects of exercise: mechanisms and implications for the prevention and treatment of disease. Nat Rev Immunol (Internet) (cited 2023 Jan 31);11(9):607–615. https://www.nature.com/articles/nri3041

Gomez-Pinilla F, Zhuang Y, Feng J, Ying Z, Fan G (2011) Exercise impacts brain-derived neurotrophic factor plasticity by engaging mechanisms of epigenetic regulation. Eur J Neurosci 33(3):383–390

Griñan-Ferré C, Puigoriol-Illamola D, Palomera-Ávalos V, Pérez-Cáceres D, Companys-Alemany J, Camins A et al (2016) Environmental enrichment modified epigenetic mechanisms in SAMP8 mouse hippocampus by reducing oxidative stress and inflammaging and achieving neuroprotection. Front Aging Neurosci 8:241

Hadad N, Unnikrishnan A, Jackson JA, Masser DR, Otalora L, Stanford DR et al (2018) Caloric restriction mitigates age-associated hippocampal differential CG and non-CG methylation. Neurobiol Aging 67:53–66

Haigis MC, Yankner BA (2010) The aging stress response. Mol Cell 40(2):333–344

Hamilton MT, Hamilton DG, Zderic TW (2014) Sedentary behavior as a mediator of type 2 diabetes. Med Sport Sci (Internet) (cited 2023 Jan 31);60:11. /pmc/articles/PMC4364419/

Harkess KN, Ryan J, Delfabbro PH, Cohen-Woods S (2016a) Preliminary indications of the effect of a brief *Yoga* intervention on markers of inflammation and DNA methylation in chronically stressed women. Transl Psychiatry (Internet) (cited 2023 Jan 31);6(11):e965. https://www.nature.com/articles/tp2016234

Harkess KN, Ryan J, Delfabbro PH, Cohen-Woods S (2016b) Preliminary indications of the effect of a brief *Yoga* intervention on markers of inflammation and DNA methylation in chronically stressed women. Transl Psychiatry 6(11):e965–e969

Horvath S, Zhang Y, Langfelder P, Kahn RS, Boks MPM, van Eijk K et al (2012) Aging effects on DNA methylation modules in human brain and blood tissue. Genome Biol 13(10):R97

Hughes DC, Ellefsen S, Baar K (2018) Adaptations to endurance and strength training. Cold Spring Harb Perspect Med (Internet) (cited 2023 Jan 31);8(6). https://pubmed.ncbi.nlm.nih.gov/28490537/

Jerath R, Barnes VA, Crawford MW (2014) Mind-body response and neurophysiological changes during stress and meditation: central role of homeostasis. J Biol Regul Homeost Agents 28(4):545–554

Jessop P, Toledo-Rodriguez M (2018) Hippocampal TET1 and TET2 expression and DNA hydroxymethylation are affected by physical exercise in aged mice. Front Cell Dev Biol 6:45

Kaminsky DA, Guntupalli KK, Lippmann J, Burns SM, Brock MA, Skelly J et al (2017) Effect of *Yoga* breathing (*Pranayama*) on exercise tolerance in patients with chronic obstructive pulmonary disease: a randomized, controlled trial. J Altern Complement Med 23(9):696–704

Kiecolt-Glaser JK, Christian L, Preston H, Houts CR, Malarkey WB, Emery CF et al (2010) Stress, inflammation, and *Yoga* practice. Psychosom Med 72(2):113–121

Kumar M, Srivastava S, Muhammad T (2022) Relationship between physical activity and cognitive functioning among older Indian adults. Sci Rep (Internet) (cited 2023 Jan 31);12(1). https://www.researchgate.net/publication/358667984_Relationship_between_physical_activity_and_cognitive_functioning_among_older_Indian_adults

Kuppusamy M, Kamaldeen D, Pittani R, Amaldas J, Ramasamy P, Shanmugam P et al (2020) Effects of *Yoga* breathing practice on heart rate variability in healthy adolescents: a randomized controlled trial. Integr Med Res 9(1):28–32

Landgrave-Gómez J, Mercado-Gómez O, Guevara-Guzmán R (2015) Epigenetic mechanisms in neurological and neurodegenerative diseases. Front Cell Neurosci 9:1–11

Laufs U, Wassmann S, Czech T, Münzel T, Eisenhauer M, Böhm M et al (2005) Physical inactivity increases oxidative stress, endothelial dysfunction, and atherosclerosis. Arterioscler Thromb Vasc Biol (Internet) (cited 2023 Jan 31);25(4):809–14. https://www.ahajournals.org/doi/abs/10.1161/01.atv.0000158311.24443.af

Le François B, Soo J, Millar AM, Daigle M, Le Guisquet AM, Leman S et al (2015) Chronic mild stress and antidepressant treatment alter 5-HT1A receptor expression by modifying DNA methylation of a conserved Sp4 site. Neurobiol Dis 82:332–341

Lee ES, Koh HL, Ho EQY, Teo SH, Wong FY, Ryan BL et al (2021) Systematic review on the instruments used for measuring the association of the level of multimorbidity and clinically important outcomes. BMJ Open (Internet) (cited 2023 Jan 31);11(5). https://pubmed.ncbi.nlm.nih.gov/33952533/

Levenson JM, O'Riordan KJ, Brown KD, Trinh MA, Molfese DL, Sweatt JD (2004) Regulation of histone acetylation during memory formation in the hippocampus. J Biol Chem 279(39):40545–40559

Lu Y, Dong Y, Tucker D, Wang R, Ahmed ME, Brann D et al (2017) Treadmill exercise exerts streptozotocin-induced rat model of sporadic Alzheimer's disease. J Alzheimers Dis 56(4):1469–1484

Lugert S, Basak O, Knuckles P, Haussler U, Fabel K, Götz M et al (2010) Quiescent and active hippocampal neural stem cells with distinct morphologies respond selectively to physiological and pathological stimuli and aging. Cell Stem Cell 6(5):445–456

Mandolesi L, Polverino A, Montuori S, Foti F, Ferraioli G, Sorrentino P, et al (2018) Effects of physical exercise on cognitive functioning and wellbeing: biological and psychological benefits. Front Psychol (Internet) (cited 2023 Jan 31);9:509. /pmc/articles/PMC5934999/

Martínez-Iglesias O, Naidoo V, Cacabelos N, Cacabelos R (2022) Epigenetic biomarkers as diagnostic tools for neurodegenerative disorders. Int J Mol Sci 23(1)

Martinowich K, Hattori D, Wu H, Fouse S, He F, Hu Y et al (2013) DNA chromatin methylation remodeling related in dependent BDNF gene regulation. Science 302:890–894

McCarthy L, Fuller J, Davidson G, Crump A, Positano S, Alderman C (2017) Assessment of *Yoga* as an adjuvant treatment for combat-related posttraumatic stress disorder. Australas Psychiatry 25(4):354–357

McGowan PO, Kato T (2008) Epigenetics in mood disorders. Environ Health Prev Med 13(1):16–24

McGowan PO, Sasaki A, D'Alessio AC, Dymov S, Labonté B, Szyf M et al (2009) Epigenetic regulation of the glucocorticoid receptor in human brain associates with childhood abuse. Nat Neurosci 12(3):342–348

Mela V, Mota BC, Milner M, McGinley A, Mills KHG, Kelly ÁM et al (2020) Exercise-induced re-programming of age-related metabolic changes in microglia is accompanied by a reduction in senescent cells. Brain Behav Immun 87:413–428

Miranda M, Morici JF, Zanoni MB, Bekinschtein P (2019) Brain-derived neurotrophic factor: a key molecule for memory in the healthy and the pathological brain. Front Cell Neurosci 13:1–25

Misztak P, Pańczyszyn-Trzewik P, Sowa-Kućma M (2018) Histone deacetylases (HDACs) as therapeutic target for depressive disorders. Pharmacol Rep 70(2):398–408

Moreno-Camacho CA, Montoya-Torres JR, Jaegler A, Gondran N (2019) Sustainability metrics for real case applications of the supply chain network design problem: a systematic literature review. J Clean Prod 231:600–618

Nascimento CMC, Pereira JR, Pires De Andrade L, Garuffi M, Ayan C, Kerr DS et al (2015) Physical exercise improves peripheral BDNF levels and cognitive functions in mild cognitive impairment elderly with different BDNF Val66Met genotypes. J Alzheimers Dis 43(1):81–91

Naveen GH, Thirthalli J, Rao MG, Varambally S, Christopher R, Gangadhar BN (2013) Positive therapeutic and neurotropic effects of *Yoga* in depression: a comparative study. Indian J Psychiatry 55(7):400–405

Naveen GH, Varambally S, Thirthalli J, Rao M, Christopher R, Gangadhar BN (2016) Serum cortisol and BDNF in patients with major depression—effect of *Yoga*. Int Rev Psychiatry 28(3):273–278

Nay K, Smiles WJ, Kaiser J, McAloon LM, Loh K, Galic S et al (2021) Molecular mechanisms underlying the beneficial effects of exercise on brain function and neurological disorders. Int J Mol Sci (Internet) (cited 2023 Jan 31);22(8):4052. https://www.mdpi.com/1422-0067/22/8/4052/htm

Neeper SA, Góauctemez-Pinilla F, Choi J, Cotman C (1995) Exercise and brain neurotrophins. Nature 373(6510):109

Nirwan M, Halder K, Saha M, Pathak A, Balakrishnan R, Ganju L (2021) Improvement in resilience and stress-related blood markers following ten months *Yoga* practice in Antarctica. J Complement Integr Med 18(1):201–207

Oh H, Lewis DA, Sibille E (2016) The role of BDNF in age-dependent changes of excitatory and inhibitory synaptic markers in the human prefrontal cortex. Neuropsychopharmacology 41(13):3080–3091

Ornish D, Lin J, Chan JM, Epel E, Kemp C, Weidner G et al (2013) Effect of comprehensive lifestyle changes on telomerase activity and telomere length in men with biopsy-proven low-risk prostate cancer: 5-year follow-up of a descriptive pilot study. Lancet Oncol 14(11):1112–1120

Pareja-Galeano H, Sanchis-Gomar F, García-Giménez JL (2014) Physical exercise and epigenetic modulation: elucidating intricate mechanisms. Sports Med 44(4):429–436

Peters R (2006) Ageing and the brain. Postgrad Med J 82(964):84–88

Pozuelo-Carrascosa DP, Martínez-Vizcaíno V, Sánchez-López M, Bartolomé-Gutiérrez R, Rodríguez-Martín B, Notario-Pacheco B (2017) Resilience as a mediator between cardiorespiratory fitness and mental health-related quality of life: a cross-sectional study. Nurs Health Sci 19(3):316–321

Provençal N, Suderman MJ, Caramaschi D, Wang D, Hallett M, Vitaro F et al (2013) Differential DNA methylation regions in cytokine and transcription factor genomic loci associate with childhood physical aggression. PLoS One 8(8):13–16

Qu S, Olafsrud SM, Meza-Zepeda LA, Saatcioglu F (2013) Rapid gene expression changes in peripheral blood lymphocytes upon practice of a comprehensive *Yoga* program. PLoS One (Internet) (cited 2023 Jan 31);8(4):e61910. https://journals.plos.org/plosone/article?id=10.1371/journal.pone.0061910

Radom-Aizik S, Zaldivar FP, Haddad F, Cooper DM (2014) Impact of brief exercise on circulating monocyte gene and microRNA expression: implications for atherosclerotic vascular disease. Brain Behav Immun 39:121–129

Raichlen DA, Klimentidis YC, Katherine Sayre M, Bharadwaj PK, Lai MHC, Wilcox RR et al (2022) Leisure-time sedentary behaviors are differentially associated with all-cause dementia regardless of engagement in physical activity. Proc Natl Acad Sci U S A (Internet) (cited 2023 Jan 31);119 (35):e2206931119. https://www.pnas.org/doi/abs/10.1073/pnas.2206931119

Rezapour S, Shiravand M, Mardani M (2018) Epigenetic changes due to physical activity. Biotechnol Appl Biochem (Internet) (cited 2023 Jan 31);65(6):761–7. https://onlinelibrary.wiley.com/doi/full/10.1002/bab.1689

Rolland Y, van Kan GA, Vellas B (2010) Healthy brain aging: role of exercise and physical activity. Clin Geriatr Med (Internet) (cited 2023 Jan 31);26(1):75–87. https://pubmed.ncbi.nlm.nih.gov/20176294/

Roth TL, Sweatt JD (2011) Epigenetic marking of the BDNF gene by early-life adverse experiences. Horm Behav 59(3):315–320

Roth TL, Zoladz PR, Sweatt JD, Diamond DM (2011) Epigenetic modification of hippocampal Bdnf DNA in adult rats in an animal model of post-traumatic stress disorder. J Psychiatr Res 45(7):919–926

Sailani MR, Halling JF, Møller HD, Lee H, Plomgaard P, Pilegaard H et al (2019) Lifelong physical activity is associated with promoter hypomethylation of genes involved in metabolism, myogenesis, contractile properties and oxidative stress resistance in aged human skeletal muscle. Sci Rep (Internet) (cited 2023 Jan 31);9(1):1–11. https://www.nature.com/articles/s41598-018-37895-8

Salim S (2017) Oxidative stress and the central nervous system. J Pharmacol Exp Ther 360(1):201–205

Salim S, Asghar M, Taneja M, Hovatta I, Chugh G, Vollert C et al (2011) Potential contribution of oxidative stress and inflammation to anxiety and hypertension. Brain Res 1404:63–71

Sánchez-Rodríguez MA, Zacarías-Flores M, Correa-Muñoz E, Arronte-Rosales A, Mendoza-Núñez VM (2021) Oxidative stress risk is increased with a sedentary lifestyle during aging in Mexican women (cited 2023 Jan 31). https://doi.org/10.1155/2021/9971765

Scartoni FR, de Oliveira Sant'Ana L, Murillo-Rodriguez E, Yamamoto T, Imperatori C, Budde H et al (2020) Physical exercise and immune system in the elderly: implications and importance in COVID-19 pandemic period. Front Psychol 11:3215

Sedentary Lifestyles Associated With Accelerated Aging Process—ScienceDaily (Internet). (cited 2023 Jan 31). https://www.sciencedaily.com/releases/2008/01/080128165734.htm

Shah SZA, Karam JA, Zeb A, Ullah R, Shah A, Haq IU et al (2021) Movement is improvement: the therapeutic effects of exercise and general physical activity on glycemic control in patients with type 2 diabetes mellitus: a systematic review and meta-analysis of randomized controlled trials. Diabetes Ther (Internet) (cited 2023 Jan 31);12(3):707–32. https://link.springer.com/article/10.1007/s13300-021-01005-1

Silverman MN, Deuster PA (2014) Biological mechanisms underlying the role of physical fitness in health and resilience. Interface Focus 4(5):20140040

Simpson RJ, Kunz H, Agha N, Graff R (2015) Exercise and the regulation of immune functions. In: Progress in molecular biology and translational science, vol 135, 1st edn. Elsevier Inc., 355–380 p

Sølvsten CAE, de Paoli F, Christensen JH, Nielsen AL (2018) Voluntary physical exercise induces expression and epigenetic remodeling of VegfA in the rat hippocampus. Mol Neurobiol 55(1):567–582

Song MS, Learman CR, Ahn KC, Baker GB, Kippe J, Field EM et al (2015) In vitro validation of effects of BDNF-expressing mesenchymal stem cells on neurodegeneration in primary cultured neurons of APP/PS1 mice. Neuroscience 307:37–50

Spielman LJ, Little JP, Klegeris A (2016) Physical activity and exercise attenuate neuroinflammation in neurological diseases. Brain Res Bull 125:19–29

Su M, Hong J, Zhao Y, Liu S, Xue X (2015) MeCP2 controls hippocampal brain-derived neurotrophic factor expression via homeostatic interactions with microRNA-132 in rats with depression. Mol Med Rep 12(4):5399–5406

Sui L, Wang Y, Ju LH, Chen M (2012) Epigenetic regulation of reelin and brain-derived neurotrophic factor genes in long-term potentiation in rat medial prefrontal cortex. Neurobiol Learn Mem 97(4):425–440

Szyf M (2012) Mind-body interrelationship in DNA methylation. Chem Immunol Allergy 98:85–99

Tolahunase M, Sagar R, Dada R (2017a) Impact of *Yoga* and meditation on cellular aging in apparently healthy individuals: a prospective, open-label single-arm exploratory study. Oxid Med Cell Longev (Internet) (cited 2023 Jan 31);2017. https://pubmed.ncbi.nlm.nih.gov/28191278/

Tolahunase M, Sagar R, Dada R (2017b) Impact of *Yoga* and meditation on cellular aging in apparently healthy individuals: a prospective, open-label single-arm exploratory study. Oxid Med Cell Longev 2017:7928981

Tolahunase MR, Sagar R, Faiq M, Dada R (2018) *Yoga*- and meditation-based lifestyle intervention increases neuroplasticity and reduces severity of major depressive disorder: a randomized controlled trial. Restor Neurol Neurosci 36(3):423–442

Tomizawa H, Matsuzawa D, Ishii D, Matsuda S, Kawai K, Mashimo Y et al (2015) Methyl-donor deficiency in adolescence affects memory and epigenetic status in the mouse hippocampus. Genes Brain Behav 14(3):301–309

Tripathi A (2012) New cellular and molecular approaches to ageing brain. Ann Neurosci 19(4):177–182

Turecki G, Meaney MJ (2016) Effects of the social environment and stress on glucocorticoid receptor gene methylation: a systematic review. Biol Psychiatry 79(2):87–96

Wahbeh H, Elsas SM, Oken BS (2008) Mind-body interventions: applications in neurology. Neurology 70(24):2321–2328

Walsh JJ, Tschakovsky ME (2018) Exercise and circulating BDNF: mechanisms of release and implications for the design of exercise interventions. Appl Physiol Nutr Metab 43(11):1095–1104

Walsh EI, Smith L, Northey J, Rattray B, Cherbuin N (2020) Towards an understanding of the physical activity-BDNF-cognition triumvirate: a review of associations and dosage. Ageing Res Rev 60:101044

Weible AP, Piscopo DM, Rothbart MK, Posner MI, Niell CM (2017) Rhythmic brain stimulation reduces anxiety-related behavior in a mouse model based on meditation training. Proc Natl Acad Sci U S A (Internet) (cited 2023 Jan 31);114(10):2532–7. https://pubmed.ncbi.nlm.nih.gov/28223484/

Winterbourn CC, Gutteridge JMC, Halliwell B (1985) Doxorubicin-dependent lipid peroxidation at low partial pressures of O2. J Free Radic Biol Med 1(1):43–49

Wu L, Zhao X, Shen Y, Huang G, Zhang M, Yan Y et al (2015) Influence of lifestyle on the FAIM2 promoter methylation between obese and lean children: a cohort study. BMJ Open (Internet) (cited 2023 Jan 31);5(4). https://pubmed.ncbi.nlm.nih.gov/25922107/

Wu C, Yang L, Tucker D, Dong Y, Zhu L, Duan R et al (2018) Beneficial effects of exercise pre-treatment in a sporadic Alzheimer's rat model. Med Sci Sports Exerc 50(5):945–956

Yadav R, Yadav RK, Khadgawat R, Pandey RM (2019) Comparative efficacy of a 12 week *Yoga*-based lifestyle intervention and dietary intervention on adipokines, inflammation, and oxidative stress in adults with metabolic syndrome: a randomized controlled trial. Transl Behav Med 9(4):594–604

Zeng Y, Tan M, Kohyama J, Sneddon M, Watson JB, Sun YE et al (2011) Epigenetic enhancement of BDNF signaling rescues synaptic plasticity in aging. J Neurosci 31(49):17800–17810

Zhang J, Li J, Zhu Y, Miao Z, Tian Y (2020) Forced running exercise mitigates radiation-induced cognitive deficits via regulated DNA hydroxymethylation. Epigenomics 12(5):385–396

Chapter 10
The Impact of *Yoga*, *Pranayama*, and Mindfulness Training on Neurocognitive, Neurophysiological, and Neuropsychological Changes in High Altitude

Gurudutta Gangenahalli, Vikas Chander, Yogesh Kumar Verma, and Shweta Singh

Contents

Abstract Our brain constitutes only 2% of the body weight, yet it is most sensitive to changes in partial oxygen pressure. It accounts for ≥20% of total oxygen metabolism, with neurons alone guzzling up around 80% of energy to maintain membrane potential and other neuronal functions. People shifting to high altitude (low-oxygen environment) experience neurological dysfunction, cognitive disorders, insomnia, dizziness, lightheadedness, etc. These cerebral syndromes were attributed to improper energy metabolism and neuronal transmission due to insufficient oxygen supply to the brain. The Indian civilization was the first to identify the positive effects of *Yoga* and *Pranayama* on the overall development of the body. The regimen combines various *Asana* (*Vyamaya* or physical exercises), *Pranayama* (breath manipulation), and *Dhyana* (meditation). It was identified that *Yoga* can also help in

G. Gangenahalli (✉) · V. Chander · Y. K. Verma · S. Singh
Division of Stem Cell and Gene Therapy Research, Institute of Nuclear Medicine and Allied Sciences (INMAS), Defence Research and Development Organization, New Delhi, India

A. Anand (ed.), *Neuroscience of Yoga*,
https://doi.org/10.1007/978-981-97-2851-0_10

improving the neurophysiological, neurophysiological, and neurocognitive function of the brain. In this chapter, we will discuss the negative effects of change in partial oxygen pressure at high altitudes on brain function. We will also discuss how *Yoga* and *Pranayama* can help in alleviating these negative effects by inducing stress erythropoiesis and neurogenesis. The ongoing COVID-19 pandemic caused by Severe Acute Respiratory Syndrome Coronavirus-2 (SARS-CoV-2) is also characterized by low oxygen levels in the blood and results in similar neurological symptoms. Hence, we will also discuss COVID-19-related neurological disorders in brief detail.

Keywords Hypoxia · Neurological disorders · Hypoxic preconditioning · *Yoga* · Meditation

10.1 Introduction

The human brain is a complex organ that controls the cognitive, psychological, and physiological functions in the body. Along with the spinal cord, it constitutes the central nervous system (CNS) (Miller and Zachary 2017). It weighs around three pounds and contains nearly 60% fat and 40% other constituents like water, protein, carbohydrates, salts, etc. It has three regions: the forebrain or prosencephalon containing the cerebrum, midbrain or mesencephalon, and the hindbrain or rhombencephalon containing the cerebellum. The midbrain pons and medulla together constitute the brainstem which controls various cognitive functions of the body. The brain is furnished with blood vessels to supply nutrients through the blood–brain barrier (BBB), nerves for transmission of electrical impulses, and cells like neurons and glial cells. Neurons are the actual workhorse of the nervous system. They receive input from the external environment, convert them into an electrical impulse, and order our muscles to react. There are around 100 billion neurons that closely interact with glial cells to transmit electrical impulses to other parts of the body through synapses. Based on the number of neurons, the brain is divided into gray matter (containing a large number of neurons and other cells) and white matter (containing myelinated axons). The color difference is due to the presence of myelin (Herbet and Duffau 2020).

Since the brain controls vital functions like our thought, memory, emotion, touch, vision, level of intelligence, breathing, hunger, etc., any damage can result in severe neurological dysfunction. A similar condition is observed in people who shift to a high-altitude environment and experience hypoxic stress leading to cerebral hypoxia and neurological dysfunction with symptoms like amnesia, dementia, delirium, insomnia, dizziness, haziness, lightheadedness, depression, etc. (Rolan 2014). In this chapter, we will discuss how *Yoga* and *Pranayama* help reduce high-altitude-induced neurological dysfunction. We will also discuss how these simple exercises help in the differentiation of neural stem cells (NSC) for the regeneration of lost neurons and induction of stress erythropoiesis in hematopoietic stem cells (HSC) for the re-establishment of the oxygen supply to the brain.

10.2 Therapeutic Role of *Yoga* and *Pranayama* in the Neurological Disorders

Yoga is an ancient Indian way of life, which includes changes in mental attitude, diet, and the practice of specific techniques such as *Yoga Asana* (postures), *Pranayama* (breathing practices), and *Dhyana* (meditation). *Yoga* originated in the Indian subcontinent around 5000 years ago as a part of *ayurvedic* medicine. The name is derived from the Sanskrit word yukti meaning union. The union of consciousness (our spirit) with almighty God (the super-spirit). *Yoga* utilizes a combination of posture-holding physical exercises and specific breathing patterns to stabilize our physiological system. On the other hand, *Pranayama* (breathing exercises) and *Dhyana* (meditation) are inclined more toward stabilizing psychological and cognitive processes. *Pranayama* or breath regulation exercises is the prime component of *Yoga*. In Sanskrit, *"prana"* means life and *"Yama"* means to control. Hence, *Pranayama* involves inhalation, exhalation, and holding/controlling the breath in between these two sequences. In short, the process involves complete silencing of the breathing activity by self-control. All *Pranayama* exercises depend on the same fundamentals that can be divided into three different phases: *Puraka* (inhalation phase), *Kumbhaka* (retention phase) *Rechaka* (exhalation phase). Some common *Pranayama* exercises are *Anuloma-Viloma* (Alternate Nostril Breathing continuously in a circle), *Nisshesha Rechaka* (Prolong Exhalation-Inhalation), *Kapalbhati* (Breath of fire), *Bhastrika*, *Nadi Shodhana* (Alternate Nostril Breathing), and *Bhramari* (Humming Bee Breath).

Yoga and *Pranayama* have been used as alternative treatment modalities for various neurological disorders in combination with normal medication. Patients with epilepsy who do not respond to anti-seizure medications like oxcarbazepine, valproic acid, etc. can find fruitful results through the regular practice of *Yoga* and meditation. In a randomized controlled study, *Panjwani* et al. demonstrated the effect of *Sahaja Yoga* meditation on 32 patients suffering from primary idiopathic epilepsy. Patients were divided into three groups: group I practiced *Sahaja Yoga* meditation twice daily for 6 months, group II practiced *Yoga* postures mimicking the meditation for 6 months; and group III was taken as control. It was found that *Sahaja Yoga* meditation brings about changes in some of the electrophysiological responses and reduced levels of stress in epileptic patients (Panjwani et al. 2000). An open non-randomized trial using a specific *Yoga* meditation technique (YMT) was evaluated for its efficacy in reducing seizures among epileptic patients with complex partial seizures (CPS). The trial consisted of 20 patients (14 males and 6 females, age range 15–47 years) diagnosed with epilepsy. In YMT intervention, patients were instructed to perform *sukhasana* and *Nadi Shodhana*. This twice-daily meditation with a weekly *Yoga* regimen demonstrated significant reductions (>50%) in seizures within 3 months in more than 50% of patients (Rajesh et al. 2006). In a randomized controlled trial conducted by *Lundgren* et al. 18 patients with EEG-diagnosed epilepsy, decreasing seizure index along with an improvement in quality of life was found after practicing *Yoga*. The treatment protocols consisted of 12 h of professional therapy distributed in two individual sessions followed by booster sessions for 6 and 12 months post-treatment (Lundgren et al. 2008).

Stress and anxiety are been increasing worldwide generally in younger individuals, which have brought about memory and cognitive decline due to an increase in levels of cortisol in the body. Cortisol is known to harm hippocampal cells. Meditation has great potential for preventing cognitive and memory decline due to its ability to reduce stress and anxiety (Lupien et al. 1998; Newcomer 1999). Furthermore, it has been suggested that *Yoga Asana* also modulate stress-induced autonomic regulatory reflex and inhibits the production of adrenocorticotropic hormone from the anterior pituitary gland, resulting in decreased production of cortisol from the adrenal gland (Katuri et al. 2016). Magnetic resonance imaging (MRI) analysis has revealed that *Yoga* and meditation modulate the activity of brain regions involved in emotion regulation, e.g., the superior parietal lobule and supramarginal gyrus leading to a diminished sympathetic response to chronic stress (Wadden et al. 2018). A meta-analysis performed by *Hoffman* et al. revealed that *Yoga* imparts greater benefits, particularly when subjects were suffering from higher values of anxiety at the early stages (Hofmann et al. 2016). Another meta-analysis of eight trials with 319 participants with anxiety disorders revealed that *Yoga* training was a safe and effective intervention to reduce the extent of anxiety (Cramer et al. 2018).

Dopamine is a neurotransmitter, which plays a role in movement, memory, and emotions. Extremely high or extremely low levels of dopamine can lead to several cognitive and neurological disorders. It was found that *Yoga* and meditation can maintain plasma dopamine levels as confirmed in studies conducted on 67 healthy participants who perform mind-body training regularly (Jung et al. 2010). *Yoga* interventions have been suggested as an effective therapy for the treatment of major depressive disorder (MDD). Few studies highlight the significance of *Yoga* and mindfulness as adjuvant treatment in MDD. It was found that *Sahaj Yoga* practices in combination with conventional antidepressant medication significantly improves depression symptoms and reduced the remission rate in patients with MDD compared to control patients (Sharma et al. 2005). It has been found that *Yoga* can regulate the GABAergic system and exert beneficial effects on MDD through cortical inhibition (Jakhar et al. 2019). γ-Aminobutyric acid (GABA) is the chief inhibitory neurotransmitter involved in the maturation of CNS. An increase in GABA levels in the brain during multiple *Yoga* sessions can lead to significant improvement of depression symptoms and improvement of mood in patients with MDD (Bhargav et al. 2020). Alternatively, *Yoga* and mindfulness training can also lead to improvement in depression symptoms in MDD by enhancing the serum levels of brain-derived neurotrophic factor (BDNF) (Tolahunase et al. 2018).

Bipolar disorder (BD) is a chronic illness characterized by extreme mood swings including hypomania (emotional highs) and depression (emotional lows). *Yoga* and meditation have been found to decrease negative emotions in patients with BD. *Hanssen* et al. conducted a randomized controlled study on 160 adults suffering from BD. It was shown that mindfulness-based cognitive therapy combined with *Yoga* practices promotes mental health and improves symptoms like anxiety and depression (Hanssen et al. 2019). *Yoga* and mindfulness training has also been found fruitful in the treatment of attention-deficit/hyperactivity disorder (ADHD) in children. Children suffering from ADHD may have trouble paying attention, controlling impulsive behavior, or might be overly active. Reports on the effect of *Yoga* and meditation on ADHD have shown the positive impact of *Sahaj Yoga* in reducing hyperactivity, inattention, and anxiety resulting in improvement of peer relationships, quality, and sleep patterns (Harrison et al. 2004).

10.3 Neurological Complications at Higher Altitude Environment

Oxygen is considered one of the most important factors influencing the evolution of organisms like eukaryotes. At first, it was toxic to living organisms, but through several mechanistic adaptations, these organisms were able to detoxify it. Soon through further adaptations, most eukaryotes started to depend on oxygen rather than detoxifying it. In more complex eukaryotes like humans, the oxygen concentrations vary from tissue to tissue for cellular metabolism, e.g., the concentration of oxygen significantly varies between lung parenchyma and blood circulation. In the brain tissue, the oxygen concentration is even lower, whereas, in bone marrow, it is around 0–4% (An and Lin 2000; Spencer et al. 2014). Although oxygen concentrations are considerably low (hypoxia) under in vivo conditions, some cells need an almost anoxic environment for maintenance, e.g., stem cells. Oxygen levels in the blood control vital functions in the human body. A low oxygen concentration in the blood and tissue leads to hypoxemia and hypoxia, respectively. The brain, which comprises only 2% of the body weight, consumes 20% of oxygen. Neurons, the functional units of the brain, consume around 80% of the oxygen brain receives from the blood (Herculano-Houzel 2011; Watts et al. 2018). It highlights the amount of energy the brain requires to operate. The energy produced is consumed for regulating various vital functions, e.g., heartbeat and contraction, hunger, vision, cognitive function, sensation on skin, memory, temperature, etc. The brain receives oxygen from the blood through a semipermeable BBB, made of a network of blood capillaries and tissue comprising closely spaced glial cells (astrocytes) (Daneman and Prat 2015). The partial oxygen pressure varies among different regions of the brain, with neurons being the most sensitive of all brain cells. However, below-normal supply to the brain for an extended period can lead to cerebral hypoxia which can cause neurological dysfunction and result in cognitive, physiological, and psychological failure. Excessive smoking, carbon monoxide poisoning, choking, pressure on the trachea, strangulation, heart attacks, strokes, accidental injury to the head (traumatic brain injury/concussion), and diseases like amyotrophic lateral sclerosis (ALS) are generally considered the main reason behind cerebral hypoxia (Busl and Greer 2010; Lu-Emerson and Khot 2010; Park et al. 2013).

Another unique factor that can cause hypoxia is exposure to a high-altitude environment, where the partial pressure of oxygen in the atmosphere is low and decreases further on moving upward, e.g., at sea level, the partial oxygen pressure is 159 mmHg and at Mt. Everest, which is 8848 m above sea level, the partial oxygen pressure drops to 53 mmHg (West 2006). People who are dwelling in high altitudes for years have become well suited for these harsh conditions through molecular and physiological adaptions in their lungs and circulatory system that allows the transport of higher amounts of oxygen to the vital organs. On the other hand, for people who have recently shifted to these places, it is a different scenario. A sudden drop in oxygen levels gives rise to hypoxemia and tissue hypoxia that may lead to organ damage, including the brain.

People traveling at higher altitudes experience hypoxic stress due to sudden drop in oxygen pressure. A miniature shift in oxygen pressure can result in mitochondrial dysfunction and a change in oxygen metabolism leading to the degeneration of neurons. These changes result in ischemic stroke and other neurological complications due to vascular edema resulting from disruption of the BBB. Cerebral hypoxia damages neurons, glial cells, and other endothelial cells as brain cells start dying less than 5 min after oxygen deprivation. If this lasts for an extended period, it can lead to severe neurodegenerative disorders (Banasiak and Haddad 1998; Banasiak et al. 2000). Our body reacts to low oxygen pressure by increasing oxygen transport to the brain through stress or emergency erythropoiesis in HSC. It leads to the formation of a high number of red blood cells (RBCs) in circulation for oxygen transport and depot to the brain. But if there is a sudden drop in oxygen levels, this phenomenon does not work efficiently, because our body needs time to analyze the situation first and respond accordingly through gradual adaptation (Bapat et al. 2021). At first, the high mountain dwellers also experience acute mountain sickness (AMS) but later adapt to these conditions. A failure to do this may lead to death within 48 h due to conditions like high-altitude cerebral edema (HACE). Given in Table 10.1 are a few neurological disorders arising due to the effects of high altitude.

Literature studies have indicated that exposure to a 4200 m high-altitude environment for 3 days can severely impair learning and memory abilities (Zhang et al. 2018). In one study, 18 healthy young men were exposed to 300 m (control), 2800 m, 3600 m, and 4400 m altitudes for an hour, respectively, in a hypobaric chamber. Results indicated that short-term memory performance decreases with increased altitude in comparison to the control group (Du et al. 1999). Previous evidence from pilots and aircrews showed that high-altitude causes altered cognition and interferes with new memory coding but not with memory retrieval (Nation et al. 2016). Exposure to short-term hypoxia hinders brain function in humans by affecting short-term memory performance and the symptoms become worse with higher altitudes. Acute and chronic hypoxia can increase oxidative stress in cerebral cortical cells that can enhance autophagy, neuronal excitability toxicity reaction, and apoptosis leading to the deterioration of brain function. Different subregions of

Table 10.1 Listing a few neurological disorders/syndromes which arise due to the effect of high altitude

S. no.	Neurological disorder	Cause	Symptoms
1.	High-altitude cerebral edema (HACE)	Prolonged hypoxia leads to hypocapnia in the blood causing the brain to swell with fluid, resulting in severe impairment	Confusion, loss of consciousness, fever, ataxia, photophobia, and altered mental state
2.	Transient ischemic attacks (TIA)	Hypoxemia may lead to failure of vascular endothelial control resulting in arterial vasospasm in the cerebral circulation and vasoconstriction	Slurred or garbled speech, blinded vision (one or both eyes), vertigo, unsteady gait, aphasia
3.	Migraine	Oxygen deprivation and cerebral hypoxia	Severe headache, dizziness, vertigo
4.	Cerebral venous sinus thrombosis	Oxygen deprivation and cerebral hypoxia	Seizure, severe headache, visual obstruction due to papilledema (swelling of the optic disk), aphasia, depression
5.	Subarachnoid hemorrhage	Oxygen deprivation and cerebral hypoxia	Thunderclap headache, seizures, confusion, cognitive impairment
6.	High-altitude syncope	Peripheral vasodilatation from acute hypoxemia followed by cerebral vasoconstriction from hyperventilation-induced hypocapnia	Blacking out, lightheaded, dizziness, drowsiness, fainting, and tunnel vision
8.	Transient global amnesia	At high altitudes, hypoxemia results in hypocapnic cerebral vasoconstriction leading to restricted blood flow to the limbic cortex of the brain	Memory loss, limb paralysis, impaired sign, and word recognition
9.	Organic brain syndrome (delirium)	Hypoxia-induced cerebral vasoconstriction. Hypoxia-induced tissue alkalosis is another contributory factor	Poor memory, inability to focus and think properly, frequent distraction, loss of focus, disoriented movement of the body, and slurred speech

the brain tolerate different degrees of hypoxia. Under hypoxia conditions, a reduced expression of glutamate decarboxylase (GAD) protease in hippocampal neurons leads to changes in inhibitory synaptic density (Hwang et al. 2018). Moreover, others studies have found that the decrease in dendritic spines in hippocampal CA1 neurons after hypoxic exposure leads to apoptosis in these neurons (Zhuravin et al. 2019).

10.4 Mechanisms Involved in Adaptation to High Altitude

In response to hypoxia, our body initiates stress erythropoiesis in HSC to produce a large number of RBCs to normalize oxygen supply to all organs. Increased oxygen transport to the brain counters cerebral hypoxia and improves neurological function. Alternatively, in response to hypoxia, the brain also enhances neurogenesis to improve cerebral function.

10.5 Induction of Stress Erythropoiesis

The maintenance of tissue oxygenation is an important goal during the treatment of several diseases. In hypoxia, cells are exposed to low oxygen supply resulting in mitochondrial dysfunction, acidosis, altered membrane permeability, and lower ATP synthesis. But a certain moderate amount of hypoxia at periodic intervals can lead to an adaptive response as seen in people living in higher altitudes. Erythropoiesis under hypoxic conditions (called stress erythropoiesis) is regulated by hypoxia-inducible factor(s) (HIFs) which increases the synthesis of kidney-derived cytokine, the erythropoietin (EPO). It is a 165-amino-acid-long highly glycosylated protein with a molecular weight of 30 kDa, which plays a vital role in erythrocyte formation. EPO stimulates the proliferation and differentiation in colony-forming unit-erythroid (CFU-E) progenitors and blocks the apoptosis in late progenitors. Kidney EPO produced by peritubular fibroblast-like type-1 interstitial cells in the renal cortex and outer medulla primarily contributes to erythropoiesis. Liver EPO, on the other hand, contributes only 10% of the plasma EPO. In the brain, neurons and astrocytes in the cortex region and the hippocampus also produce cerebral EPO, but its role in erythropoiesis is unclear due to the presence of the BBB. Besides, several other tissues also produce EPO to normalize oxygen supply, e.g., the spleen, lung, testis, and even erythroid progenitors.

In response to hypoxia, hypoxia-inducible factors (HIFs) regulate the expression of genes involved in erythropoiesis. HIFs also stimulate the genes essential for angiogenesis and glucose metabolism. Hypoxia stimulate HIFs to trigger the synthesis of EPO and promote the differentiation of HSC and progenitors into functional RBCs. The role of HIFs under hypoxia was proven in studies by *Sheldon* et al. where hypoxia-mediated adaptive response was eliminated in HIF1-α knockout mice (Sheldon et al. 2014). HIF is a helix–loop–helix PAS (Per–Arnt–Sim) family transcription factor, first identified by *Semenza* et al. (Semenza and Wang 1992). Under hypoxic stress, HIFs increase the expression of the EPO gene by tenfold in the kidney and liver (Fandrey 2004). HIF contains a hypoxia-regulated α subunit and a constitutively expressed β subunit (also called aryl hydrocarbon receptor nuclear translocator, ARNT). There are three different isoforms of the α subunit, namely, HIF-1α, HIF-2α, and HIF-3α. The β subunit, on the other hand, has only one isoform: HIF-1β. The three different oxygen-dependent enzymes, prolyl hydroxylase domain proteins, namely, PHD1, PHD2, and PHD3, regulate the availability of HIFα subunits (Semenza 2001).

Under normoxic conditions, PHD-2 inactivates HIF-1α and HIF-2α hydroxylating two specific proline residues (Pro402/Pro564 in HIF-1α and Pro405/Pro531 in HIF-2α) in the oxygen-dependent degradation domain (Strowitzki et al. 2019). The hydroxylated forms then interact with von Hippel Lindau (VHL) tumor suppression protein, leading to their ubiquitination and destruction in proteasomes. On the other hand, in hypoxic conditions, HIF-1α and HIF-2α subunits escape hydroxylation and degradation. The α subunits translocate to the nucleus and attach to the β subunit to form a functionally stable HIF heterodimer that binds to the target genes and modulates the expression. The hypoxia-responsive element (HRE) located immediately after the polyadenylation site of the human and mouse *EPO* gene was found responsible for the expression of the EPO gene (Orlando et al. 2019). The formation of erythrocytes starts with the binding of EPO to its erythropoietin receptor (EPOR), an oxygen-sensitive receptor that lacks intrinsic catalytic activity. Upon binding, the EPOR is phosphorylated by Janus Kinase-2 (JAK2), leading to rapid activation of the transcription factors: signal transducer and activator of transcription5 (Stat5), phosphoinositide-3 kinase/Akt, and p42/44 mitogen-activated protein kinase. A rapid signal transduction cascade that follows regulates the anti-apoptotic effect and expansion in liver erythroid precursors. Besides, hypoxia also controls the expression of EPOR, hemoglobin synthesis, and regulation of stem cell maintenance.

Induction of stress erythropoiesis thus results in a large number of RBCs for oxygen supply to the brain and helps in normalizing the functions previously hampered due to hypoxic stress. It was also found that in clinical settings, recombinant EPO has emerged as a remarkable treatment option for the improvement of cognition. High-dose administration of recombinant EPO was found to amplify auto/paracrine EPO/EPOR signaling, which prompted the generation of new hippocampal CA1 neurons with dendritic spine densities (Wakhloo et al. 2020). This observation further confirms the importance of erythropoiesis in enhancing brain function.

10.6 Induction of Neurogenesis

For a stem cell, the decision of self-renewal or differentiation depends upon various microenvironmental cues like growth factors, extracellular matrix, and mechanical forces. The most vital force which governs the fate of stem cells is oxygen. Although oxygen concentrations are crucial under in vivo conditions, some cells need an almost anoxic environment for maintenance, e.g., stem cells. According to the "generation-age hypothesis," the self-renewal and differentiation process of the stem cells is related to actual oxygen availability (the oxygen-stem cell paradigm). In adulthood, stem cells reside in specific regions within the bone marrow called niches. Niches are of two types: the vascular niche which is well supplied with blood vessels and contains actively proliferating and differentiating stem cells, and the endosteal niche, which is devoid of the blood vessels and contains quiescent cells which maintain their primitiveness/stemness. But despite having an ample supply of blood vessels, vascular niches are still hypoxic since it contains very few afferent arterioles for oxygen delivery, and only 1–2% oxygen is present. Therefore, stem cells are "anaerobic cells" maintained at a certain oxygen tension. A small change in oxygen concentration may promote the differentiation of stem cells toward a specific lineage.

In the brain also, a defined zone is present in which NSC are retained after embryonic development. This zone (s) is called the NSC niche which helps in the production of neurons for the replenishment of the nervous system (Conover and Notti 2007). NSC has been found at high densities in the subventricular zone (SVZ) and the dentate gyrus (DG) of the forebrain. They are also found in the hippocampal subgranular zone (SGZ). For years it was assumed that once a neuron is lost it cannot be replaced. However, recent research findings have proven that neurons too can regrow to improve brain function. NSC from niches readily differentiate and give rise to juvenile neurons, and non-neuronal glial cells (namely, oligodendrocytes and astrocytes) through neurogenesis and gliogenesis, respectively. The functionally mature cells migrate and integrate themselves into the nervous system.

Adult neurogenesis is a complicated process, beginning with the differentiation of NSC into neural progenitor cells, finally leading to the functional integration of newly formed neurons into a pre-existing neural circuit. Neurogenesis occurs primarily in the olfactory bulb (OB) and the hippocampus. Two types of interneurons are found in the OB, generated from a population of progenitor cells present in the SVZ. The continuous addition of new interneurons modulates the spatial and temporal coding of olfactory information and helps in adaption to environmental changes. Among various intrinsic and extrinsic molecular signals that regulate neurogenesis in the CNS, hypoxic damage plays a vital role in the maintenance and function of NSC. Although HIF-1α normally undergoes proteasomal degradation under non-hypoxic conditions, it is constitutive stabilized in multiple stem cell types in vivo, including bone marrow HSC, MSC, and NSC of adult SVZ, and hippocampal SGZ (Horie et al. 2008).

Hypoxia promotes the differentiation of NSC into particularly dopaminergic neurons. It was also found that in the hippocampus, neurogenesis helps in the regulation of learning and memory, the repair of brain injuries, and the fight against depression. The treatment of hypoxia promotes angiogenesis, gliogenesis, and neurogenesis in vivo by enhancing the expression of HIF-1α in neurogenic regions of the brain including the DG and the SVZ (Yuen et al. 2014; Jablonska et al. 2012). However, it is still unclear whether external hypoxia affects the actual oxygen concentrations in the SVZ and DG for promoting neurogenesis. The reports on hypoxia have been indirect and reveal that hypoxia damage is immense on the brain's cognitive functions. Long-term operations carried out at high altitudes by military personnel, pilots, and astronauts may trigger health complications. In particular, chronic exposure to high altitude has been associated with deficits in cognitive function (Sharma et al. 2019).

It can be concluded from the above discussion that sudden hypoxia can lead to impairment of brain function, but periodic hypoxia in form of hypoxia training or preconditioning can result in better brain function through the adaptive mechanism.

10.7 Benefits of *Yoga*, *Pranayama*, and Meditation

Physical exercises help in coping with hypoxic stress by triggering a state of hypoxia in the body. Now the question arises, how does induced hypoxia fight against hypoxic stress? Continuous physical exercises induce a state of hypoxia in the body and in doing so make our body resistant to any subsequent hypoxic event. The role of physical exercises in acclimatization at higher altitudes deals with gradual adaptation to climate and oxygen pressure changes. Upon first exposure to a low-oxygen environment, hypoxia is induced in the body followed by a feeling of uneasiness. Repetitive exposure to the same environments makes the person more tolerant to these oxygen levels due to adaptation. More importantly, the brain senses the hypoxia as a "warning signal" and responds by inducing the transcription of various genes through the formation and activation of the complete HIF heterodimeric complex. A second-round exposure to the same or similar hypoxia results in a multifold faster response. Hence hypoxic stress could be beneficial when utilized in a controlled manner as a part of a training program. It is being used for patients suffering from chronic obstructive pulmonary disease (COPD), diabetes, coronary heart disease, heart failure, hypertension, etc.

The concept of using intermittent hypoxia as a therapy for avoiding hypoxic stress-related tissue damage was first introduced by Lu (1963). He defined it as an induced tolerance developed in tissue and cells after exposure to short bouts of hypoxia in between normoxia. The procedure was named "Hypoxic Preconditioning" (HP). It generally involves exposure to a short period of moderate hypoxia, followed by normoxia. The protocol involves three phases that include the initial tolerance phase, which provides immediate tolerance to hypoxia (first few minutes of exposure), the long-term hypoxic tolerance phase, and the third phase in which the expression of hypoxic tolerance genes and other pro-adaptive genes takes place (generally after 24 h) (Rybnikova and Samoilov 2015). Under hypoxia, HIF first induces the expression of genes involved in the maintenance of homeostasis. HIFs have a very distinct role in the hypoxia-affected brain. In earlier stages, it is involved in cell death, and later, after 4–5 days, it promotes the survival of the cells (Barteczek et al. 2016), and later it allows cytokine production that either directly or indirectly helps in neurogenesis through the differentiation of NSC (Leu et al. 2021). Studies conducted by *Bin-Jaliah* et al. showed that 10 rounds of 6 mins each containing 10% O_2 and 21% O_2 could stimulate EPO and RBC production in male Wistar rats (Bin-Jaliah et al. 2009). Another study conducted in 2019 showed that HP increased EPO levels both centrally and peripherally under the same hypoxia conditions (Törpel et al. 2019).

Currently, there is no standardized plan to follow for the administration of HP. The parameters vary for each patient and their sensitivity to hypoxia. Thus, in these situations, clinicians may sometimes alter the rounds of exposure to hypoxia by increasing the ratio of hypoxia to normoxia or increasing the period of exposure, which may sometimes be confusing. Breathing exercises can be beneficial under these circumstances as the rate of inhalation or exhalation is under the control of the individual itself and through routine practice, it can be developed as a strategy for self-administration of hypoxia. Because physical exercises are not feasible for all ages and body types, alternatively, *Yoga* and *Pranayama* can be used to subdue any negative impacts of hypoxia on the brain. It is the easiest way by which HP can be self-administered.

Pranayama exposes the tissue to periodic hypoxia and stimulates HSC to form a large number of functional RBCs through the activation of HIFs. In a study conducted by *Singh* et al. it was found that during the retention (Kumbhaka) phase in *Nadi Shodhana*, a state of hypoxia was created in the body, which caused a noticeable increase of EPO in circulation, enhancing the RBC production by 10 times which lasted until the exhalation (Puraka) phase (Singh 2013). Nadi Shodhana also increased hemoglobin levels in patients suffering from anemia, thereby increasing the oxygen-delivering capacity of the RBC (Kumar et al. 2018). Similarly, *Nisshesha Rechaka* produce brief, recurrent hypoxia that induced similar effects (Malshe 2011).

Tiwari et al. showed that hypoxia can also be self-administered by combining *Yoga* with *Pranayama*. They found that *Pranayama*, when done along with *Yoga* exercises like *Pawanmuktasana* (wind relieving pose), *Mandukasana* (frog pose), and *Mulabandh* (root lock), leads to squeezing of abdominal viscera which increases and decreases the blood flow and built a hypoxic environment leading to erythropoiesis in HSC. It was also mentioned that *Yoga* decreases the synthesis of hepcidin (Tiwari et al. 2017). Hepcidin is a protein that regulates iron homeostasis in the body by coordinating iron absorption and its distribution to the tissues. It is a 25-amino-acid-long peptide produced in the liver hepatocytes. Higher levels of hepcidin inhibit intestinal iron absorption and macrophage iron recycling, thus causing anemia due to restricted erythropoiesis. On the other hand, lower levels of hepcidin favor sufficient iron supply to the BM for hemoglobin synthesis and uninterrupted erythropoiesis (Pagani et al. 2019). Production of hepcidin is suppressed by hypoxia, and HP through *Pranayama* increases RBC formation and hemoglobin synthesis through increased iron absorption (Liu et al. 2012). Hence, hypoxia-induced suppression of hepcidin is indirectly regulated by HIFs through the synthesis of EPO. The maturing erythroblasts secrete growth differentiation factor-15 (GDF15) and transforming growth factor-β (TGF-β), which also inhibits the hepcidin production by suppressing the hepcidin gene (HAMP1) expression in hepatocytes (Serebrovskaya et al. 2011). Figure 10.1 shows how *Yoga* and *Pranayama* induce stress erythropoiesis.

Hypoxic stress at higher altitudes leads to structural and functional alterations in brain regions, resulting in psychological problems, including cognitive impairment. A few research studies have emphasized the benefits of *Yoga* and *Pranayama* in reducing the neurological dysfunction related to high-altitude travels by induction

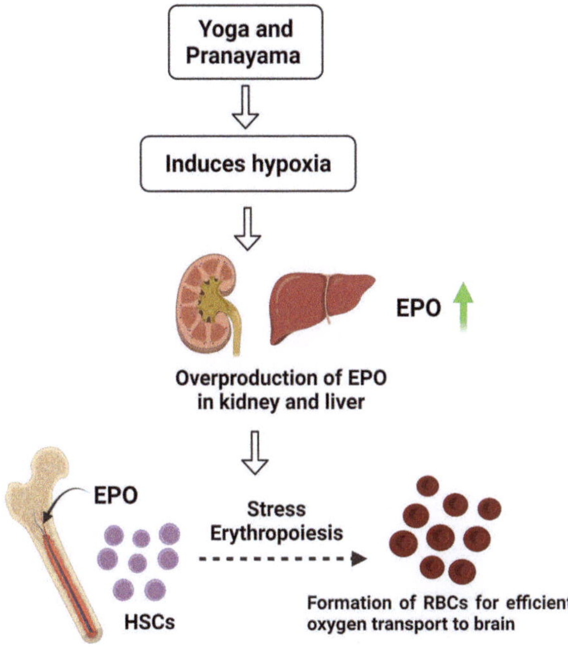

Fig. 10.1 Representing how *Yoga* and *Pranayama* help in self-administration of HP and induction of stress erythropoiesis: *Yoga* and *Pranayama* induced hypoxia, which increases the production of EPO in the kidney and liver. In the bone marrow, EPO enhances erythropoiesis in HSCs and increases the number of RBCs for oxygen transport to the brain

of neurogenesis through administering HP. Regular meditation has positive effects on several regions of the brain (frontal and parietal lobes, medulla, etc.). Deep meditation practices improve all necessary cognitive functions like memory, attention, perception, observation ability, processing speed, neural activity, intelligence function, etc., which energizes the brain to focus on a task. Meditation may also be able to offset age-related cognitive decline and perhaps even increase cognitive capabilities, thereby preventing mental illness. Deep breathing during *Pranayama* sessions affects several receptors in the alveoli, including the stretch receptors, baroreceptors, and chemoreceptors, which in turn pass the information to the CNS, resulting in specific responses like perception, cognition, emotional behavior, etc. *Yoga* and *Pranayama* may also stimulate hippocampal neurogenesis through HP.

In a study conducted by *Bhanushali* et al. it was found that an advanced meditation program (AMP) including *Sudarshan Kriya*, *Nadi Shodhana*, *Anuloma-Viloma*, *Bhastrika*, and 18 other *Yoga* poses was able to confer biochemical and neurocognitive benefits to the participants (Bhanushali and Tyagi 2020). *Gupta* et al. observed a sudden weight loss and decrease in body mass index (BMI) due to hypoxia followed by an increase in blood pressure due to activation of the sympathetic nervous system. They also investigated the role of *Sudarshan kriya* and observed positive psychological changes in the participants like cognitive performance, mood, sleeping pattern, happiness quotient, etc. (Gupta et al. 2022). *Himashree* et al. explored the effects of *Yoga*, *Pranayama*, and meditation on the Indian soldiers living at Karu, Leh (altitude >3000 m). The soldiers were divided into two groups. One group followed a comprehensive *Yoga* package that consisted of poses *Padmasana* (2 min), *Suptapavanmuktasana* (2 min), *Pavanmuktasna* (2 min), *Vajarasana* (2 min), *Gomukhasana* (2 min), *Makarasana* (2 min), *Shalabhasana* (2 min), *Shavasana* (6 min); *Pranayama*/breathing-*Anuloma-Viloma* (5 min), *Kapalbhati* (1–2 min), *Brahmari* (2–3 min), and Meditation *Omkar* (2 min), guided meditation (8 min). It was found that the soldiers from the *Yoga* group showed improvement in neural health indices at high altitudes in comparison to the other group that followed a regimen of only routine physical exercises (Himashree et al. 2016).

Xue et al. investigated the effects of mindful breathing on rapid hypoxia acclimatization training at high altitudes and reducing the symptoms of HACE. Their Hypoxic Preacclimatization Training Scheme (HPTS) at 3600 m and 3900 m successfully helped volunteers adapt to the high altitude in a short time using the Mindful Breathing Training Scheme (MBTS). This induced significant gain effects by increasing the acclimatization speed by obtaining higher SpO_2. MBTS can enable beginners without any experience in mindfulness and breathing training to initially stimulate the trend of "energy-saving" acclimatization in 2–3 days (Xue et al. 2020). *Yoga*, *Pranayama*, and meditation are ideal exercises for brain cell regeneration. *Gothe* et al. reviewed 11 clinical studies and found that *Yoga* positively impacts the hippocampus, amygdala, prefrontal cortex, cingulate cortex, and brain networks structurally and functionally (Gothe et al. 2019). A meditation exercise called *kirtan kriya* was found to increase the blood flow toward the brain regions that control memory, thereby improving the symptoms of dementia in Alzheimer's disease (Khalsa and Perry 2017). In a different study, it was found that *Bhastrika* was helpful in decreasing anxiety by modulating the activity of emotion-processing regions of the brain, which was confirmed by resting-state functional MRI (Novaes et al. 2020).

Overall brain growth is known to depend upon several neurotrophic factors. High expression of BDNF in the brain plays a vital role in the differentiation of NSC into neurons, and synaptic connectivity, which forms the basis of learning, cognition, and memory. The crucial role of BDNF was proven by *Ernfors* et al. where excessive apoptosis and minimal formation of new neurons in the BDNF knockout mice that displayed a lack of coordinated movement and balance in the body was observed (Ernfors et al. 1995). *Yoga*, *Pranayama*, and meditation were found to increase the serum levels of BDNF together with the reduction in cortisol levels in patients suffering from depression and anxiety (Cahn et al. 2017). Similarly, *Yoga*, *Pranayama*, and meditation also increase the level of the glial cell line-derived neurotrophic factor (GDNF), which is known to regulate the functioning of the brain's dopaminergic system. Insulin-like growth factor-1 (IGF-1) and insulin-like growth factor-2 (IGF2) regulate neurogenesis and survival of neurons, angiogenesis, neurotransmissions, and cognitive behavior. *Gallegos* et al. performed experiments to study the effects of mindfulness-based stress reduction (MBSR) activities like *Yoga* and meditation on circulating IGF-1 concentrations. A committed practice of *Yoga* and meditation in sitting posture was found to increase IGF-1 levels leading to an improvement in cognitive behavior (Gallegos et al. 2013).

Studies conducted by *Zhu* et al. showed that HP promotes the proliferation of neuronal progenitors, leading to the formation of neurons in the hippocampus in adult rats. The effect was due to the enhancement of BDNF expression after HP (Zhu et al. 2010). *Garner* et al. also concluded that hippocampal gray matter density increased in the group who performed *Hatha Yoga* for HP (Garner et al. 2019). The breathing pattern during *Pranayama* was shown to impart significant cardiovascular and psychological health benefits. It was found that fast breathing (120 breaths/min) followed by slow breathing (6 breaths/min) results in an increase in both the right and left prefrontal cortex, a change in hemodynamics, and an increase in brain oxygenation in participants (Bellissimo et al. 2020).

Streeter et al. found a 12-week *Yoga* intervention causes improvement in the mood and anxiety of the practitioners due to an increase in thalamic levels of neurotransmitter GABA in comparison to normal walking exercise (Streeter et al. 2010). Higher GABA levels in the thalamus may be due to enhanced cerebrovascular blood flow in the prefrontal cortex, which can lead to the activation of the reticular nucleus in the thalamus resulting in higher GABA synthesis. Higher GABA levels were earlier observed through magnetic resonance spectroscopy (MRS) in *Yoga* practitioners after a 60 min session of *Yoga* training (sun salutation, standing poses, sitting poses, twists, supine and prone poses, inverted poses, balancing poses, and backbend) compared to controls (Streeter et al. 2007). In addition, enhanced levels of dopamine were also observed in the ventral striatum of *Yoga* practitioners (Kjaer et al. 2002).

Echoencephalograph (EEG) studies were conducted to examine the effects of *Yoga* on the brain. Brain waves revealed that breathing, meditation, and posture-based *Yoga* and meditation practice increase overall brain activity by structural alteration in the amygdala, frontal cortex, hippocampus, insula, and anterior cingulate cortex (Cahn and Polich 2006). The gray matter volume in the *Yoga* practitioners was also found high in different areas of the dominant hemisphere, including the ventromedial orbitofrontal, ventrolateral prefrontal, and inferior temporal and parietal cortices as well as the left insula in practitioners of *Sahaja Yoga* meditation. Experienced *Yoga* practitioners have shown greater neocortical thickness in the left prefrontal complex cluster, which includes part of the lateral middle frontal gyrus, dorsal superior frontal gyrus, and anterior superior frontal gyrus compared to non-practitioners (Hernández et al. 2016, 2018, 2020). Figure 10.2 shows how *Yoga* and *Pranayama* induce neurogenesis.

10.8 COVID-19: Neurological Complications

Apart from high altitude, conditions like pulmonary fibrosis, anemia, cardiovascular diseases, asthma, bronchitis, and pneumonia can also induce hypoxia in your body. The pandemic COVID-19 is also characterized by the slow development of

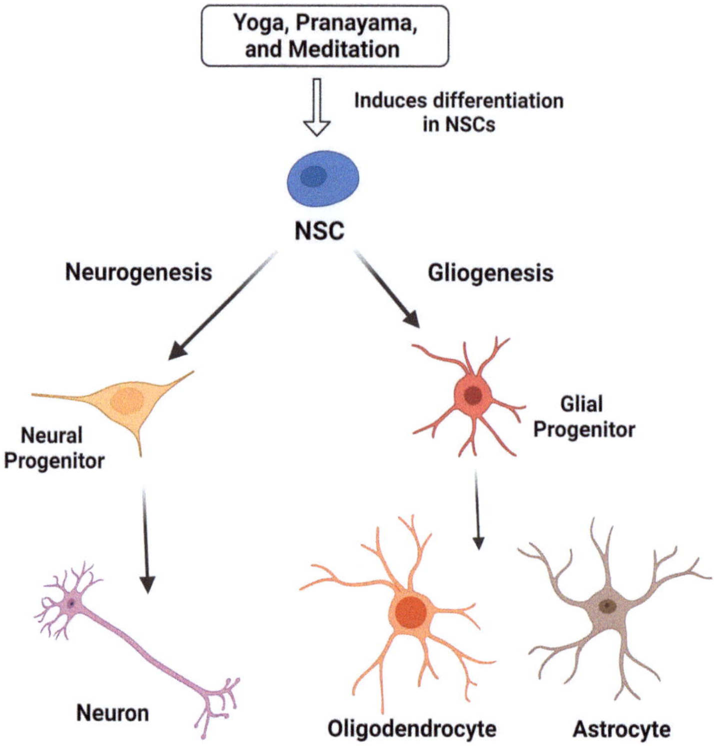

Fig. 10.2 Representing how *Yoga*, *Pranayama*, and meditation improve brain function: *Yoga*, *Pranayama*, and meditation induce neurogenesis and gliogenesis in neural stem cells and help in the regeneration of lost tissue in the brain

hypoxemia and tissue hypoxia. Silent or happy hypoxemia/hypoxia leads to organ damage which is the prime reason for high mortality rates. Chronic respiratory failure leads to hypoxia in the brain giving rise to neurological dysfunction, e.g., strokes, epilepsy, nerve damage, dementia, etc. In the section below, we will discuss COVID-19 and how it leads to the failure of neurological dysfunction.

COVID-19 is an infectious disease caused by Severe Acute Respiratory Syndrome Coronavirus-2 (SARS-CoV-2). SARS-CoV-2 is an enveloped, single-stranded RNA virus having a large genome of around 28.9–30 kb that codes for 9860 amino acids (29 proteins). The genome comprises 11 open reading frames coding for all essential proteins for virus entry, assembly, release, and other non-essential proteins (Srinivasan et al. 2020). Angiotensin-converting enzyme 2 (ACE2) expressed on alveoli type-2 (AT2) and capillary endothelium is recognized by spike (S) protein (1272 amino acid; 180–200 kb; 20–40 nm length) of SARS-CoV-2. S-protein is coded by ORF2 and has two subunits S1 and S2. The S1 subunit binds to the ACE2 receptor through its receptor-binding domain (RBD), facilitating the attachment of the virus to the target cells. Another protein known as transmembrane protease serine-2 (TMPRSS2) assists in the entry of the virus by cleaving the

S-protein at S1/S2 and S2′ sites by the protease activity, allowing the fusion of virus and host cell membranes. After fusion, the virus enters the host cells by an endocytic pathway to release the virus particles into the cells (Hoffmann et al. 2020). Remaining ORFs code for other essential structural and non-structural proteins. ORF1a and ORF1b code for 16 non-structural proteins essential for the replication and transcription of the viral genome. ORF4 codes for envelope protein and an ion-channel protein, Viroporin. ORF5 codes for membrane protein. ORF9 codes for nucleocapsid protein (Schoeman and Fielding 2019; Thomas 2020).

After infection, a sudden increase in proinflammatory cytokines results in a cytokine storm (Mehta et al. 2020). High inflammation results in increased levels of D-dimer, C-reactive protein (CRP), and ferritin (Yu et al. 2020; Ali 2020; Lin et al. 2020). In the lungs, it impairs oxygen transport and gaseous exchange, leading to systemic hypoxia. Altogether, these factors lead to intravascular blood coagulation and multiple organ failures due to oxygen shortage. Besides pulmonary symptoms, several hematological abnormalities such as reduced hemoglobin levels and a non-significant but gradual decline in RBC number with structural abnormalities have also been seen. SARS-CoV-2 uses a very different receptor called the BAND3 protein for entry into RBC which causes cell death leading to even more severe hypoxemia (Cosic et al. 2020).

10.8.1 Neurological Complications and Mechanisms

The most common symptoms of COVID-19-induced neurological complications are unstable gait, ataxia, headache, dizziness, loss of taste and smell, vision impairment, nerve pain, epileptic strokes, and malaise. Besides, acute disseminated encephalomyelitis (ADEM), encephalopathy, viral meningitis/encephalitis, polyneuritis, Guillain–Barré syndrome (GBS), Miller Fisher syndrome (MSF), cerebral hemorrhage, ischemic strokes, hyposmia and hypogeusia, and infractions are also reported (Balcom et al. 2021).

The BBB and cerebrospinal fluid (CSF) prevent the entry of pathogens and toxic molecules into the brain. However, due to its semipermeable nature, viruses and a few pathogens have developed multiple mechanisms to traverse the BBB (Kim 2008). The most common route followed by coronavirus is the hematological route or vascular route, which starts by entering the bloodstream, causing viremia. SARS-CoV-2 infects brain endothelial cells, which express ACE2 and induce endothelial damage leading to cerebral vascular thrombosis and disruption of the BBB that increases its permeability. Alternately, vascular damage may also be caused by infection-induced vasculitis or vasculopathy (Krasemann et al. 2022). Other routes include the dynein- and kinesin-mediated transport of the virus into the CNS using an infected motor or sensory nerves in the lungs (Li et al. 2020a) and through olfactory sensory neurons (retrograde axonal transport) that leads to the olfactory deficit (Meinhardt et al. 2020).

Hypoxemia can also induce brain damage by inducing ischemic encephalopathy, microhemorrhages, and oligodendroglia cell injury. Prolonged hypoxia also causes BBB disruption leading to leaky capillaries resulting in microhemorrhages. Aberrant immune response due to the presence of SARS-CoV-2 can cause cranial nerve dysfunction resulting in neurological disorders such as Miller Fisher syndrome and polyneuritis cranialis (Popescu 2021). In addition, cytokine storm also disrupts BBB and results in ischemic strokes and other thromboembolic events (Erickson et al. 2021). Additionally, hypokalemia and hyponatremia in severe COVID-19 patients can lead to brain edema and demyelination in entire nervous systems (Pourfridoni et al. 2021).

Taquet et al. studied the neurological complication among 2,36,379 COVID-19 patients. Out of the neurological and psychological incidences which were observed, 0.56% of patients suffered from intracranial hemorrhage, 2.10% suffered an ischemic stroke, 0.11% developed Parkinsonism, 0.67% had dementia, 17.39% have an anxiety disorder, and 1.40% other psychotic disorders. In the group inside the intensive therapy unit (ITU), estimated incidences were 2.66% patients with intracranial hemorrhage, 6.92% with ischemic stroke, 0.26% with developed Parkinson's, 1.74% have dementia, 19.15% have an anxiety disorder, and 2.77% have other psychotic disorders. Most diagnostic categories were more common in patients who had COVID-19 than in those who had other diseases such as influenza. Hence, the risk of neurological complications was found high in severe COVID-19 (Taquet et al. 2021).

Yang et al. profiled around 65,309 single-nucleus transcriptomes from 30 frontal cortex and choroid plexus samples of COVID-19 patients and observed broad cellular perturbations in barrier cells resulting in peripheral inflammation in the brain. Further, damage to microglia and astrocytes was also found, a pathological symptom that is similar to various human neurodegenerative diseases (Yang et al. 2021). Out of the 214 patients studied by *Mao* et al. 78 (36%) developed neurological complications. They found that the chances of developing neurological complications are more in older patients and patients with severe COVID-19 infection as they display symptoms such as ataxia, acute cerebrovascular disease, unconsciousness, etc. Non-severe patients, on the other hand, developed underlying symptoms such as hypertension, headache, etc. (Mao et al. 2020). In a UK-based study, elderly patients were examined for having neurological manifestations. Complete clinical datasets were available for 125 out of 153 patients. Seventy-seven of these patients had an ischemic stroke intracerebral and CNS vasculitis, 39 patients had altered mental status, and 23 patients with altered mental status fulfilled the clinical case definitions for psychiatric diagnoses (Varatharaj et al. 2020). In 509 COVID-19 patients examined by *Liotta* et al. nearly 82% were having neurological complications like myalgias, headaches, encephalopathy, dizziness, dysgeusia, anosmia, strokes, movement disorders, motor and sensory deficits, ataxia, and seizures with around 63% of patients having complications at hospitalization (Liotta et al. 2020).

Post-hypoxic myoclonus has been observed in patients, which can lead to movement disorders. *Ros-Castello* et al. studied a 72-year-old female patient suffering from severe COVID-19. After the patient was discharged from the intensive care unit, she was diagnosed with progressively disabling myoclonus in the upper limbs and negative myoclonus in the lower limbs leading to movement impairment. Myoclonus may be caused by metabolic abnormalities such as liver failure, renal failure, or hypercapnia or due to medications such as cephalosporins, quinolones, imipenem, or hypoxia. In the patient, MRI showed cortical and brainstem ischemic lesions leading to the conclusion that myoclonus can be attributed to CNS hypoxia (Ros-Castelló et al. 2020). *Serafini* et al. highlighted a few other consequences of the infection that included stress, anxiety, and panic symptoms arising due to fear, loneliness, and separation from society resulting in a depressive state in the individual and cognitive impairment (Serafini et al. 2020). Hypoxia induces an excessive accumulation of anaerobic metabolites in the mitochondria and acid metabolites in the brain, leading to edema of the brain cells and obstruction of cerebral blood flow. Hyposmia in patients was observed which may be due to "conductive" loss or "neural" loss.

As mentioned earlier, SARS-CoV-2 infects humans by targeting ACE2 via its spike protein through the receptor-binding domain. ACE2 is widely expressed in the lungs, kidneys, guts, cardiovascular system, and also in CNS. In the brain, it is expressed in the neurons, astroglia cells, microglia cells, and endothelial cells. Alternative receptors are neuropilin-1 (NRP1) (Cantuti-Castelvetri et al. 2020) and basigin (BSG)/CD147 (Wang et al. 2020) which are expressed highly in the CNS. Alternatively, proteases-furin and cathepsin might permit viral entry in cells with low levels of TMPRSS2 expression (Shang et al. 2020). Another potential mechanism that induces neurological changes is neuro-inflammation triggered by the expression of viral antigens into the CNS. Glial cell activation and occasional leucocyte infiltrate were observed in patients, although the associated molecular pathways remain unclear. CSF studies suggest activation of innate immune responses characterized by elevated levels of β2-microglobulin and neopterin and the presence of dedifferentiated monocytes and increased levels of neurofilament in the CNS suggesting neuronal injury (Heming et al. 2021).

The olfactory epithelium (OE) is another route for entry of SARS-COV-2 in the brain. The nasal cavity contains olfactory sensory neurons (OSNs). The horizontal basal cells (HBC) of the OE are progenitors to OSNs. Although OSNs do not express ACE2; HBC show a surface expression of ACE2 which becomes the target of SARS-COV-2. Infected HBC mature into bipolar unmyelinated OSNs and gain access to OB by a synaptic path. Subsequently, the virus could infect mitral cells which are connected to several parts of the brain and facilitate infection of the cortex, the mesolimbic cortex, the hippocampus, the amygdala, and eventually the brainstem and spinal cord via a trans-synaptic pathway, using endocytosis and exocytosis (Li et al. 2020b). The neurotoxic effect of SARS-CoV-2 like smell or taste disorder might be due to changes in the phosphorylation pattern of proteins associated with axons and synapses in olfactory or gustatory neurons or injuries to any cranial nerves (Berth et al. 2009). In addition, the virus could use other peripheral nerves to reach the CNS including the pulmonary network and enteric nervous system (Briguglio et al. 2020; Esposito et al. 2020).

10.8.2 Yoga *and* Pranayama *as a Possible Treatment Option*

The symptoms of high-altitude and COVID-19-related neurological complications are similar, although the mechanism of the manifestation may differ slightly (later is caused by viral infection). Cerebral hypoxia in both cases arises due to a low supply of oxygen to the brain. Interestingly, a few reports suggest that high-altitude inhabitants were found comparatively less prone to develop respiratory and neurological complications of COVID-19 (Arias-Reyes et al. 2020). People living at high altitudes are constantly exposed to hypoxic conditions which make them physiologically adapted to the hypoxic environment and more resilient to the hypoxic manifestation of COVID-19 (Millet et al. 2020). A few potential mechanisms have been proposed: first, the altitude-related decrease of the ACE2 expression may provide a protective effect against COVID-19-related respiratory distress (Arias-Reyes et al. 2020); second, the enhanced expression of EPO at high-altitude leads to high RBC numbers in residents for increasing the oxygen transport to organs including the brain (Soliz et al. 2020); third, the higher level of UV radiation at high altitude may hamper the survival of the virus; and fourth, the lower barometric pressure leads to a reduced air density and flow that would reduce the viral dissemination between people (Arias-Reyes et al. 2020). Data analysis of the actual geographical region of the COVID-19 epidemic was carried out, and results from 70 cities with altitudes ranging between 1 and 3180 m showed a negative correlation between altitude and COVID-19, concluding that higher altitude has lesser infection and fatality rate (Cano-Pérez et al. 2020). The capital of the Central South American country Bolivia, La Paz, is a classic example. La Paz lies at a high altitude ranging between 2400 and 4000 m above sea level. The first case was reported on 10 March 2020 and the total number was 54 in the Bolivian region located in the highland. The low rate of transmission of infection was remarkable and did not show the expected rapid infection rates (Accinelli and Leon-Abarca 2020).

HP through *Yoga* and *Pranayama* can be used to prevent or treat COVID-19-related respiratory and neurological complications. From this perspective, *Rain* et al. emphasized the importance of *Simha Kriya* and *Isha kriya*, in combating COVID-19-related depressive states and how *Yoga Asana* leads to psychological and physiological improvement in patients (Rain et al. 2021). Similarly, *Kulkarni* et al. mentioned that *Yoga* and *Pranayama* have a moderate effect on mental health when used as adjunct therapy which can be beneficial to improve the psychological health of people during the COVID-19 pandemic (Kulkarni et al. 2022). *Nagarathna* et al. performed a pan-India cross-sectional survey study, during the COVID-19 lockdown. The population under study was divided into two groups: one who practices *Yoga* during lockdown (*Yoga* group) and the other who doesn't practice *Yoga* (non-*Yoga* group). It was found that in comparison to the non-*Yoga* group, the *Yoga* group displayed a better ability to cope with stress and anxiety during the lockdown (Nagarathna et al. 2021). *Jerrin* et al. performed a quasi-experimental study on 130 patients. *Yoga* and naturopathic intervention were given for 60 min a day for 2 weeks in the morning between 7.30 a.m. and 8.30 a.m. which includes *Yoga* poses (*tadasana*, *chakrasana*), *Pranayama* (*Bhramari*, *Nadi shuddhi*), stretches, steam inhalation, and sun exposure. Anxiety and depression levels were significantly reduced in these patients (Jenefer Jerrin et al. 2021).

This quasi-randomized clinical trial was conducted to assess the efficacy of *Pranayama* in preventing infection in healthcare professionals dealing with COVID-19 patients. In total, 280 participants were assigned to intervention and control groups. Participants practiced *Vaata neti*, *Ujjaayee*, *Bhramari*, *Kapalbhati*, *Nadi Shodhana*, and meditation except for the control group. All the participants were tested for antibodies to COVID-19 on the 28th day of the intervention. At the end of 28 days of intervention, nine participants in the control group and one in the intervention group developed COVID-19. The finding suggests that the daily practice of the *Pranayama* module for 28 days in healthcare professionals may have helped in preventing COVID-19 infection by improving lung capacity, cardio-respiratory functions, neuro-endocrine activity, and mental activity (Sarwal et al. 2022).

10.9 Future Direction for Research

Scientists, physicians, and clinicians have tried to utilize hypoxia to their advantage by developing a therapeutic approach that involves building a state of resistance through intermittent doses of hypoxia in a planned manner. The build-up of resistance or protective mechanism against hypoxia depends on the individual's age, immune system, and how quickly the body responds and adapts. A strategy to incorporate short doses of hypoxia in between normoxia to build resistance against any subsequent hypoxic event leading to an even faster response is called HP. HP can be self-administered through easy physical exercises, especially *Yoga* and breathing exercises such as *Pranayama*. *Yoga* and *Pranayama* in combination with meditation can also be used to prevent neurological dysfunction at high altitudes by increasing neurogenesis and safeguarding brain function. Similarly, in COVID-19, *Yoga*, *Pranayama*, and meditation can be used to reduce respiratory and neurological complications. HP through *Yoga* and *Pranayama* is perfect for patients with mild to moderate cases. It is also beneficial to people who are vulnerable to the infection or old-age people because, for them, an adaptive response due to HP can act as a sort of vaccine by providing them with long-term benefits against breathing and neurological complications.

Although successful, the practical utility of *Yoga*, *Pranayama*, and mediation in rescuing from high altitude-related neurophysiological, neuropsychological, and neurocognitive dysfunction has been marred by a lack of a large number of clinical studies providing satisfactory pieces of evidence, especially under these harsh conditions. Hence, more clinical phase trials are needed in the future to successfully establish a protocol for self-administration of hypoxia through *Yoga* and *Pranayama* for neurological benefits. Lastly, more intricate details about the mechanism involved directly in neurogenesis are still to be elucidated.

10.10 Clinical Implications and Conclusion

Due to low oxygen pressure, first exposure to high altitude causes a massive drop in blood oxygen levels leading to hypoxemia. It is an alarming situation as it results in organ failure due to low tissue oxygenation. Mountaineers, tourists, and people who have recently shifted to high altitude display neurological complications due to cerebral hypoxia such as headache, anxiety, dizziness, lightheadedness, and depression (epilepsy, seizures, hemorrhage in worst cases). On the other hand, people who have been living for years do not display such symptoms. The gradual adaptation has led these people to tolerate these harsh conditions. The adaption to hypoxia is primarily through increased oxygen transport to CNS through enhanced erythropoiesis and neurogenesis. The phenomenon of gradual adaptation has been utilized in the form of HP in which intermittent doses of hypoxia are administered in between normoxia to make an individual resistant to any subsequent and even harsher hypoxic event.

In this chapter, we have tried to highlight the effectiveness of *Yoga* and *Pranayama* in the self-administration of hypoxia. *Yoga* and *Pranayama* can be used to alleviate the neurophysiological, neuropsychological, and neurocognitive dysfunctions arising due to hypoxia at high altitudes by enhancing erythropoiesis for better oxygen supply to the brain. Combined with mindfulness training, *Yoga* and *Pranayama* can also help in enhancing the rate of neurogenesis in the brain to improve brain function. Although effective, there are doubts about its potential and more preference is given to allopathic medication due to lack of knowledge. But these medicines have some serious side effects on the brain. Hence, it is suggested that *Yoga*, *Pranayama*, and meditation should be practiced together with medication for faster recovery. Regular practice may also reduce the cost of extended medical treatment.

References

Accinelli RA, Leon-Abarca JA (2020) En la altura la covid-19 ES menos frecuente: La Experiencia del Perú. Arch Bronconeumol 56:760–761

Ali N (2020) Elevated level of c-reactive protein may be an early marker to predict risk for severity of COVID-19. J Med Virol 92:2409–2411

An H, Lin W (2000) Quantitative measurements of cerebral blood oxygen saturation using magnetic resonance imaging. J Cereb Blood Flow Metab 20:1225–1236

Arias-Reyes C, Zubieta-DeUrioste N, Poma-Machicao L, Aliaga-Raduan F, Carvajal-Rodriguez F, Dutschmann M, Schneider-Gasser EM, Zubieta-Calleja G, Soliz J (2020) Does the pathogenesis of SARS-COV-2 virus decrease at high-altitude? Respir Physiol Neurobiol 277:103443

Balcom EF, Nath A, Power C (2021) Acute and chronic neurological disorders in COVID-19: Potential mechanisms of disease. Brain 144:3576–3588

Banasiak KJ, Haddad GG (1998) Hypoxia-induced apoptosis: effect of hypoxic severity and role of p53 in neuronal cell death. Brain Res 797:295–304

Banasiak KJ, Xia Y, Haddad GG (2000) Mechanisms underlying hypoxia-induced neuronal apoptosis. Prog Neurobiol 62:215–249

Bapat A, Schippel N, Shi X, Jasbi P, Gu H, Kala M, Sertil A, Sharma S (2021) Hypoxia promotes erythroid differentiation through the development of progenitors and proerythroblasts. Exp Hematol 97:32–36

Barteczek P, Li L, Ernst A-S, Böhler L-I, Marti HH, Kunze R (2016) Neuronal HIF-1α and HIF-2A deficiency improves neuronal survival and sensorimotor function in the early acute phase after ischemic stroke. J Cereb Blood Flow Metab 37:291–306

Bellissimo G, Leslie E, Maestas V, Zuhl M (2020) The effects of fast and slow *Yoga* breathing on cerebral and central hemodynamics. Int J Yoga 13:207

Berth SH, Leopard PL, Morfini GN (2009) Virus-induced neuronal dysfunction and degeneration. Front Biosci 14:5239–5259

Bhanushali D, Tyagi R, Limaye N, Anand A (2020) Effect of mindfulness meditation protocol in subjects with various psychometric characteristics at high altitude. Brain Behav 10:e01604

Bhargav PH, Reddy PV, Govindaraj R et al (2020) Impact of a course of add-on supervised *Yoga* on cortical inhibition in major depressive disorder: a randomized controlled trial. Can J Psychiatry 66:179–181

Bin-Jaliah I, Ammar HI, Mikhailidis DP, Dallak MA, Al-Hashem FH, Haidara MA, Yassin HZ, Bahnasi AA, Rashed LA, Isenovic ER (2009) Cardiac adaptive responses after hypoxia in an experimental model. Angiology 61:145–156

Briguglio M, Bona A, Porta M, Dell'Osso B, Pregliasco FE, Banfi G (2020) Disentangling the hypothesis of host dysosmia and SARS-COV-2: the bait symptom that hides neglected neurophysiological routes. Front Physiol 11:671

Busl KM, Greer DM (2010) Hypoxic–ischemic brain injury: pathophysiology, neuropathology and mechanisms. NeuroRehabilitation 26:5–13

Cahn BR, Polich J (2006) Meditation states and traits: EEG, ERP, and neuroimaging studies. Psychol Bull 132:180–211

Cahn BR, Goodman MS, Peterson CT, Maturi R, Mills PJ (2017) *Yoga*, meditation and mind-body health: increased BDNF, cortisol awakening response, and altered inflammatory marker expression after a 3-month *Yoga* and meditation retreat. Front Hum Neurosci 11:229690

Cano-Pérez E, Torres-Pacheco J, Fragozo-Ramos MC, García-Díaz G, Montalvo-Varela E, Pozo-Palacios JC (2020) Negative correlation between altitude and Covid-19 pandemic in Colombia: a preliminary report. Am J Trop Med Hyg 103:2347–2349

Cantuti-Castelvetri L, Ojha R, Pedro LD et al (2020) Neuropilin-1 facilitates SARS-COV-2 cell entry and infectivity. Science 370:856–860

Conover JC, Notti RQ (2007) The neural stem cell niche. Cell Tissue Res 331:211–224

Cosic I, Cosic D, Loncarevic I (2020) RRM prediction of erythrocyte BAND3 protein as alternative receptor for SARS-COV-2 virus. Appl Sci 10:4053

Cramer H, Lauche R, Anheyer D, Pilkington K, de Manincor M, Dobos G, Ward L (2018) *Yoga* for anxiety: a systematic review and meta-analysis of randomized controlled trials. Depress Anxiety 35:830–843

Daneman R, Prat A (2015) The blood–brain barrier. Cold Spring Harbor Perspect Biol 7:a020412

Du JY, Li XY, Zhuang Y, Wu XY, Wang T (1999) Effects of acute mild and moderate hypoxia on human short memory. Space Med Med Eng (Beijing) 12:270–273

Erickson MA, Rhea EM, Knopp RC, Banks WA (2021) Interactions of SARS-COV-2 with the blood–brain barrier. Int J Mol Sci 22:2681

Ernfors P, Kucera J, Lee K, Loring J, Jaenisch R (1995) Studies on the physiological role of brain-derived neurotrophic factor and neurotrophin-3 in knockout mice. Int J Dev Biol 39:799–807

Esposito G, Pesce M, Seguella L, Sanseverino W, Lu J, Sarnelli G (2020) Can the enteric nervous system be an alternative entrance door in SARS-CoV2 neuroinvasion? Brain Behav Immun 87:93–94

Fandrey J (2004) Oxygen-dependent and tissue-specific regulation of erythropoietin gene expression. Am J Physiol Regul Integr Comp Physiol 286:977–988

Gallegos AM, Hoerger M, Talbot NL, Krasner MS, Knight JM, Moynihan JA, Duberstein PR (2013) Toward identifying the effects of the specific components of mindfulness-based stress reduction on biologic and emotional outcomes among older adults. J Altern Complement Med 19:787–792

Garner M, Reith W, Krick C (2019) 10-week Hatha *Yoga* increases right hippocampal density compared to active and passive control groups: a controlled structural cMRI study. J Neuroimaging Psychiatry Neurol 4:1. https://doi.org/10.17756/jnpn.2019-027

Gothe NP, Khan I, Hayes J, Erlenbach E, Damoiseaux JS (2019) *Yoga* effects on brain health: a systematic review of the current literature. Brain Plast 5:105–122

Gupta SJ, Pradhan B, Bhanushali D, Nityapragya R, Limaye N (2022) Positive psychological changes at high altitude shown by the low landers after the *Yoga* intervention—Sudarshan Kriya *Yoga* (sky). Ann Neurosci 30:20–25

Hanssen I, Huijbers MJ, Lochmann-van Bennekom MW, Regeer EJ, Stevens AW, Evers SM, Wensing M, Kupka RW, Speckens AE (2019) Study protocol of a multicenter randomized controlled trial of mindfulness-based cognitive therapy and treatment as usual in bipolar disorder. BMC Psychiatry 19:1–10

Harrison LJ, Manocha R, Rubia K (2004) Sahaja *Yoga* meditation as a family treatment programme for children with attention deficit-hyperactivity disorder. Clin Child Psychol Psychiatry 9:479–497

Heming M, Li X, Räuber S et al (2021) Neurological manifestations of Covid-19 feature T cell exhaustion and dedifferentiated monocytes in cerebrospinal fluid. Immunity 54:164–175.e6

Herbet G, Duffau H (2020) Revisiting the functional anatomy of the human brain: toward a meta-networking theory of cerebral functions. Physiol Rev 100:1181–1228

Herculano-Houzel S (2011) Scaling of brain metabolism with a fixed energy budget per neuron: implications for neuronal activity, plasticity and evolution. PLoS One 6:e17514

Hernández SE, Suero J, Barros A, González-Mora JL, Rubia K (2016) Increased grey matter associated with long-term Sahaja *Yoga* meditation: a voxel-based morphometry study. PLoS One 11:e0150757

Hernández SE, Barros-Loscertales A, Xiao Y, González-Mora JL, Rubia K (2018) Gray matter and functional connectivity in anterior cingulate cortex are associated with the state of mental silence during Sahaja *Yoga* meditation. Neuroscience 371:395–406

Hernández SE, Dorta R, Suero J, Barros-Loscertales A, González-Mora JL, Rubia K (2020) Larger whole brain grey matter associated with long-term Sahaja *Yoga* meditation: a detailed area by area comparison. PLoS One 15:e0237552

Himashree G, Mohan L, Singh Y (2016) *Yoga* practice improves physiological and biochemical status at high altitudes: a prospective case-control study. Altern Ther 22:53–59

Hoffmann M, Kleine-Weber H, Schroeder S et al (2020) SARS-COV-2 cell entry depends on ACE2 and TMPRSS2 and is blocked by a clinically proven protease inhibitor. Cell 181:271–280

Hofmann SG, Andreoli G, Carpenter JK, Curtiss J (2016) Effect of hatha *Yoga* on anxiety: a meta-analysis. J Evid Based Med 9:116–124

Horie N, So K, Moriya T, Kitagawa N, Tsutsumi K, Nagata I, Shinohara K (2008) Effects of oxygen concentration on the proliferation and differentiation of mouse neural stem cells in vitro. Cell Mol Neurobiol 28:833–845

Hwang S, Ham S, Lee S-E, Lee Y, Lee GH (2018) Hypoxia regulates the level of glutamic acid decarboxylase enzymes and interrupts inhibitory synapse stability in primary cultured neurons. Neurotoxicology 65:221–230

Jablonska B, Scafidi J, Aguirre A, Vaccarino F, Nguyen V, Borok E, Horvath TL, Rowitch DH, Gallo V (2012) Oligodendrocyte regeneration after neonatal hypoxia requires FoxO1-mediated p27Kip1 expression. J Neurosci 32:14775–14793

Jakhar J, Mehta UM, Ektare A, Vidyasagar PD, Varambally S, Thirthalli J, Gangadhar BN (2019) Cortical inhibition in major depression: investigating the acute effect of single-session *Yoga* versus walking. Brain Stimul 12:1597–1599

Jenefer Jerrin R, Theebika S, Panneerselvam P, Venkateswaran ST, Manavalan N, Maheshkumar K (2021) *Yoga* and naturopathy intervention for reducing anxiety and depression of COVID-19 patients—a pilot study. Clin Epidemiol Glob Health 11:100800

Jung Y-H, Kang D-H, Jang JH, Park HY, Byun MS, Kwon SJ, Jang G-E, Lee US, An SC, Kwon JS (2010) The effects of mind–body training on stress reduction, positive affect, and plasma catecholamines. Neurosci Lett 479:138–142

Katuri KK, Dasari AB, Kurapati S, Vinnakota NR, Bollepalli AC, Dhulipalla R (2016) Association of *Yoga* practice and serum cortisol levels in chronic periodontitis patients with stress-related anxiety and depression. J Int Soc Prev Community Dent 6:7–14

Khalsa DS, Perry G (2017) The four pillars of Alzheimer's prevention. Cerebrum 2017:03–17

Kim KS (2008) Mechanisms of microbial traversal of the blood–brain barrier. Nat Rev Microbiol 6:625–634

Kjaer TW, Bertelsen C, Piccini P, Brooks D, Alving J, Lou HC (2002) Increased dopamine tone during meditation-induced change of consciousness. Cognit Brain Res 13:255–259

Krasemann S, Haferkamp U, Pfefferle S et al (2022) The blood–brain barrier is dysregulated in Covid-19 and serves as a CNS entry route for SARS-COV-2. Stem Cell Rep 17:307–320

Kulkarni MS, Kakodkar P, Nesari TM, Dubewar AP (2022) Combating the psychological impact of Covid-19 pandemic through *Yoga*: recommendation from an overview. J Ayurveda Integr Med 13:100433

Kumar S, Kumari P, Priyadarshi A (2018) Role of Nadi Shodhana *Pranayama* in the management of hyperventilation syndrome. Int J Res Anal Rev 5:1691–1697

Leu T, Fandrey J, Schreiber T (2021) (H) If applicable: promotion of neurogenesis by induced HIF-2 signalling after ischaemia. Pflügers Arch Eur J Physiol 473:1287–1299

Li YC, Bai WZ, Hashikawa T (2020a) The neuroinvasive potential of SARS-COV2 may play a role in the respiratory failure of COVID-19 patients. J Med Virol 92:552–555

Li Z, Liu T, Yang N, Han D, Mi X, Li Y, Liu K, Vuylsteke A, Xiang H, Guo X (2020b) Neurological manifestations of patients with COVID-19: potential routes of SARS-COV-2 neuroinvasion from the periphery to the brain. Front Med 14:533–541

Lin Z, Long F, Yang Y, Chen X, Xu L, Yang M (2020) Serum ferritin as an independent risk factor for severity in COVID-19 patients. J Infect 81:647–679

Liotta EM, Batra A, Clark JR, Shlobin NA, Hoffman SC, Orban ZS, Koralnik IJ (2020) Frequent neurologic manifestations and encephalopathy-associated morbidity in COVID-19 patients. Ann Clin Transl Neurol 7:2221–2230

Liu Q, Davidoff O, Niss K, Haase VH (2012) Hypoxia-inducible factor regulates hepcidin via erythropoietin-induced erythropoiesis. J Clin Investig 122:4635–4644

Lu G (1963) Tissue-cell adaptation to hypoxia. Adv Pathophysiol 1:197

Lu-Emerson C, Khot S (2010) Neurological sequelae of hypoxic–ischemic brain injury. NeuroRehabilitation 26:35–45

Lundgren T, Dahl JA, Yardi N, Melin L (2008) Acceptance and commitment therapy and *Yoga* for drug-refractory epilepsy: a randomized controlled trial. Epilepsy Behav 13:102–108

Lupien S, de Leon M, de Santi S, Convit A, Tarshish C, Nair NPV, Thakur M, McEwen B, Hauger R, Meaney M (1998) Correction: Cortisol levels during human aging predict hippocampal atrophy and memory deficits. Nat Neurosci 1:329–329

Malshe PC (2011) Nisshesha rechaka *Pranayama* offers benefits through brief intermittent hypoxia. AYU 32:451

Mao L, Jin H, Wang M et al (2020) Neurologic manifestations of hospitalized patients with Coronavirus disease 2019 in Wuhan, China. JAMA Neurol 77:683

Mehta P, McAuley DF, Brown M, Sanchez E, Tattersall RS, Manson JJ (2020) Covid-19: consider cytokine storm syndromes and immunosuppression. Lancet 395:1033–1034

Meinhardt J, Radke J, Dittmayer C et al (2020) Olfactory transmucosal SARS-COV-2 invasion as a port of central nervous system entry in individuals with Covid-19. Nat Neurosci 24:168–175

Miller AD, Zachary JF (2017) Nervous system. In: Pathologic basis of veterinary disease. Elsevier

Millet GP, Debevec T, Brocherie F, Burtscher M, Burtscher J (2020) Altitude and Covid-19: friend or foe? A narrative review. Physiol Rep 8:e14615

Nagarathna R, Anand A, Rain M, Srivastava V, Sivapuram MS, Kulkarni R, Ilavarasu J, Sharma MN, Singh A, Nagendra HR (2021) *Yoga* practice is beneficial for maintaining healthy lifestyle and endurance under restrictions and stress imposed by lockdown during COVID-19 pandemic. Front Psychiatry 12:613762

Nation DA, Bondi MW, Gayles E, Delis DC (2016) Mechanisms of memory dysfunction during high altitude hypoxia training in military aircrew. J Int Neuropsychol Soc 23:1–10

Newcomer JW (1999) Decreased memory performance in healthy humans induced by stress-level cortisol treatment. Arch Gen Psychiatry 56:527–533

Novaes MM, Palhano-Fontes F, Onias H, Andrade KC, Lobão-Soares B, Arruda-Sanchez T, Kozasa EH, Santaella DF, de Araujo DB (2020) Effects of *Yoga* respiratory practice (bhastrika *Pranayama*) on anxiety, affect, and brain functional connectivity and activity: a randomized controlled trial. Front Psychiatry 11:492687

Orlando IMC, Lafleur VN, Storti F, Spielmann P, Crowther L, Santambrogio S, Schödel J, Hoogewijs D, Mole DR, Wenger RH (2019) Distal and proximal hypoxia response elements cooperate to regulate organ-specific erythropoietin gene expression. Haematologica 105:2774–2784

Pagani A, Nai A, Silvestri L, Camaschella C (2019) Hepcidin and anemia: a tight relationship. Front Physiol 10:1294

Panjwani U, Selvamurthy W, Singh S, Gupta H, Mukhopadhyay S, Thakur L (2000) Applied psychophysiology and biofeedback, vol 25. Springer, pp 1–12

Park S-Y, Kim S-M, Sung J-J, Lee K-M, Park K-S, Kim S-Y, H-woo N, Lee K-W (2013) Nocturnal hypoxia in ALS is related to cognitive dysfunction and can occur as clusters of desaturations. PLoS One 8:e75324

Popescu C (2021) Hypoxic–ischemic injury of basal ganglia associated with the COVID-19 infection: case report. Case Rep Neurol 13:668–671

Pourfridoni M, Abbasnia SM, Shafaei F, Razaviyan J, Heidari-Soureshjani R (2021) Fluid and electrolyte disturbances in COVID-19 and their complications. Biomed Res Int 2021:1–5

Rain M, Subramaniam B, Avti P, Mahajan P, Anand A (2021) Can yogic breathing techniques like Simha Kriya and Isha Kriya regulate Covid-19-related stress? Front Psychol 12:635816

Rajesh B, Jayachandran D, Mohandas G, Radhakrishnan K (2006) A pilot study of a *Yoga* meditation protocol for patients with medically refractory epilepsy. J Altern Complement Med 12:367–371

Rolan T (2014) Neurology and altitude illness. Neurol Clin Pract 5:102–107

Ros-Castelló V, Quereda C, López-Sendón J, Corral I (2020) Post-hypoxic myoclonus after COVID-19 infection recovery. Mov Disord Clin Pract 7:983–984

Rybnikova E, Samoilov M (2015) Current insights into the molecular mechanisms of hypoxic pre- and postconditioning using hypobaric hypoxia. Front Neurosci 9:151094

Sarwal R, Dhamija RK, Jain K, Basavaraddi IV (2022) Efficacy of *Pranayama* in preventing COVID-19 in exposed healthcare professionals: a quasi-randomized clinical trial. J Ayurveda Integr Med 100:586

Schoeman D, Fielding BC (2019) Coronavirus envelope protein: current knowledge. Virol J 16:1–22

Semenza GL (2001) HIF-1, O2, and the 3 PHDs: how animal cells signal hypoxia to the nucleus. Cell 107:1–3

Semenza GL, Wang GL (1992) A nuclear factor induced by hypoxia via de novo protein synthesis binds to the human erythropoietin gene enhancer at a site required for transcriptional activation. Mol Cell Biol 12:5447–5454

Serafini G, Parmigiani B, Amerio A, Aguglia A, Sher L, Amore M (2020) The psychological impact of Covid-19 on the mental health in the general population. QJM Int J Med 113:531–537

Serebrovskaya TV, Nikolsky IS, Nikolska VV, Mallet RT, Ishchuk VA (2011) Intermittent hypoxia mobilizes hematopoietic progenitors and augments cellular and humoral elements of innate immunity in adult men. High Altitude Med Biol 12:243–252

Shang J, Wan Y, Luo C, Ye G, Geng Q, Auerbach A, Li F (2020) Cell entry mechanisms of SARS-COV-2. Proc Natl Acad Sci 117(21):11727–11734

Sharma VK, Das S, Mondal S, Goswampi U, Gandhi A (2005) Effect of Sahaj *Yoga* on depressive disorders. Indian J Physiol Pharmacol 49:462–468
Sharma R, Cramer NP, Perry B et al (2019) Chronic exposure to high altitude: synaptic, astroglial and memory changes. Sci Rep 9:16406
Sheldon RA, Lee CL, Jiang X, Knox RN, Ferriero DM (2014) Hypoxic preconditioning protection is eliminated in HIF-1α knockout mice subjected to neonatal hypoxia–ischemia. Pediatr Res 76:46–53
Singh VK (2013) The Effect of Nadisodhan *Pranayama* on blood hemoglobin among healthy volunteers. Int J Yoga Allied Sci 1:46–51
Soliz J, Schneider-Gasser EM, Arias-Reyes C, Aliaga-Raduan F, Poma-Machicao L, Zubieta-Calleja G, Furuya WI, Trevizan-Baú P, Dhingra RR, Dutschmann M (2020) Coping with hypoxemia: could erythropoietin (EPO) be an adjuvant treatment of Covid-19? Respir Physiol Neurobiol 279:103476
Spencer JA, Ferraro F, Roussakis E et al (2014) Direct measurement of local oxygen concentration in the bone marrow of live animals. Nature 508:269–273
Srinivasan S, Cui H, Gao Z, Liu M, Lu S, Mkandawire W, Narykov O, Sun M, Korkin D (2020) Structural genomics of SARS-COV-2 indicates evolutionary conserved functional regions of viral proteins. Viruses 12:360
Streeter CC, Jensen JE, Perlmutter RM, Cabral HJ, Tian H, Terhune DB, Ciraulo DA, Renshaw PF (2007) *Yoga Asana* sessions increase brain GABA levels: a pilot study. J Altern Complement Med 13:419–426
Streeter CC, Whitfield TH, Owen L et al (2010) Effects of *Yoga* versus walking on mood, anxiety, and brain GABA levels: a randomized controlled MRS study. J Altern Complement Med 16:1145–1152
Strowitzki M, Cummins E, Taylor C (2019) Protein hydroxylation by hypoxia-inducible factor (HIF) hydroxylases: unique or ubiquitous? Cells 8:384
Taquet M, Geddes JR, Husain M, Luciano S, Harrison PJ (2021) 6-month neurological and psychiatric outcomes in 236,379 survivors of Covid-19: a retrospective cohort study using electronic health records. Lancet Psychiatry 8:416–427
Thomas S (2020) The structure of the membrane protein of SARS-COV-2 resembles the sugar transporter semisweet. Pathog Immun 5:342–363
Tiwari V, Mishra R, Chaudhary S, Pandey H (2017) *Yoga* enhances bioavailability and assimilation of oral iron in young female patients of iron deficiency anemia. Int J Contemp Med Res 4:1475–1477
Tolahunase MR, Sagar R, Faiq M, Dada R (2018) *Yoga* and meditation-based lifestyle intervention increases neuroplasticity and reduces severity of major depressive disorder: a randomized controlled trial. Restor Neurol Neurosci 36:423–442
Törpel A, Peter B, Hamacher D, Schega L (2019) Dose–response relationship of intermittent normobaric hypoxia to stimulate erythropoietin in the context of health promotion in young and old people. Eur J Appl Physiol 119:1065–1074
Varatharaj A, Thomas N, Ellul MA et al (2020) Neurological and neuropsychiatric complications of COVID-19 in 153 patients: a UK-wide surveillance study. Lancet Psychiatry 7:875–882
Wadden KP, Snow NJ, Sande P, Slawson S, Waller T, Boyd LA (2018) *Yoga* practitioners uniquely activate the superior parietal lobule and supramarginal gyrus during emotion regulation. Front Integr Neurosci 12:60
Wakhloo D, Scharkowski F, Curto Y et al (2020) Functional hypoxia drives neuroplasticity and neurogenesis via brain erythropoietin. Nat Commun 11:1313
Wang K, Chen W, Zhang Z et al (2020) CD147-spike protein is a novel route for SARS-COV-2 infection to host cells. Signal Transduct Target Ther 5:283
Watts ME, Pocock R, Claudianos C (2018) Brain energy and oxygen metabolism: emerging role in normal function and disease. Front Mol Neurosci 11:216
West JB (2006) Human responses to extreme altitudes. Integr Comp Biol 46:25–34

Xue Y, Yang J, Feng Y, Zhou Y, Qin Y, Li Y, Li Y, Ren Q, Liu C, Cao Z (2020) Effects of mindful breathing on rapid hypoxia preacclimatization training. J Med Imaging Health Inform 10:718–723

Yang AC, Kern F, Losada PM et al (2021) Dysregulation of brain and choroid plexus cell types in severe COVID-19. Nature 595:565–571

Yu H-H, Qin C, Chen M, Wang W, Tian D-S (2020) D-Dimer level is associated with the severity of COVID-19. Thromb Res 195:219–225

Yuen TJ, Silbereis JC, Griveau A, Chang SM, Daneman R, Fancy SPJ, Zahed H, Maltepe E, Rowitch DH (2014) Oligodendrocyte-encoded HIF function couples' postnatal myelination and white matter angiogenesis. Cell 158:383–396

Zhang XY, Zhang XJ, Xv J, Jia W, Pu XY, Wang HY, Liang H, Lu DX (2018) Crocin attenuates acute hypobaric hypoxia-induced cognitive deficits of rats. Eur J Pharmacol 818:300–305

Zhu X-H, Yan H-C, Zhang J et al (2010) Intermittent hypoxia promotes hippocampal neurogenesis and produces antidepressant-like effects in adult rats. J Neurosci 30:12653–12612

Zhuravin IA, Dubrovskaya NM, Vasilev DS, Postnikova TY, Zaitsev AV (2019) Prenatal hypoxia produces memory deficits associated with impairment of long-term synaptic plasticity in young rats. Neurobiol Learn Mem 164:107066

Chapter 11
Yoga as a Primer for Cardiac Health

Pooja Nadholta and Somya Saxena

Contents

Abstract Achieving a positive health state is dependent on several factors of which the heart plays an important role. Cardiovascular disease risk is low in people who maintain ideal cardiac health over their lifetime. This indicates that a healthy heart is better than a diseased one, and therefore maintaining good cardiac health is better than treating a diseased heart. It is therefore important to promote ideal cardiac health in order to prevent the prevalence of cardiac disorders. There is considerable evidence to confirm that *Yoga* promotes physical and mental well-being by various pathways. Since cardiac health determines the health of overall body, it is important to assess the practical significance and suitability of incorporating *Yoga* into daily lifestyle for promoting a healthy heart which will lead to primary prevention of

P. Nadholta
Department of Neurology, Postgraduate Institute of Medical Education and Research, Chandigarh, India

S. Saxena (✉)
Department of Physical Medicine and Rehabilitation, Postgraduate Institute of Medical Education and Research, Chandigarh, India
e-mail: saxena.somya@pgimer.edu.in

A. Anand (ed.), *Neuroscience of Yoga*,
https://doi.org/10.1007/978-981-97-2851-0_11

several cardiovascular diseases. This chapter discusses the utility of *Yoga* as a natural health care measure for the preventive, promotive and curative aspects of various cardiovascular disorders.

Keywords Yoga · Cardiovascular health · Lifestyle modifications · Physical activity · Rehabilitation · Holistic health

11.1 Introduction

Lifestyle modification and adapting a healthy lifestyle is important and significant for promoting health of the heart. Consider yourself, running on a treadmill on a specific speed, how long will you be able to run? After a certain period, you will be exhausted, stressed and wanting to rest. Similarly, our heart functions day in and day out without taking any break. It remains active even when we are sleeping. So, isn't that enough for the heart to do or "Do we need to increase the burden by increasing workload on heart?" Obviously not; but this is what we are doing; increasing the workload on heart and not giving it time to relax and rejuvenate. According to the World Health Organization, approximately 32% people are dying due to cardiovascular disorder, most of which occur by them due to lifestyle disorders such as obesity, hypertension, diabetes, sedentary lifestyle, smoking, etc. (World Health Organization 2022a). There is no available treatment for these lifestyle disorders. The only way to avoid these disorders is prevention by incorporating healthy habits like regular physical activity and healthy eating. *Yoga* is a natural remedy which has proved to be effective in prevention, utilized for maintenance therapy and helpful in rehabilitative management of cardiovascular diseases. It does this by reducing stress, improving self-awareness and normalizing autonomic functions in the body (Sengupta 2012).

There are many factors that threaten the health of our heart. These include hereditary conditions, lifestyle choices, societal and environmental factors. Cardiac health is an umbrella term used for healthy cardiac vessels (cardiovascular health), healthy coronary arteries and strong heart muscles which pump blood efficiently to all body parts as well as to heart itself. Good cardiac health is not only necessary for heart but for all other body, since all body parts depend on heart for adequate blood supply to provide them with oxygen and nutrients. Various factors which are responsible for maintaining good cardiac health include modifiable factors, i.e. behavioural and biological, and non-modifiable factors, i.e. age, gender, family history and race (Kokane et al. 2020) (Fig. 11.1).

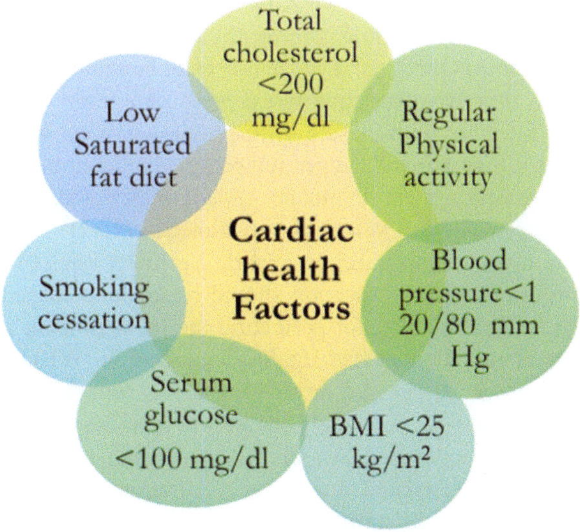

Fig. 11.1 Cardiac health parameters

11.2 Modifiable Parameters

11.2.1 Behavioural Domain

It includes the factors which are acquired by individuals during their lifetime as habit. Acute change in behaviour is not harmful but if that change persists for a longer period, it may affect cardiac health.

11.2.1.1 Smoking

Smoking causes many health conditions like cancer, lung diseases, tuberculosis. It promotes atherosclerosis, i.e. plaque build-up in the arteries causing narrowing of arteries and hence restricts blood flow (Howard et al. 1998). For chronic smokers, heart and blood vessels are the target organ. The available evidences indicate that the development of ischemic heart diseases (due to decreased blood flow caused by narrowed arteries) and myocardial infarction (lack of blood flow to the heart muscles) occurs more commonly in smokers (Leone 1993). Both active and/or passive smoking causes an impairment in cardiac performance which can worsen if an ischemic patient gets exposure to smoking (Leone 1995). Vagal-cardiac nerve activity which is responsible for maintaining physiological homeostasis by activating the arterial baroreceptor reflex response is seen to be reduced in habitual smokers, and therefore, they have impaired blood pressure control. Smoking has also been reported to interfere with muscle sympathetic nerve activity (MSNA) with impaired baroreflex function and trigger sympathetic activation, thereby stimulating sympathetic nervous system dominance (Niedermaier et al. 1993). It has been reported that patients with left ventricular dysfunction (LVD) who stop smoking before or after a myocardial infarction experience lower rates of morbidity and mortality. Therefore, smoking cessation should be considered and adopted as a highly effective medical therapy for protecting the heart (Suskin et al. 2001). In context to smoking cessation, World Health Organization (WHO) has suggested 5As (ask, advise, assess, assist and arrange) and 5Rs (relevance, risks, rewards, roadblocks and repetition) model to help smokers to increase the motivation to quit smoking. As per this model, self-awareness and finding some way to decrease sympathetic activation (regulated breathing, meditation) that curbs the urge of smoking may encourage these individuals to quit smoking (World Health Organization 2014). Furthermore, the craving for smoking and the bad mood after quitting smoking which usually become hurdles can be handled with optimum physical activity including aerobic exercise, running or physical postures combined with some relaxation practices, breathing practices and meditation (Ussher et al. 2014).

11.2.1.2 Alcohol

Alcohol is a toxic and psychoactive abusive substance with dependence producing properties affecting the proper functioning of central nervous system. It has complex effect on cardiovascular health (World Health Organization 2022b). Studies have reported both positive and negative effects of alcohol consumption on cardiac health. Short-term acute alcohol consumption causes sudden increase in heart rate and blood pressure. In long term, when alcohol consumption is regular with increased amount and frequency, it leads to continuously elevated heart rate and high blood pressure which increase the workload on heart. This predisposes the individual to early onset of conditions like cardiac wall thickening, irregular heart beat and uncontrolled hemodynamic (Husain et al. 2014). However, some reports suggest moderate use of alcohol by healthy individual is associated with reduced risk of cardiovascular diseases, but there are chances of sudden death due to ventricular arrhythmia. Some reports suggest benefits of moderate alcohol consumption with pre-existing CVD (Poli et al. 2013). High-density cholesterol (HDL) which is considered as good cholesterol and involves in the clearing of bad cholesterol is present in increased concentration in the plasma of moderate drinker (Brien et al. 2011). These evidences explain the cardioprotective role of alcohol but considering its addictive nature and detrimental effects on other systems and even on cardiac system in long term, we do not recommend the use of alcohol in any form in any age group.

11.2.1.3 Dietary Intake

Diet should be of primary concern when it comes to the management of cardiac health since it is the primary factor responsible for cardiac diseases, after excluding all biological factors. Unhealthy diet, such as excessive intake of sugar, unhealthy fat, processed food containing high sodium levels and low intake of proteins, nuts, fibres, whole grains, vegetable and fruits, slowly deteriorates normal cardiac function and increases the workload on heart by narrowing the arteries and increasing the blood pressure. The heart thus becomes more exposed to this chronically increased workload state as compared to the normal workload state to which it becomes unaccustomed and therefore finds it difficult to relax. As our body requires sleep to rejuvenate and operate at their best, in the same way heart also needs some leisure time to relax and rejuvenate. Therefore, to reduce the workload on heart and supplement cardiac function, dietary intake should be consciously chosen. In studies, it has been demonstrated that intake of high-fat diets without physical activity is linked to increased regional adiposity, whereas intake of high-quality diets (high fibre, protein, minerals, etc.) is associated with decreased regional adiposity (Casas et al. 2018). Intake of healthy diet is also associated with better control of risk factors responsible for the development of cardiac disease. These risk factors are elevated blood pressure, altered lipid profile, altered gut microbiome and abnormal glucose metabolism, etc. (Mozaffarian 2016). A very famous saying is that *"The way to man's heart is through gut"*. Giving more scientific insight into this, one can equate the "gut" here to the "gut microbiota" which is a manifestation of what we eat. Altered gut microbiota profile is associated with obesity, diabetes, inflammation, etc., all of which are risk factors for cardiac disease. High-fibre diets with a moderate amount of other supplements have been shown to promote bacterial activity in the gut, resulting in improved gut hormone levels and other physiological functions associated with cardiovascular health (Tuohy et al. 2014).

A balanced diet plays an important role in promoting overall health and can be easily practised by following the principles of "Yogic Diet" or *"Mitahar"*. *Yoga* is not only about physical postures and breathing practices, but it is a vast science of health and wellness which focus on every aspect of well-being, which includes body cleansing procedures (*Shatkriya*), social ethics (*Yama*), personal discipline (*Niyama*), postures from physical health (*Asana*), regulated breathing techniques for controlled breath (*Pranayama*), mental preparation for increasing willpower by self-assessment (*Pratyahara*), concentration (*Dharna*), meditation for calming down mental fluctuation (*Dhyana*) and being fully conscious (*samadhi*) (Gruber 2008). There is enough evidence that dietary modification with other lifestyle changes can lower serum cholesterol levels, reverse coronary lesions and improve myocardial perfusion, all of which are indicators of cardiac health (Yogendra et al.

2004). *Mitahar* emphasizes on moderate eating by filling only three-fourth part of stomach with food and leave rest empty for water and air. Vegetables, fruits, high protein and fibres, balanced fat and micro-nutrients-rich food are recommended as a wholesome diet for any person to prevent the occurrence of diseases resulting from dietary imbalances, especially by cholesterol imbalance (Nilakhe et al. 2018).

11.2.1.4 Stress

Stress is an umbrella term used for the situations when physiological demands or psychological pressures outweigh an individual's capability to cope with them. Acute stress can also be termed eustress as it prepares the body and the mind to deal with a stressful situation more efficiently (Dierolf et al. 2018). Chronic stress, on the other hand, primarily raises the catecholamine and suppressor T cells, suppressing immune system, which ultimately manifests in various illnesses. In addition, stress increases histamines, alters insulin requirements in the body and contributes to plaque formation in arteries if combined with sedentary lifestyles and poor food choices (Salleh 2008). Increased occupational stress in men is associated with higher serum cholesterol and accelerated blood clot time, whereas this is not the case in less stressful situations. This indicates that stress is a risk factor for cardiovascular diseases (Friedman et al. 1958). Stress-induced disruptions of inflammatory, homeostatic and autonomic processes may trigger acute myocardial infarction and increase the chances of coronary heart disease by 40–50%, whereas stress management may prevent these adverse effects of stress on heart and promote the cardiac health by increased quality of life (Steptoe and Kivimäki 2012).

It is impossible not to think of *Yoga* practice when we talk about stress management. Practising *Yoga* promotes self-awareness. It helps in developing healthy coping mechanisms, a positive attitude, compassion, calmness and mindfulness, all of which help reduce stress. Biologically, *Yoga* modulates the autonomic nervous system, hypothalamic–pituitary axis, release of nitric oxide and endocrine response (Riley and Park 2015). *Yoga* has been shown to decrease sympathetic response and hence regulate heart rate, blood pressure and cortisol level (Vijayalakshmi et al. 2004). With advancing age performing intense physical activities become more challenging and there is an increase in the incidence of stress, obesity, diabetes, hypertension and cardiac ailments. *Yoga* can be adopted as a substitute for intense physical activity as it is safe and less intense and provides equal benefits as other intense physical activities do (Satin et al. 2014).

Stress management just by regulating the breath is an easy and simple idea which can be adapted by every age group irrespective of race, gender, health condition or any other physical deformities. The *Pranayama* component of *Yoga*, which specifically focuses on breath regulation, helps to combat stress by decreasing heart rate and blood pressure and increasing pulse pressure (Singh et al. 2011). Resonance breathing, which is also known as coherent breathing, is a type of slow and deep breathing with only 5–7 breaths/min and helps in slowing down the heart rate,

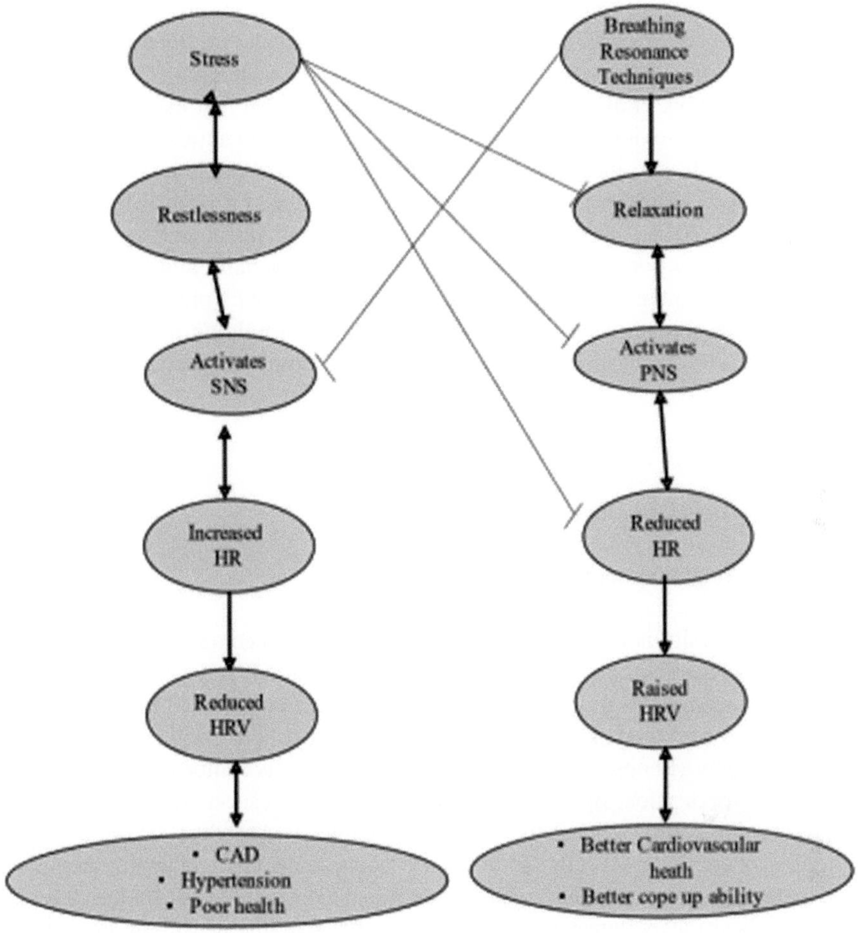

Fig. 11.2 Stress management through regulated breathing techniques

activating the Parasympathetic Nervous System (PNS) and increasing the Heart Rate Variability (HRV), which alleviates stress (Chaitanya et al. 2022) (Fig. 11.2).

11.2.1.5 Physical Activity

Sedentary lifestyle is one of the worst enemies of the heart. According to epidemiological data, people who are physically active tend to have lower rates of cardiovascular disease. Approximately 35% incidence of coronary heart disease is due to physical inactivity. In order to maintain cardiovascular fitness, the American Heart Association recommends at least 30 min of physical activity three to four times per week (Department of Health 1999). Involvement in any kind of physical activity is associated with reduced CVD risk. Not only vigorous exercise but person who reduced their sitting time had better cardiovascular profile than those with more sitting time which is also found to be associated with higher CVD mortality rates (Katzmarzyk et al. 2009). It has been demonstrated that exercise increases the level of super-oxide dismutase 2 (SOD2), an important antioxidant enzyme in mitochondria. It also increases mitochondrial ATP-sensitive potassium channels and promotes cardioprotection through the expression of mitochondrial proteins (Powers et al. 2014). *Yoga* in comparison to intense physical activity is advantageous for elderly population, pregnant women and specially abled persons who cannot perform intense practices. Apart from that, *Yoga* with physical movements focuses on the breath awareness, and longer duration holding in particular postures relaxes the body muscles and activates PNS promoting better cardiovascular health.

11.2.2 Biological Domain

11.2.2.1 Weight Management (Obesity)

In general, obesity is a condition in which there is an accumulation of unhealthy body fat, which requires an increased supply of blood, oxygen and nutrients. Increased demand puts more pressure on heart, which makes heart to pump harder and increases workload of heart which alters the heart structure and causes ventricular hypertrophy (Ebong et al. 2014). Other conditions accompanying obesity are diabetes, hypertension, dyslipidaemia, atherosclerosis, cancer, stress, etc., which are result of increased adipose tissue mass and pathogenic product secretion such as cytokines from these fat cells (Bray 2004). In obese people, endocrine, metabolic and thermodynamic mechanisms have been found to be responsible for the development of obesity-hypertensive condition by suppressing the natriuretic peptides in arteries (rapid-acting regulators of volume homeostasis and arterial pressure); increasing sympathetic adrenergic activity; leptin resistance; release of Ang II from adipocytes; chronic hyperleptinemia leptin resistance; chronic hyperleptinemia; and hyperinsulinemia. Weight management is the most important and primary strategy for obesity and related conditions (Morse et al. 2005). Few popular weight loss strategies include fad dieting, juice detoxification diet, intermittent fasting, behavioural change therapy and high-intensity training (Obert et al. 2017). Affected people are seeking treatment of obesity through medical, commercial and surgical procedures which have other side effects and should be opted only if required medically after considering risk-benefit ratio (Kushner 2014). Keeping in mind the psychological state and physical capacity, organic weight loss management by lifestyle modification and combining diet, physical activity and behavioural therapy could be the best strategy for obesity (Nemet et al. 2005). Lifestyle modification through *Yoga* is promising as well as feasible for obese or overweight people as it includes physical activity (*Asana*), diet modification (Yogic diet) and behavioural change

(*Pranayama* and meditation). *Yoga* can be easily used in managing obesity since obese individuals are less capable of participating in intense physical activity and have more mental stress and other medical conditions (Bernstein et al. 2014). Studies reported both short- and long-term benefits of *Yoga* practice. Three months *Yoga* practice showed significant reduction in body mass index (BMI) of obese male (Gadham et al. 2015). Further, 1 year practice was observed to improve exercise capacity (can help obese people to improve the stamina) and decrease body weight, serum total cholesterol, LDL cholesterol and triglycerides levels (Manchanda et al. 2000). The results of these studies indicate that *Yoga* practice possesses the potential to help obese people manage their weight, cope with psychological and physical stress and reduce the risk of cardiovascular disease and other diseases associated with obesity.

11.2.2.2 Lipid Management (Cholesterol)

Lipoproteins are precursors to atherosclerosis, so maintaining optimum levels of lipids is essential for achieving optimal cardiovascular health. Low-density lipoprotein (LDL) and other apo-protein B-containing lipoproteins transport cholesterol and lipids throughout the body and play a critical role in atherosclerosis development (Ference et al. 2017). Rate of atherosclerosis progression has been seen directly proportional to absolute plasma LDL level. Lipoprotein particles become retained on arterial wall over the time and the size of plaque keeps on increasing with the accumulation of more lipids on arterial wall (Chien 2003). Therefore, plaque burden can be determined by concentration of circulating LDLs and the duration of exposure. Untreated high level of LDL/total cholesterol can be threatening in the long run if not managed on time.

Principal approaches for dyslipidaemia management can be primary prevention by lifestyle management. Preventing and managing dyslipidaemia by lifestyle modification is the ideal intervention option when considering the side effects, economic burden and dependency on medications (Gotto Jr 2002). Diet and regular physical activity are recommended to the people with dyslipidaemia irrespective of their weight (Huffman et al. 2012). Even though these interventions are cost effective, feasible, with no side effects, people still do not regularly engage in these practices due to non-specific/avoidable reasons. Noncompliance to recommended levels of physical activity and a balanced diet is a major reason for the development of many lifestyle disorders and deterioration of cardiovascular health (Lalonde et al. 2002). An adverse physical condition affects a person's psychological state also due to the mind-body connection. So, we suggest exercising in such ways which leads to improvement of both psychological and physical well-being. This can be done by *Yoga* practice, which gives the benefits of physical exercise along with calmness and improved emotional health (Shantakumari and Sequeira 2013). Therefore, *Yoga* by mindful movement of body with rhythmic breathing practice is an excellent practice to be adopted in prevention as well as management of any lifestyle disease including imbalanced lipid/dyslipidaemia.

11.2.2.3 Blood Pressure Control

A blood pressure measurement indicates the pressure of blood flowing through arteries as the heart pumps. The heart functions normally when blood pressure is within its threshold (120/80 mmHg), but when there is any external disturbance such as exercise, volume overload or sympathetic over-activity, the heart has to pump harder and there is increase in blood pressure. Besides the non-modifiable factors (age, race, family history), other factors which are responsible for hypertension are obesity, sedentary lifestyle, tobacco use, alcohol, stress and some chronic conditions (diabetes, pregnancy, kidney diseases, etc.) (Hajjar et al. 2006). Although earlier high blood pressure and hypertension was more prevalent in elderly population, but these days due to sedentary lifestyle and wrong eating habits it is becoming more common in adults and kids also (Myoclinic 2022). Hyperactive sympathetic nervous system also plays an important role in the development of hypertension by releasing norepinephrine and increasing heart rate and blood pressure (Wyss 1993). Usually, there are no specific presenting symptoms of hypertension, therefore it often goes unnoticed/untreated. If patients are left untreated and in chronic phase, there is risk of heart failure and other associated co-morbidities affecting heart, blood vessels and other organs (Wilber 1973). Primary prevention by healthy lifestyle habits such as exercising, healthy eating and no smoking can help in blood pressure control. Cardiovascular training is known to be the most effective form of exercise for blood pressure control by lowering the blood pressure in both normotensive and hypertensive patients, therefore exercise can be a preventive as well as treatment strategy for blood pressure control (Wallace 2003).

Yogic deep breathing practices are known to slow down the sympathetic nervous system, activate the parasympathetic nervous system and decrease the blood pressure and heart rate (Mourya et al. 2009). Breathing, relaxation and meditation Yogic practices rather than physical aspect of *Yoga* seem to be more beneficial for blood pressure management as these practices can increase parasympathetic activity and downregulate sympathetic activities exerting calming effects (Cramer 2016).

11.2.2.4 Blood Glucose

High blood pressure, abnormal cholesterol, obesity, smoking and stress can be directly connected to cardiovascular health, but how blood glucose can affect the cardiac health? Data suggests that around 30% people with type 2 diabetes also have cardiac vascular disorder because elevated blood glucose level damages the blood vessels of and contribute to the formation of plaques in arteries (Einarson et al. 2018). Diabetes increases the risk of obesity, hypertension and dyslipidaemia, which are risk factors for CVD, and this is how patients of diabetes mellitus have increased risk of cardiac diseases (Leon and Maddox 2015). Hyperinsulinemia also increases the sympathetic nerve activity which is considered as one of the characteristics of CVD (Anderson and Mark 1993). Conversely, obesity and physical inactivity develop insulin resistance and hence contribute to diabetes (Grundy et al. 1999), which means all risk factors somehow are related to each other and presence of even a single condition is potent to develop another condition and combination of these risk factors make a huge base for the development of cardiac diseases. WHO suggests lifestyle measures as an effective prevention strategy for delaying the onset of type 2 diabetes mellitus. This management includes healthy body weight, more physical activity, eating healthy diet and avoiding use of tobacco (World Health Organization 2022c). The benefits of exercise in the treatment of type 1 (genetic) and type 2 (lifestyle) diabetes are known to lead to a reduction in adiposity and an increase in insulin sensitivity. Apart from treatment, onset of T2DM can also be prevented by exercise. American Diabetes Association (ADA) and the American College of Sports Medicine (ACSM) have jointly suggested exercise guideline for T2DM patients which include a gentle warm up, stretching and active cool down to allow gradual adjustment of heart rate and blood pressure (American Diabetes Association 1997).

An ideal *Yoga* protocol which starts with warm up and stretching followed by *Asana*, then *Pranayama* and finally relaxation and meditation, fulfil the requirements of suggested guidelines for T2DM patients. Malhotra et al. have suggested a series of 13 practices for T2DM patients which can be practised by every age group as per their own capacity. These include "*Surya Namaskar, Trikonasana, Tadasana, Sukhasana, Padmasana, Bhastrika Pranayama, Pashimottanasana, Ardhmatsyendrasana, Pawanmuktasana, Bhujangasana, Vajrasana, DhanurAsana and ShavAsana*". Practice of this protocol every day for 40 days (30–40 min/day) showed significant reduction in body weight and blood glucose while increased serum insulin was observed (Malhotra et al. 2005). We suggest that dietary management by following Yogic diet along with practising *Yoga* can potentially manage the blood glucose level without any side effects and putting economic burden on oneself.

11.3 Non-modifiable Parameters

Other non-modifiable parameters which affect cardiac health are age (risk increase with increase in age), gender (more prevalent in male as female sex hormones are cardioprotective), family history and genetics. These non-modifiable risk factors cannot be avoided but risk reduction can be done by lifestyle and behavioural modifications (Tabei et al. 2014). Evidence suggests that *Yoga* practice, i.e. combination of physical activity with regulated breathing technique, relaxation and meditation, is a promising holistic approach for improving cardio-metabolic health (Chu et al. 2016). Based on the available evidences as discussed in the above sections we believe that if risk factors are avoided and lifestyle management is done by a holistic health approach utilizing *Asana* for physical health, *Pranayama* for breath regulation, relaxation techniques to decrease stress, meditation for emotional and spiritual health and following a Yogic diet, one can maintain better cardiac health even if non-modifiable risk factors are present.

11.4 Management of Cardiac Health by *Yoga*: Physiological Benefits for Cardiac Health

As per the growing scientific evidences, the hypothalamus–pituitary–adrenal (HPA) axis and the sympathetic nervous system may be downregulated by certain *Yoga* techniques, which indicate that *Yoga* is an effective method to manage stress and stress-induced disorders (Riley and Park 2015). Exercise focuses on strengthening the physical body and similar is the purpose of the *Asana* component of *Yoga*. In addition to this *Asana* also acts on smooth muscles of targeted body parts and brings stability and deeper attention. Besides *Asana*, Yogic practice also comprises loosening exercise (Suksham Vyayam), cleansing practice (*Shatkriya*), relaxation techniques (relax body and mind), *Pranayama* (breath control), meditation (mental exercise), *Mantra* chanting (awareness), *Mudras, Bandhas,* etc., which make *Yoga* a holistic health approach for both mind and body.

Practising *Asana*, it is not just about maintaining a specific posture, control of the breath is also equally important. In some poses one need to inhale, in some exhalation is required and, in some postures, breath holding is required. This increases awareness, improves blood flow to the organs, strengthens the heart and prevents adverse events. Relaxation and *Pranayama* practice balance the function of sympathetic and parasympathetic nervous system and hence normalize the function of autonomic nervous system and decrease cortisol levels. This helps in managing stress/anxiety and keeps blood glucose, heart rate and blood pressure within physiologically normal levels (Sengupta 2012).

In today's digitalised and technology-driven world everybody seems to be busy with work. This has led to an epidemic of inactivity and a sedentary lifestyle. Coupled with this innumerable tempting and unhealthy food options are readily delivered at workplace/homes at the click of a button. So, in these circumstances it becomes really challenging to steer clear from lifestyle diseases. Increasing work load leads to stress, anxiety and tension which even affects the sleeping pattern. Yogic practices which help an individual to stay active and yet in a relaxed state of mind can become a tool to reduce stress which will promote relaxation with enhanced productivity.

Bhramari *Pranayama* in which a humming bee sound is produced during a longer than normal exhalation through nostrils, while keeping mouth, eyes and ears closed, is a simple and feasible technique to improve cardiac health. This simple practice produces pleiotropic benefits which include parasympathetic dominance and improved cognition. The longer exhalation period gives heart more time to relax and hence reduces the workload of heart (Kuppusamy et al. 2018). Similarly, all other *Pranayama* practices promote the cardiac health by reducing the work load on heart.

Some of the basic *Yoga* practices which can be adopted for promoting good cardiac health are represented in tabulated form below. Although there is still lack of complete scientific evidence to justify the benefits of individual practice, but physiologically, we propose the following benefits of a particular practice on the cardiac system. Just as any medication should always be taken after the prescription of specialized physician, similarly, *Yoga* practices should also be started under the guidance of experts. We suggest readers to consult expert in the field of *Yoga* before starting any practice (Table 11.1).

Table 11.1 Proposed protocol can be practised for promotion of good cardiac health after consultation of expert

Recommended Practice	**Possible Benefits**	**Common Precautions**
Surya Namaskar A sequence of 12 *Asana* (65)	• In each round 27 ml/ kg/ min oxygen consumption occurs , hence improve cardio-respiratory fitness. • Expenditure of 230 kcal during 30 min session for a 60 kg individual, hence it promotes weight management. 	Avoid- In hypertension, CAD, Myocardial Infraction and other cardiac disorders. Severe osteoporosis, Acute PIVD, Angle closure glaucoma
Standing Postures *Tadasana* *Ardhachakrashna* *Ardhakaticharasana*	Standing postures with full stretch, slight backward bending and side twisting increase the blood flow in body, stretches the muscles and produce relaxation. 	Backward bending should be avoided in case of recent surgery. Severe osteoporosis, Acute PIVD
Sitting *Gomukhasana* *Ustrasana* *Sashankasana*	These *Asana* stretch chest and shoulder, and hence increase the lungs capacity, normalize blood pressure, improve blood circulation and reduces the strain on the blood vessels. Improve blood flow to heart and reduce blood pressure. 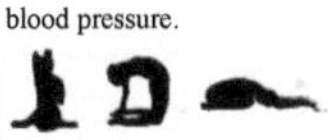	Should be avoided in case of recent surgery.
Prone position *Bhujangasana* *Dhanurasana*	Improve blood circulation to heart Relaxes heart muscles stretch the chest, shoulders and abdomen Increase lung capacity. Stretch whole body, increase circulation and relax the body. 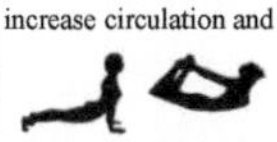	Dhanur*Asana* should be avoided in case of recent surgery.
Supine Posture *Setu Bandhasana,* *Pawanmuktasana,* *Halasana,* *Sarvangasana*	Helps in increasing the endurance and strengthen the cardiac muscles. Prepare heart for any adverse condition. Increase blood flow to heart, hence relax the heart. 	Should be avoided by patients who have undergone cardiac surgery and are hypertensive. Severe osteoporosis, Acute PIVD Glaucoma, cervical instability
***Pranayama* Practice** Deep slow breathing, *Bhastrika,* *Nadi Shodhan,*	*Pranayama* practice helps heart to relax and decrease the workload on heart by decreasing the heartbeat per minutes. Balances the Autonomic function of nervous system	Bhastrika should be avoided in hypertension and other cardiac disorders.

11.5 Potential Mechanism of Action of *Yoga* as a Preventive Tool

As a preventive treatment measure, *Yoga* is effective in managing a range of lifestyle disorders, many of which manifest as cardiac issues if not addressed at an early stage. Many cardiac disorders are manifestations of other lifestyle and behavioural disorders that can be modified. As discussed, *Yoga* helps in management of risk factors such as obesity, dyslipidaemia, hypertension and stress responsible for cardiac malfunctions. As vaccines immunize the body against infections in the same way, *Asana* practice strengthens the cardiac muscles, making it stronger to handle any adverse event.

Breathing pattern affects the emotional regulation, neural, respiratory, cardiac, digestive and all other systems working in the body. A rapid breathing rhythm can induce a state of aroused anxiety, increased heartbeat, but as breathing rhythms slow down, cardiovascular, endocrine, autonomic and parasympathetic system get activated resulting in physiological relaxation (Ma et al. 2017). When breathing rhythm is slowed down (0.1 Hz or 6 breaths/min), a cardiorespiratory synchronization occur, in which heart rate slows down, heart comes to relaxed state, blood pressure comes to normal state and interval between two heart beats, i.e. hear rate variability, increases, which is a physiological indicator of good cardiac health (Boyadzhieva and Kayhan 2021).

An important role in maintaining physiological homeostasis is played by the Vagus nerve, which regulates cardiac function also. Vagus nerve stimulation has been identified as a therapeutic tool for various disorders including cardiovascular disorders. This indicates the role of vagus nerve in maintaining the cardiac health. Vagus nerve carries parasympathetic fibers and supplies sinus node of heart which initiates cardiac rhythm. Vagal stimulation increases the parasympathetic dominance, which maintains the respiratory sinus arrhythmia (RSA) indicating a good cardiac health. RSA is normal increase in heart rate while inhaling and decrease in heart rate while exhaling, which is usually impaired/absent during stress, anxiety and depression. *Yoga* practice, specifically, deep breathing and meditation, has been reported to stimulate the vagus nerve and hence increase RSA, indicating the protective role of *Yoga* for cardiac health.

So, in summary, in this chapter we tried to evaluate various *Yoga* interventions and their effects on cardiac health via autonomic and peripheral nervous system besides generalized biopsychosocial effect on whole individual/body and would like to mention that the holistic approach to the preventive, curative and rehabilitative aspect of cardiac health must include *Yoga* as a primer for greater benefits.

11.6 Clinical Implication

Clinical implication of our limited experience and review of available literature here is to promote the utilization of our ancient tradition in the form of *Yoga* to maintain and enhance cardiac health by adopting appropriate lifestyle changes and include *Yoga* as an adjunct modality for prevention and rehabilitation of cardiac ailments.

11.7 Future Perspective

There is unmet need of evidence-based clinical trials and research in the field of *Yoga* as a preventive and probably curative/rehabilitative modality. It is strongly recommended to initiate high-quality clinical studies and closely observed findings over regular follow-ups with *Yoga* as a primary treatment approach for healthy population as well as for maintenance of good cardiac health.

11.8 Conclusion

There is limited evidence to support the exact mechanism through which *Yoga* exerts its effect but most of the studies show that even if few components of *Yoga* like *Asana*, *Pranayama* and Meditation are practised by healthy individual, there are less chances of development of cardiac and cardiovascular ailments.

References

American Diabetes Association (1997) Diabetes mellitus and exercise: joint position statement. Diabetes Care 29(12):1–6

Anderson EA, Mark AL (1993) The vasodilator action of insulin. Implications for the insulin hypothesis of hypertension. Hypertension 21(2):136–141

Bernstein AM, Bar J, Ehrman JP, Golubic M, Roizen MF (2014) *Yoga* in the management of overweight and obesity. Am J Lifestyle Med 8(1):33–41

Boyadzhieva A, Kayhan E (2021) Keeping the breath in mind: respiration, neural oscillations, and the free energy principle. Front Neurosci 15:647579

Bray GA (2004) Medical consequences of obesity. J Clin Endocrinol Metab 89(6):2583–2589

Brien SE, Ronksley PE, Turner BJ, Mukamal KJ, Ghali WA (2011) Effect of alcohol consumption on biological markers associated with risk of coronary heart disease: systematic review and meta-analysis of interventional studies. BMJ 342:d636

Casas R, Castro-Barquero S, Estruch R, Sacanella E (2018) Nutrition and cardiovascular health. Int J Mol Sci 19(12):3988

Chaitanya S, Datta A, Bhandari B, Sharma VK (2022) Effect of resonance breathing on heart rate variability and cognitive functions in young adults: a randomised controlled study. Cureus 14:2

Chien S (2003) Molecular and mechanical bases of focal lipid accumulation in arterial wall. Prog Biophys Mol Biol 83(2):131–151

Chu P, Gotink RA, Yeh GY, Goldie SJ, Hunink MM (2016) The effectiveness of *yoga* in modifying risk factors for cardiovascular disease and metabolic syndrome: a systematic review and meta-analysis of randomized controlled trials. Eur J Prev Cardiol 23(3):291–307

Cramer H (2016) The efficacy and safety of *yoga* in managing hypertension. Exp Clin Endocrinol Diabetes 124(02):65–70

Department of Health (1999) Physical inactivity and cardiovascular disease. https://www.health.ny.gov/diseases/chronic/cvd.htm

Dierolf AM, Schoofs D, Hessas E-M, Falkenstein M, Otto T, Paul M et al (2018) Good to be stressed? Improved response inhibition and error processing after acute stress in young and older men. Neuropsychologia 119:434–447

Ebong IA, Goff DC Jr, Rodriguez CJ, Chen H, Bertoni AG (2014) Mechanisms of heart failure in obesity. Obes Res Clin Pract 8(6):e540–e5e8

Einarson TR, Acs A, Ludwig C, Panton UH (2018) Prevalence of cardiovascular disease in type 2 diabetes: a systematic literature review of scientific evidence from across the world in 2007–2017. Cardiovasc Diabetol 17(1):1–19

Ference BA, Ginsberg HN, Graham I, Ray KK, Packard CJ, Bruckert E et al (2017) Low-density lipoproteins cause atherosclerotic cardiovascular disease. 1. Evidence from genetic, epidemiologic, and clinical studies. A consensus statement from the European Atherosclerosis Society Consensus Panel. Eur J Heart 38(32):2459–2472

Friedman M, Rosenman RH, Carroll V, Tat RJ (1958) Changes in the serum cholesterol and blood clotting time in men subjected to cyclic variation of occupational stress. Circulation 17(5):852–861

Gadham J, Sajja S, Rooha V (2015) Effect of *yoga* on obesity, hypertension and lipid profile. Int J Res Med Sci 3(5):1061–1065

Gotto AM Jr (2002) Management of dyslipidemia. Am J Med 112(8):10–18

Gruber K (2008) The physiological and psychological effects of Ashtanga *yoga*

Grundy SM, Benjamin IJ, Burke GL, Chait A, Eckel RH, Howard BV et al (1999) Diabetes and cardiovascular disease: a statement for healthcare professionals from the American Heart Association. Circulation 100(10):1134–1146

Hajjar I, Kotchen JM, Kotchen TA (2006) Hypertension: trends in prevalence, incidence, and control. Annu Rev Public Health 27:465–490

Howard G, Wagenknecht LE, Burke GL, Diez-Roux A, Evans GW, McGovern P et al (1998) Cigarette smoking and progression of atherosclerosis: the Atherosclerosis Risk in Communities (ARIC) Study. JAMA 279(2):119–124

Huffman KM, Hawk VH, Henes ST, Ocampo CI, Orenduff MC, Slentz CA et al (2012) Exercise effects on lipids in persons with varying dietary patterns—does diet matter if they exercise? Responses in studies of a targeted risk reduction intervention through defined exercise I. Am Heart J 164(1):117–124

Husain K, Ansari RA, Ferder L (2014) Alcohol-induced hypertension: mechanism and prevention. World J Cardiol 6(5):245

Katzmarzyk PT, Church TS, Craig CL, Bouchard C (2009) Sitting time and mortality from all causes, cardiovascular disease, and cancer. Med Sci Sports Exerc 41(5):998–1005

Kokane AM, Joshi R, Kotnis A, Chatterjee A, Yadav K, Revadi G et al (2020) Determinants of behavioural and biological risk factors for cardiovascular diseases from state level STEPS survey (2017–2019) in Madhya Pradesh. PeerJ 8:e10476

Kuppusamy M, Kamaldeen D, Pittani R, Amaldas J, Shanmugam P (2018) Effects of *Bhramari Pranayama* on health—a systematic review. J Tradit Complement Med 8(1):11–16

Kushner RF (2014) Weight loss strategies for treatment of obesity. Prog Cardiovasc Dis 56(4):465–472

Lalonde L, Gray-Donald K, Lowensteyn I, Marchand S, Dorais M, Michaels G et al (2002) Comparing the benefits of diet and exercise in the treatment of dyslipidemia. Prev Med 35(1):16–24

Leon BM, Maddox TM (2015) Diabetes and cardiovascular disease: epidemiology, biological mechanisms, treatment recommendations and future research. World J Diabetes 6(13):1246

Leone A (1993) Cardiovascular damage from smoking: a fact or belief? Int J Cardiol 38(2):113–117

Leone A (1995) Cigarette smoking and health of the heart. J R Soc Health 115(6):354–355

Ma X, Yue Z-Q, Gong Z-Q, Zhang H, Duan N-Y, Shi Y-T et al (2017) The effect of diaphragmatic breathing on attention, negative affect and stress in healthy adults. Front Neurosci 8:874

Malhotra V, Singh S, Tandon OP, Sharma SB (2005) The beneficial effect of *yoga* in diabetes. Nepal Med Coll J 7(2):145–147

Manchanda S, Narang R, Reddy K, Sachdeva U, Prabhakaran D, Dharmanand S et al (2000) Retardation of coronary atherosclerosis with *yoga* lifestyle intervention. J Assoc Physicians India 48(7):687–694

Mody BS (2011) Acute effects of Surya Namaskar on the cardiovascular and metabolic system. J Bodyw Mov Ther 15(3):343–347

Morse SA, Bravo PE, Morse MC, Reisin E (2005) The heart in obesity-hypertension. Expert Rev Cardiovasc Ther 3(4):647–658

Mourya M, Mahajan AS, Singh NP, Jain AK (2009) Effect of slow-and fast-breathing exercises on autonomic functions in patients with essential hypertension. J Altern Complement Med 15(7):711–717

Mozaffarian D (2016) Dietary and policy priorities for cardiovascular disease, diabetes, and obesity: a comprehensive review. Circulation 133(2):187–225

Myoclinic (2022) High blood pressure 2022. https://www.mayoclinic.org/diseases-conditions/high-blood-pressure/symptoms-causes/syc-20373410#:~:text=High%20levels%20of%20stress%20can,Certain%20chronic%20conditions

Nemet D, Barkan S, Epstein Y, Friedland O, Kowen G, Eliakim A (2005) Short-and long-term beneficial effects of a combined dietary–behavioral–physical activity intervention for the treatment of childhood obesity. Pediatrics 115(4):e443–e4e9

Niedermaier ON, Smith ML, Beightol LA, Zukowska-Grojec Z, Goldstein DS, Eckberg DL (1993) Influence of cigarette smoking on human autonomic function. Circulation 88(2):562–571

Nilakhe SS, Joshi YV, Kulkarni SS (2018) Understanding dietary do's and don'ts from hatha*yoga* with ayurvedic principles—a review. Int J Health Sci Res 8(9):255–262

Obert J, Pearlman M, Obert L, Chapin S (2017) Popular weight loss strategies: a review of four weight loss techniques. Curr Gastroenterol Rep 19(12):1–4

Poli A, Marangoni F, Avogaro A, Barba G, Bellentani S, Bucci M et al (2013) Moderate alcohol use and health: a consensus document. Nutr Metab Cardiovasc Dis 23(6):487–504

Powers SK, Smuder AJ, Kavazis AN, Quindry JC (2014) Mechanisms of exercise-induced cardioprotection. Physiology (Bethesda) 29(1):27–38

Riley KE, Park CL (2015) How does *Yoga* reduce stress? A systematic review of mechanisms of change and guide to future inquiry. Health Psychol Rev 9(3):379–396

Salleh MR (2008) Life event, stress and illness. Malays J Med Sci 15(4):9

Satin JR, Linden W, Millman RD (2014) *Yoga* and psychophysiological determinants of cardiovascular health: comparing *yoga* practitioners, runners, and sedentary individuals. Ann Behav Med 47(2):231–241

Sengupta P (2012) Health impacts of *Yoga* and *Pranayama*: a state-of-the-art review. Int J Prev Med 3(7):444

Shantakumari N, Sequeira S (2013) Effects of a *yoga* intervention on lipid profiles of diabetes patients with dyslipidemia. Indian Heart J 65(2):127–131

Singh S, Gaurav V, Parkash V (2011) Effects of a 6-week nadi-shodhana *Pranayama* training on cardio-pulmonary parameters. J Phys Educ Sports Manag 2(4):44–47

Steptoe A, Kivimäki M (2012) Stress and cardiovascular disease. Nat Rev Cardiol 9(6):360–370

Suskin N, Sheth T, Negassa A, Yusuf S (2001) Relationship of current and past smoking to mortality and morbidity in patients with left ventricular dysfunction. J Am Coll Cardiol 37(6):1677–1682

Tabei SMB, Senemar S, Saffari B, Ahmadi Z, Haqparast S (2014) Non-modifiable factors of coronary artery stenosis in late onset patients with coronary artery disease in Southern Iranian population. J Cardiovasc Thorac Res 6(1):51

Tuohy KM, Fava F, Viola R (2014) 'The way to a man's heart is through his gut microbiota'—dietary pro-and prebiotics for the management of cardiovascular risk. Proc Nutr Soc 73(2):172–185

Ussher MH, Taylor AH, Faulkner GEJ (2014) Exercise interventions for smoking cessation. Cochrane Database Syst Rev 8:CD002295

Vijayalakshmi P, Madanmohan BA, Patil A, Babu K (2004) Modulation of stress induced by isometric handgrip test in hypertensive patients following yogic relaxation training. Indian J Physiol Pharmacol 48(1):59–64

Wallace JP (2003) Exercise in hypertension. Sports Med 33(8):585–598

Wilber JA (1973) The problem of undetected and untreated hypertension in the community. Bull N Y Acad Med 49(6):510

World Health Organization (2014) Toolkit for delivering the 5A's and 5R's brief tobacco interventions in primary care 2014. https://apps.who.int/iris/bitstream/han-

dle/10665/112835/9789241506953_eng.pdf;jsessionid=596DC4BDD0AF43E6622287BB3F19E516?sequence=1
World Health Organization (2022a) Cardiovasular diseases. https://www.who.int/health-topics/cardiovascular-diseases#tab=tab_1
World Health Organization (2022b) Health topic/alcohol 2022. https://www.who.int/health-topics/alcohol#tab=tab_1
World Health Organization (2022c) Diabetes 2022. https://www.who.int/news-room/fact-sheets/detail/diabetes
Wyss JM (1993) The role of the sympathetic nervous system in hypertension. Curr Opin Nephrol Hypertens 2(2):265–273
Yogendra J, Yogendra HJ, Ambardekar S, Lele R, Shetty S, Dave M et al (2004) Beneficial effects of *yoga* lifestyle on reversibility of ischaemic heart disease: caring heart project of International Board of *Yoga*. JAPI 52:283–289

Chapter 12
Effect of Meditation on Neuropsychological Profile of Prisoners

Kalyan Maity, Vijaya Majumdar, Parth Lal, Sanjib Patra, Pramod Avti, and Krishan Kumar

Contents

K. Maity
Division of Yoga and Life Sciences, Swami Vivekananda Yoga Anusandhana Samsthana (S-VYASA), Bengaluru, Karnataka, India

Neuroscience Research Laboratory, Department of Neurology, Postgraduate Institute of Medical Education and Research, Chandigarh, India

V. Majumdar
Division of Yoga and Life Sciences, Swami Vivekananda Yoga Anusandhana Samsthana (S-VYASA), Bengaluru, Karnataka, India

P. Lal
Advanced Pediatric Centre, Postgraduate Institute of Medical Education and Research, Chandigarh, India

S. Patra
Department of Yoga, Central University of Rajasthan, Ajmer, Rajasthan, India

P. Avti
Department of Biophysics, Postgraduate Institute of Medical Education and Research, Chandigarh, India

K. Kumar (✉)
Department of Psychiatry, Postgraduate Institute of Medical Education and Research, Chandigarh, India
e-mail: kumar.krishan@pgimer.edu.in

A. Anand (ed.), *Neuroscience of Yoga*,
https://doi.org/10.1007/978-981-97-2851-0_12

Abstract Prisoners live their life in a very stressful situation and are prone to have neuropsychological abnormalities such as impaired cognition, attention deficit, mood disorders, anxiety, migraine/headache, eating disorders, addictions, metabolic issues, etc. Even though in some cases causes are unknown but these problems may be caused by environmental factors, side effects of drugs, stressful life events, and finally affect their daily activities and damage their quality of life. As of now there is a paucity of treatment modalities to mitigate these issues in prisoners. Few studies have documented the positive effect of interventions such as *Yoga*, exercise, music therapy, cognitive behavioral therapy, and third-wave psychotherapy on prisoners. On the other hand, various studies showed the positive effect of different types of meditation like mindfulness meditation, cyclic meditation, *mantra* meditation, transcendental meditation, and heartfulness meditation on various populations to reduce mental conflicts, stress, and tranquilizing the mind. Meditation can change the structure and function of the brain and also reduce the default mode network activity and activate the brain region which is related to cognition, attention, working memory, conflict monitoring, and emotional control. Even though meditation has remarkable neuropsychological effects, but there is very little scientific evidence that has examined the effect of meditation in prison settings. In this chapter, we have highlighted the scientific evidence of meditation in the different neuropsychological profiles of prisoners.

Keywords Prison · *Yoga* · Meditation · Neuropsychology · Mental health

12.1 Introduction

The current estimates indicated that around 11 million people were incarcerated in prison globally and more worrisome is the fact that this figure is growing day by day, and the rate can be more than the normal population growth rate (Kinner and Young 2018). The study suggests that more than 50% of prisoners suffer from poor mental health (James and Glaze 2006). However, there remains a lack of available data on the state of health care in prisons in most of the countries and they are mostly not integrated with health coverage.

Apart from personal factors and imported vulnerabilities like previous episodes of suicide attempts, self-harm, and psychiatric illnesses, the isolated and impoverished environment of the prisons and the associated pain and traumas (Sparks et al. 1996) contribute significantly to the vulnerability of reduced overall health and particularly mental disorders. These mental health issues are major social concerns and increase the risk of re-offending which is known as criminal recidivism. Prisoners experience very serious mental health issues and are under the treatment of psychiatric medicine (Wilper et al. 2009). These mental health problems can also lead to multiple harmful health outcomes even during imprisonment and also after imprisonment (Wallace and Wang 2020). Most of the studies on the prison population have been associated with poor mental health but they have not focused on the variety of environmental experiences in prison, which may have played a potential role in the poor mental conditions (Liebling 1995). Prisoners face significant mental pressure in prisons, which may have been aggravated by the COVID pandemic.

12.2 Neuropsychological Profiles of Prisoners

Prisoners show a higher prevalence of psychosis, affective disorder, anxiety, suicidal thoughts, substance use, personality disorders, and neurasthenia. The isolated and impoverished environment are crucial contributors to reduced life expectancy and the overall health of prisoners across the globe. In particular, mental health conditions, if left untreated, could increase the risk for treatment failure and recidivism rate in prisoners. Prisoners showed a higher prevalence of psychosis (male 3.6%, female 3.9%), depression (male 10.2%, female 14.1%), and post-traumatic stress disorder (PTSD) (male 17.8%, female 40.1%). Females are more prone to get anxiety, depression, and have a higher suicide rate compared to males (Belet et al. 2020). Prisoners with a history of domestic violence during childhood are having more stress, anxiety, and depression levels compared to those who do not have such a history. Also, young prisoners with a history of imprisonment have elevated anger levels compared to other prisoners (Unver et al. 2013).

Age and residential environment can also count as causative factors for anxiety in prisoners. Prisoners with age from 50 to 59 years have been found to be more vulnerable to getting generalized anxiety disorder (GAD). On the other hand, prisoners who stayed in closed-type prisons are having three times higher chance to get GAD compared to half-open or open services prisoners (Stawinska-Witoszynska et al. 2021). Longer period of locked time due to a lack of prison staff leads to anxiety and anger in prisoners (Nurse et al. 2003). A positive correlation of stress with cortisol levels and imprisonment time has been seen (Fenol et al. 2017). PTSD, anxiety, depression, self-harm, and suicide have also been reported to be highly prevalent in prisoners (Sen et al. 2014). Especially young prisoners with age ranging from 16 to 21 years have been reported to be having more suicidal rates and mortality (Hales et al. 2015). A meta-analysis showed suicide and aggressive behaviors which were significantly associated with PTSD symptoms (Facer-Irwin et al. 2019). Childhood abuse also plays an important role in PTSD. Around 40% of men was having a history of childhood abuse in that emotional abuse was 24.5%, physical abuse 25%, and sexual abuse 10% among prisoners with PTSD symptoms. All types of childhood abuse prisoners have a higher prevalence of anxiety, depression, near-lethal suicide attempts, and stress disorder in comparison to prisoners who have not been exposed (Caravaca Sánchez et al. 2019).

Women inmates have higher mental illnesses compared to women in the community (Tye and Mullen 2006). Studies showed that adverse childhood experiences and lack of social support can lead to higher stress, anxiety, and depression levels in victimized inmates compared to non-victimized inmates (Caravaca-Sánchez et al. 2019). These psychological factors are also hurdles for the motherhood of a prisoner. Mental illness like depression (80%) is much higher in pregnant prisoners which may lead to an adverse effect on the fetus' health. Also, stressors in prison and the separation of newborn babies from the mother lead to various mental illnesses such as PTSD, anxiety, depression, and insomnia in female prisoners (Ferszt et al. 2013). On the other hand, female prisoners with PTSD have a history of higher rape and sexual abuse (Pérez-Pedrogo et al. 2018). Female prisoners have higher mortality and comorbidity rates. Around 26.1% of female prisoners are having mental illnesses in that 14.7% are having depression, 6.4% PTSD, and 1.8% are having psychotic symptoms.

12.3 Available Treatment Modalities to Improve Neuropsychological Deficit in Prisoners

Though many interventions had targeted prisoner mental health, they were mostly limited by small sample size, heterogeneity of prisoners, and the settings that limit firm conclusive interpretation (Fazel et al. 2016). Unfortunately, there still exists perceptive demarcation between prisoner health and general public health which is a major obstacle preventing the optimization of healthcare in prison settings. Therefore, there is limited availability of the implementation details of mental health treatments in real circumstances of prison. However, this issue has been lately recognized and numerous psychosomatic interventions have been introduced in prison to improve health conditions and reduce the recidivism rate. A recent systematic review by Thekkumkara et al. (2022), compiled the effect of different interventions such as "Motivation enhancement Therapy, Seeking Safety Program, Cognitive Behavior Therapy, Interpersonal Therapy, Acceptance, and Commitment Therapy, Cognitive-behavioral Suicidal Prevention Therapy, Group Interpersonal Psychotherapy, Music Therapy, Art Therapy, Group Psychological Intervention, and Mindfulness-Based Relapse Prevention Intervention" on prisoners. Most of the interventions focused on substance use disorder in high-income countries but there were very less studies focused on estimating effect of psychosocial interventions (Thekkumkara et al. 2022).

12.4 Mind-Body Intervention, as a Treatment Modality for Prisoners

12.4.1 Yoga

Yoga is a well-known alternative therapy useful for treating various psychological symptoms such as stress, depression, novelty-seeking, harm avoidance, improved self-awareness, and self-directedness in inmates (Danielly and Silverthorne 2017).

Yoga is also helpful for the self-regulation, physical, emotional, and metabolic well-being of inmates (Tibbitts et al. 2021). Bartels et al. (2019), showed significant improvement in anxiety, depression, goal direction, self-esteem, negative affect, non-acceptance, flexibility, pain reduction, sleep, relaxation, and mental well-being after *Yoga* practice (Bartels et al. 2019). *Yoga* may also help to reduce substance use by coping with stress (Wimberly and Engstrom 2018). A systematic review of 37 randomized trials showed promising evidence of *Yoga* to improve cardio-metabolic health (Chu et al. 2016). Another form of yogic practice called *Sudarshan Kriya* can reduce depressive symptoms and anxiety and increase self-control, vitality, and positive well-being (Sureka et al. 2014). A randomized control trial (RCT) in Sweden concluded that *Yoga* has a positive impact on factors associated with recidivism like impulsivity and antisocial behavior (Kerekes et al. 2017). *Yoga* also has a longitudinal outcome on the psychological profile of prisoners. Lower recidivism has been noticed even 5 years after release from prison. A study by Kovalsky et al. (2021), showed a reduction in recidivism rates (33.82%) in the *Yoga* group in comparison to those who did not practice *Yoga* (40.74%) during their imprisonment (Kovalsky et al. 2021). A study conducted over a period of 10 weeks in prison inmates showed that *Yoga* substantially reduced obsessive-compulsive paranoid ideations as well as somatization symptoms (Sfendla et al. 2018). A 10-week RCT conducted in Britain by Bilderbeck and colleagues showed that *Yoga* increased self-reported positive affect and reduced psychological distress among prison inmates (Bilderbeck et al. 2013).

12.4.2 Exercise

Many studies showed exercise can reduce stress, anxiety, depression, and other psychological and metabolic variables. A 2006 study showed that regular moderate physical activity showed beneficial results on the physical as well as mental well-being of prisoners (Nelson et al. 2006). A systematic review suggests that exercise can improve the physical as well as mental well-being of prisoners (Sanchez-Lastra et al. 2019). A randomized control trial showed that a combination of aerobic and resistance training caused a significant decrease in BMI, BP, and coronary heart disease (Battaglia et al. 2013). A multicenter study from Italy also concluded the positive impact of exercise on the mental health of prisoners (Mipatrini et al. 2015). A study by Legrand et al. (2020), showed that 6-week interval exercise training can significantly reduce anxiety, stress, depression, anger, and improvement in self-esteem levels in prisoners (Legrand et al. 2020). A 2021 German study found that functional high-intensity interval training improved coordination muscular as well as aerobic endurance in prison inmates (Dransmann et al. 2021).

12.4.3 Music Therapy (MT)

Relaxing music therapy is effective to reduce anxiety and anger in jail inmates (Bensimon et al. 2015). Two weeks of MT showed a significant reduction of state anxiety in prisoners (Gold et al. 2014). MT can reduce anxiety and depression in prisoners. Greater improvement in anxiety levels was found in younger prisoners and prisoners with low literacy.

12.4.4 Cognitive Behavioral Therapy (CBT)

CBT is a psychological approach designed by Back. CBT uses to remove inhibited thought processes and suppressed negative filling. On the other hand, changing self-criticism and automatic thinking pattern to develop a healthy personality is the major goal of CBT (Beck 2020; Matusiewicz et al. 2010). CBT is an effective intervention to deal with many psychological problems. Studies showed that CBT can reduce depressive symptoms, anxiety (Leigh-Hunt and Perry 2015), psychological distress (Amoke et al. 2020), insomnia (Randall et al. 2019), violent sexual behaviors (Nwokeoma et al. 2019), and helps in suicide prevention (Pratt et al. 2016), alcohol, and drug use in the prison population (Doyle et al. 2019). Some reviews have indicated that CBT programs are more effective compared to other psychological interventions. Also, CBT has shown different effects for different psychological aspects. A meta-analysis showed a reduction in recidivism risk (20–30%) after CBT (Beaudry et al. 2021), whereas a systematic review showed limited effects of CBT programs in prisoners for reducing violent behavior (Auty et al. 2017a). However, the longitudinal effect of CBT was not studied effectively (Johnsen and Friborg 2015).

12.4.5 Third-Wave Psychotherapy

Third-wave behavior therapy is a new way of psychotherapy that generally includes "Acceptance and commitment therapy (ACT), dialectical behavior therapy (DBT), mindfulness-based intervention (MBI), compassion-focused therapy (CFT), and schema therapy (ST)" to address various physical and psychological experiences such as personality disorder, depression, generalized anxiety disorder, and to bring mindful changes and self-compassion (Kahl et al. 2012). Prisoners come under highly stigmatized groups and have several restrictions in the community, such as housing, voting rights, financial aid, employment, etc. Third-wave psychotherapy does not focus on the content itself but it is highlighting the context and function of these internal unfriendly events. That is the major difference between CBT and third-wave psychotherapy. A study showed improvement in abstinence rate, psychological flexibility, and reduction of anxiety sensitivity after ACT in female prisoners (Lanza and Menéndez 2013). A comparative study between ACT and CBT among female prisoners showed ACT was more effective to improve mental health compared to CBT (Lanza et al. 2014). A study by Riley et al. (2019), showed the practice of acceptance-based group therapy and mindfulness reduced anxiety, depression, and somatoform symptoms and improved acceptance and mindfulness in female prisoners (Riley et al. 2019).

12.5 Meditation

Meditation is a part of *Ashtanga Yoga* described by *Maharshi Patanjali.* Meditation is defined as an "uninterrupted flow of consciousness" and has a potential role on body, mind, and emotion (Dunlop 2015). Many types of meditation such as Mindfulness meditation, Cyclic meditation, *Mantra* meditation, Transcendental meditation, and Heartfulness meditation have shown beneficial effects on the different psychological and neurological aspects. A meta-analysis by Goyal et al. showed the positive effect of meditation on anxiety, depression, and pain (Goyal et al. 2014).

12.6 Effect of Different Types of Meditation in Different Populations

Very importantly, a neurophenomenological account could be presented based on the scientific evidence of different physiological and neurological markers of meditation and MBIs that aid in substantiating the claims of effectiveness. For example, studies have reported improvement in functional activity in the executive control network such as attention, working memory, conflict monitoring, and inhibition (Taren et al. 2017). A study by Afonso et al. (2020), showed that meditation can change the structure and function of the brain and also reduce the default mode network activity and activate the brain region which is related to cognition and emotional control (Afonso et al. 2020). Different meditation techniques are having various psychological and neurological effects as follows.

12.6.1 Mindfulness Meditation

Mindfulness-Based Stress Reduction (MBSR) was first established by Jon Kabat-Zinn in 1979 (Kabat-Zinn 2003). Later on, several similar mindfulness meditation techniques arose like Mindfulness Meditation (MM) and Mindfulness-Based Cognitive Therapy (MBCT) which is mainly characterized by compassion and acceptance (Crane 2017). In the field of psychological and behavioral research MM has immense contributions. The study showed significant improvement in mindfulness, cognitive function (Wong et al. 2017), spirituality, and reduction of perceived stress, depression, insomnia, and pain severity after MM in older adults. Another review of RCTs showed the improvement of sleep quality, positive affect, memory, and executive function along with a reduction of anxiety, depression, and insomnia symptoms after MM in older adults (Hazlett-Stevens et al. 2019).

A study by Bostock and Crosswell (2019), showed improvement in general well-being, positive affect, and reduction of anxiety and depression after 8 weeks of mindfulness meditation in employees (Bostock et al. 2019). A 2017 pilot RCT aimed at value-based mindfulness for prison inmates, post release risk behavior, showed statistically significantly reduced recidivism rates (Malouf et al. 2017). A 2019 study found that mindfulness-based stress reduction improved psychological outcomes significantly in prison inmates (Bouw et al. 2019). A recent study by Youngs et al. (2021), indicated improvement in visual short-term memory after 8 min of MM in college students (Youngs et al. 2021).

12.6.2 Cyclic Meditation

Cyclic meditation (CM) is a moving meditation technique developed by H.R. Nagendra, Chancellor of *Swami Vivekananda Yoga Anusandhana Samsthana*, Bengaluru, India. The principle of cyclic meditation is "alternate stimulation and relaxation." The main concept of cyclic meditation was taken from the ancient Indian scripture called *Mandukya Upanishad (*Subramanya and Telles 2009*)*. Many studies showed the effect of cyclic meditation on different neuropsychological variables. Studies showed improved sustained attention after CM, and prolonged latencies in the cerebral cortex after CM indicate cortical inhibition. Another study by Subramanya and Telles (2009), showed the reduction of anxiety and increased memory functions immediately after practicing CM compared to supine rest. There was also improved performance in tasks requiring selective attention, concentration, and visual scanning abilities. A 2021 article discussed the possible benefits of cyclic mediation on the regulation of the gut-brain axis (Ningthoujam et al. 2021).

12.6.3 **Mantra** *Meditation*

Mantra meditation is a technique in which we mentally repeat or chant the spiritual or non-spiritual syllables. However, a spiritually meaningful *mantra* makes a positive impact on the mind and body (Bormann and Carrico 2009). Pradhan and Derle (2012), showed an improvement in digit-letter substitution tasks after chanting of *Gayatri Mantra* (Pradhan and Derle 2012*)*. A meta-analysis by Alvarez-Perez et al. (2022), showed a significant reduction in stress, anxiety, depressive symptoms, post-traumatic stress, and improvement in mental health-related quality of life

(QOL) after mantra meditation; however, religion can act as a limitation in this case (Alvarez-Perez et al. 2022).

12.6.4 Transcendental Meditation (TM)

TM originated from the ancient yogic practice of South Asia. It comprises a mental repetition of a *mantra* (Sanskrit syllables). TM improves concentration and helps to calm down the mind. Also one can experience the subtle and transcendental state after TM practice (Roth 1994). The study showed that TM is beneficial for stress reduction, improved relaxation, and leads to higher states of consciousness (Mason et al. 1997). The study showed decreased PTSD (Herron and Rees 2018), depression, and increased mindfulness and QOL after TM practice (Kang et al. 2018). A narrative review showed TM has neuropsychological and cognitive-behavioral benefits (Mosini et al. 2019). A 2017 randomized trial found that transcendental meditation was effective in reducing trauma symptoms in prison inmates (Nidich et al. 2017).

12.6.5 Heartfulness Meditation

Heartfulness Meditation (HM) system is also called *Sahaj Marg*. During HM, meditators focus on *Pranahuti* or yogic transmission. Many research studies showed the different effects of HM. HM reduced stress, depression, anger, and improved overall mood and health-related quality of life (Yadav et al. 2021; Thimmapuram et al. 2022). A study in the United States showed a reduction of perceived stress and improvement in sleep quality after HM during the COVID-19 pandemic. Another study showed the reduction of burnout and improved emotional wellness and telomere length after HM practice in medical professionals (Thimmapuram et al. 2017).

12.7 Meditation as a Treatment Modality for Prisoners

There are several lines of research that support meditation-based programs as rehabilitative for prison populations. Even though mindfulness intervention is extensively used for reducing stress and anxiety but in recent years it has been introduced to prisoners. It serves as an effective addition to behavior-based interventions that were already in practice in prison settings for longer periods of time. In 2017, Auty et al. reported the meta-analysis reports of the available literature indicating the benefits of meditation programs in prisoners to improve behavioral functioning and psychological well-being. They also emphasized the better results of a longer duration of practice compared to the more intensive programs (Auty et al. 2017b).

Violence underlies important negative psychological consequences in prisoners. Mindfulness meditation training for prison inmates has been proposed to be the most effective intervention, toward the prevention of violence, and reducing recidivism (Cooley 2019). Mind-body relaxation techniques were also found to be a beneficial treatment modality to reduce stress and anxiety in prisoners. Dominique et al. showed the reduction of anxiety, physical tension, and improvement of sleep quality and physical fitness after relaxation techniques in prisoners and a significant reduction of anxiety, depression, somatoform symptoms, and improvement in acceptance and mindfulness in female prisoners after 10 mindfulness and acceptance-based group programs (Riley et al. 2019). Nidich et al. showed significant improvement in perceived stress, trauma, anxiety, depression, and sleep after Transcendental Meditation in male prisoners (Nidich et al. 2016). A study by An et al. showed the mindfulness training program has a positive effect on mindfulness, aggression, anxiety, depression, mood, and perceived stress in prisoners with long-term imprisonment compared to prisoners with short-term imprisonment. Meditation also reduces the recidivism rate (Lyons and Cantrell 2016). Duration of the practice and intensity of the practice are most important in *Yoga* practice. Longer duration with low-intensity *Yoga* and meditation practice has more psychological well-being compared to short-term high-intensity practice in prisoners. Meditation reduces substance abuse by reducing anxiety and strengthening mental resolve. Another study showed the reduction of trauma symptoms after TM practice (Nidich et al. 2017). Also, another meditation technique called *Vipassana* Meditation (VM) can be used for traumatized prisoners (Phillips and Hopper 2015) and is effective for reducing anxiety and depression (Chandiramani et al. 1998). Studies showed the effects on good perception, overcoming ordeal behavior, positive relationship, and the social atmosphere immediately after 10 days of VM and even after 3–4 months. VM practice is also able to reduce mood disturbance and improve emotional intelligence and mindfulness in prisoners.

12.8 Clinical Implications

Different meditation techniques may be used as an alternative therapy for prisoners to manage neuropsychological deficits of prisoners. Also, meditation techniques can be implemented in society during the quarantine and lockdown era for their neuropsychological health.

12.9 Future Directions for Research

Many studies have shown the positive effect of meditation on various neuropsychological profiles such as cognition, stress, anxiety, depression, etc. But there were very less meditation techniques used in the prison population for treating various neuropsychological issues. Also, the number of RCTs was less in the prison population. Hence, more future studies are required to identify the role of various specific meditation techniques on the prison population and implement those meditation techniques to treat neuropsychological issues of prisoners. Very importantly, the world has recently witnessed the Covid-19 pandemic and its associated quarantine, lockdown, and restricted social interactions. Even though the nature of isolation is different, but Bursky et al. have recently proposed the beneficial effect of meditation and mindfulness-based intervention for maintaining psychological health during quarantine and lockdown.

12.10 Conclusion

Meditation techniques may be beneficial for managing different neuropsychological aspects such as stress, anxiety, sleep quality, depression, trauma symptoms, mood, aggression, recidivism rate, and can help to improve the overall well-being of prisoners. A collaboration must be there between policy makers, physicians, and police personnel for a radical change in the current system. We need to set up special clinics/teaching classes for prison inmates and most importantly these people need to be treated as human beings who deserve the right to a better environment where they can heal in all aspects.

References

Afonso RF, Kraft I, Aratanha MA, Kozasa EH (2020) Neural correlates of meditation: a review of structural and functional MRI studies. Front Biosci (Schol Ed) 12(1):92–115

Alvarez-Perez Y, Rivero-Santana A, Perestelo-Perez L, Duarte-Diaz A, Ramos-Garcia V, Toledo-Chavarri A, Serrano-Aguilar P (2022) Effectiveness of mantra-based meditation on mental health: a systematic review and meta-analysis. International Journal of Environmental Research and Public Health 19(6):3380

Amoke CV, Ede MO, Nwokeoma BN, Onah SO, Ikechukwu-Ilomuanya AB, Albi-Oparaocha FC, Nweke ML, Amoke CV, Amadi KC, Aloh HE, Anyaegbunam EN (2020) Effects of group cognitive-behavioral therapy on psychological distress of awaiting-trial prison inmates. Medicine 99(17):e18034

Auty KM, Cope A, Liebling A (2017a) Psychoeducational programs for reducing prison violence: a systematic review. Aggress Violent Behav 33:126–143

Auty KM, Cope A, Liebling A (2017b) A systematic review and meta-analysis of *Yoga* and mindfulness meditation in prison: effects on psychological well-being and behavioural functioning. Int J Offender Ther Comp Criminol 61(6):689–710

Bartels L, Oxman LN, Hopkins A (2019) "I would just feel really relaxed and at peace": findings from a pilot prison *Yoga* program in Australia. Int J Offender Ther Comp Criminol 63(15–16):2531–2549

Battaglia C, Di Cagno A, Fiorilli G, Giombini A, Fagnani F, Borrione P, Marchetti M, Pigozzi F (2013) Benefits of selected physical exercise programs in detention: a randomized controlled study. Int J Environ Res Public Health 10(11):5683–5696

Beaudry G, Yu R, Perry AE, Fazel S (2021) Effectiveness of psychological interventions in prison to reduce recidivism: a systematic review and meta-analysis of randomised controlled trials. Lancet Psychiatry 8(9):759–773

Beck JS (2020) Cognitive behavior therapy: basics and beyond. Guilford Publications, New York

Belet B, D'Hondt F, Horn M, Amad A, Carton F, Thomas P, Vaiva G, Fovet T (2020) Post-traumatic stress disorder in prison. L'encephale 46(6):493–499

Bensimon M, Einat T, Gilboa A (2015) The impact of relaxing music on prisoners' levels of anxiety and anger. Int J Offender Ther Comp Criminol 59(4):406–423

Bilderbeck AC, Farias M, Brazil IA, Jakobowitz S, Wikholm C (2013) Participation in a 10-week course of *Yoga* improves behavioural control and decreases psychological distress in a prison population. J Psychiatr Res 47(10):1438–1445

Bormann JE, Carrico AW (2009) Increases in positive reappraisal coping during a group-based mantram intervention mediate sustained reductions in anger in HIV-positive persons. Int J Behav Med 16(1):74–80

Bostock S, Crosswell AD, Prather AA, Steptoe A (2019) Mindfulness on-the-go: effects of a mindfulness meditation app on work stress and well-being. J Occup Health Psychol 24(1):127

Bouw N, Huijbregts SCJ, Scholte E, Swaab H (2019) Mindfulness-based stress reduction in prison: experiences of inmates, instructors, and prison staff. Int J Offender Ther Comp Criminol 63(15–16):2550–2571

Caravaca Sánchez F, Ignatyev Y, Mundt AP (2019) Associations between childhood abuse, mental health problems, and suicide risk among male prison populations in Spain. Crim Behav Ment Health 29(1):18–30

Caravaca-Sánchez F, Fearn NE, Vidovic KR, Vaughn MG (2019) Female prisoners in Spain: adverse childhood experiences, negative emotional states, and social support. Health Soc Work 44(3):157–166

Chandiramani K, Dhar PL, Verma SK (1998) Psychological effects of *Vipassana* on Tihar jail inmates

Chu P, Gotink RA, Yeh GY, Goldie SJ, Hunink MM (2016) The effectiveness of *Yoga* in modifying risk factors for cardiovascular disease and metabolic syndrome: a systematic review and meta-analysis of randomized controlled trials. Eur J Prev Cardiol 23(3):291–307

Cooley C (2019) Escaping the prison of mind: meditation as violence prevention for the incarcerated. Health Promot Pract 20(6):798–800

Crane R (2017) In: Dryden W (ed) Mindfulness-based cognitive therapy: distinctive features, 2nd edn. Taylor & Francis Group, London, 1–173 p

Danielly Y, Silverthorne C (2017) Psychological benefits of *Yoga* for female inmates. Int J Yoga Therap 27(1):9–14

Doyle MF, Shakeshaft A, Guthrie J, Snijder M, Butler T (2019) A systematic review of evaluations of prison-based alcohol and other drug use behavioural treatment for men. Aust N Z J Public Health 43(2):120–130

Dransmann M, Koddebusch M, Gröben B, Wicker P (2021) Functional high-intensity interval training lowers body mass and improves coordination, strength, muscular endurance, and aerobic endurance of inmates in a German prison. Front Physiol 12:733774

Dunlop J (2015) Meditation, stress relief, and well-being. Radiol Technol 86(5):535–555

Facer-Irwin E, Blackwood NJ, Bird A, Dickson H, McGlade D, Alves-Costa F, MacManus D (2019) PTSD in prison settings: a systematic review and meta-analysis of comorbid mental disorders and problematic behaviours. PLoS One 14(9):e0222407

Fazel S, Hayes AJ, Bartellas K, Clerici M, Trestman R (2016) Mental health of prisoners: prevalence, adverse outcomes, and interventions. Lancet Psychiatry 3(9):871–881

Fenol A, Jebi S, Krishnan S, Perayil J, Vyloppillil R, Bhaskar A, Menon SM, Mohandas A (2017) Association of stress, salivary cortisol level, and periodontitis among the inmates of a central prison in Kerala. Dent Res J (Isfahan) 14(4):288

Ferszt GG, Hickey JE, Seleyman K (2013) Advocating for pregnant women in prison: the role of the correctional nurse. J Forensic Nurs 9(2):105–110

Gold C, Assmus J, Hjørnevik K, Qvale LG, Brown FK, Hansen AL, Waage L, Stige B (2014) Music therapy for prisoners: pilot randomised controlled trial and implications for evaluating psychosocial interventions. Int J Offender Ther Comp Criminol 58(12):1520–1539

Goyal M, Singh S, Sibinga EM, Gould NF, Rowland-Seymour A, Sharma R, Berger Z, Sleicher D, Maron DD, Shihab HM, Ranasinghe PD (2014) Meditation programs for psychological stress and well-being: a systematic review and meta-analysis. JAMA Intern Med 174(3):357–368

Hales H, Edmondson A, Davison S, Maughan B, Taylor PJ (2015) The impact of contact with suicide-related behavior in prison on young offenders. Crisis 36(1):21–30

Hazlett-Stevens H, Singer J, Chong A (2019) Mindfulness-based stress reduction and mindfulness-based cognitive therapy with older adults: a qualitative review of randomized controlled outcome research. Clin Gerontol 42(4):347–358

Herron RE, Rees B (2018) The transcendental meditation program's impact on the symptoms of post-traumatic stress disorder of veterans: an uncontrolled pilot study. Mil Med 183(1–2):e144–e150

James DJ, Glaze LE (2006) Mental health problems of prison and jail inmates

Johnsen TJ, Friborg O (2015) The effects of cognitive behavioral therapy as an anti-depressive treatment is falling: a meta-analysis. Psychol Bull 141(4):747

Kabat-Zinn J (2003) Mindfulness-based stress reduction (MBSR). Constructivism in the Human Sciences 8(2):73

Kahl KG, Winter L, Schweiger U (2012) The third wave of cognitive behavioural therapies: what is new and what is effective? Curr Opin Psychiatry 25(6):522–528

Kang SS, Erbes CR, Lamberty GJ, Thuras P, Sponheim SR, Polusny MA, Moran AC, Van Voorhis AC, Lim KO (2018) Transcendental meditation for veterans with post-traumatic stress disorder. Psychol Trauma Theory Res Pract Policy 10(6):675–680

Kerekes N, Fielding C, Apelqvist S (2017) *Yoga* in correctional settings: a randomized controlled study. Front Psychiatry 8:204

Kinner SA, Young JT (2018) Understanding and improving the health of people who experience incarceration: an overview and synthesis. Epidemiol Rev 40(1):4–11

Kovalsky S, Hasisi B, Haviv N, Elisha E (2021) Can *Yoga* overcome criminality? The impact of *Yoga* on recidivism in Israeli prisons. Int J Offender Ther Comp Criminol 65(6–7):671–691

Lanza PV, Menéndez AG (2013) Acceptance and commitment therapy for drug abuse in incarcerated women. Psicothema 25(3):307–312

Lanza PV, Garcia PF, Lamelas FR, González-Menéndez A (2014) Acceptance and commitment therapy versus cognitive behavioral therapy in the treatment of substance use disorder with incarcerated women. J Clin Psychol 70(7):644–657

Legrand FD, Ory E, Herring MP (2020) Evaluation of a brief interval exercise training (IET) intervention for first-time prisoners with elevated anxiety symptoms. Anxiety, Stress, & Coping 33(5):581–589

Leigh-Hunt N, Perry A (2015) A systematic review of interventions for anxiety, depression, and PTSD in adult offenders. Int J Offender Ther Comp Criminol 59(7):701–725

Liebling A (1995) Vulnerability and prison suicide. Br J Criminol 35(2):173–187

Lyons T, Cantrell WD (2016) Prison meditation movements and mass incarceration. Int J Offender Ther Comp Criminol 60(12):1363–1375

Malouf ET, Youman K, Stuewig J, Witt EA, Tangney JP (2017) A pilot RCT of a values-based mindfulness group intervention with jail inmates: evidence for reduction in post-release risk behavior. Mindfulness (N Y) 8(3):603–614

Mason LI, Alexander CN, Travis FT, Marsh G, Orme-Johnson DW, Gackenbach J, Mason DC, Rainforth M, Walton KG (1997) Electrophysiological correlates of higher states of consciousness during sleep in long-term: practitioners of the transcendental meditation program. Sleep 20(2):102–110

Matusiewicz AK, Hopwood CJ, Banducci AN, Lejuez CW (2010) The effectiveness of cognitive behavioral therapy for personality disorders. Psychiatr Clin 33(3):657–685

Mipatrini D, Mannocci A, Masala D, Di Thiene D, Rizzo J, Meggiolaro S, La Torre G (2015) The impact of physical activity on prisoners' quality of life in Italy: a multicenter study: Daniele Mipatrini. Eur J Public Health 25(suppl_3):ckv175–ckv258

Mosini AC, Saad M, Braghetta CC, Medeiros RD, Peres MF, Leão FC (2019) Neurophysiological, cognitive-behavioral and neurochemical effects in practitioners of transcendental meditation—a literature review. Rev Assoc Med Bras 65(5):706–713

Nelson M, Specian VL, Tracy NC, DeMello JJ (2006) The effects of moderate physical activity on offenders in a rehabilitative program. J Correct Educ 57:276–285

Nidich S, O'Connor T, Rutledge T, Duncan J, Compton B, Seng A, Nidich R (2016) Reduced trauma symptoms and perceived stress in male prison inmates through the transcendental meditation program: a randomized controlled trial. Perm J 20(4):16–007

Nidich S, Seng A, Compton B, O'Connor T, Salerno JW, Nidich R (2017) Transcendental meditation and reduced trauma symptoms in female inmates: a randomized controlled study. Perm J 21:16–008

Ningthoujam DS, Singh N, Mukherjee S (2021) Possible roles of cyclic meditation in regulation of the gut-brain axis. Front Psychol 12:768031

Nurse J, Woodcock P, Ormsby J (2003) Influence of environmental factors on mental health within prisons: focus group study. BMJ 327(7413):480

Nwokeoma BN, Ede MO, Ugwuanyi C, Mezieobi D, Ugwoezuonu AU, Amoke C, Egenti NT, Nwosu N, Oforka TO, Victor-Aigbodion V, Offordile EE (2019) Efficacy of prison-based cognitive behavioral rehabilitation intervention on violent sexual behaviors among sex offenders in Nigerian prisons. Medicine (Baltimore) 98(29):e16103

Pérez-Pedrogo C, Martínez-Taboas A, González RA, Caraballo JN, Albizu-García CE (2018) Sex differences in traumatic events and psychiatric morbidity associated to probable posttraumatic stress disorder among Latino prisoners. Psychiatry Res 265:208–214

Phillips J, Hopper JW (2015) Intensive *Vipassana* meditation practice for traumatized prisoners

Pradhan B, Derle SG (2012) Comparison of effect of Gayatri Mantra and poem chanting on digit letter substitution task. Anc Sci Life 32(2):89–92

Pratt D, Gooding P, Awenat Y, Eccles S, Tarrier N (2016) Cognitive behavioral suicide prevention for male prisoners: case examples. Cogn Behav Pract 23(4):485–501

Randall C, Nowakowski S, Ellis JG (2019) Managing acute insomnia in prison: evaluation of a "one-shot" cognitive behavioral therapy for insomnia (CBT-I) intervention. Behav Sleep Med 17(6):827–836

Riley BJ, Smith D, Baigent MF (2019) Mindfulness and acceptance–based group therapy: an uncontrolled pragmatic pre–post pilot study in heterogeneous population of female prisoners. Int J Offender Ther Comp Criminol 63(15–16):2572–2585

Roth R (1994) Maharishi Mahesh Yogi's transcendental meditation. Primus, Washington, DC

Sanchez-Lastra MA, de Dios Álvarez V, Pérez CA (2019) Effectiveness of prison-based exercise training programs: a systematic review. J Phys Act Health 16(12):1196–1209

Sen P, Exworthy T, Forrester A (2014) Mental health care for foreign national prisoners in England and Wales. J Ment Health 23(6):333–339

Sfendla A, Malmström P, Torstensson S, Kerekes N (2018) Yoga practice reduces the psychological distress levels of prison inmates. Front Psychiatry 9:407

Sparks R, Bottoms A, Hay W (1996) Prisons and the problem of order. Clarendon Press

Stawinska-Witoszynska B, Czechowska K, Moryson W, Wieckowska B (2021) The prevalence of generalised anxiety disorder among prisoners of the penitentiary institution in north-eastern Poland. Front Psychiatry 12:671019

Subramanya P, Telles S (2009) A review of the scientific studies on cyclic meditation. Int J Yoga 2(2):46

Sureka P, Govil S, Dash D, Dash C, Kumar M, Singhal V (2014) Effect of Sudarshan Kriya on male prisoners with non psychotic psychiatric disorders: a randomized control trial. Asian J Psychiatr 12:43–49

Taren AA, Gianaros PJ, Greco CM, Lindsay EK, Fairgrieve A, Brown KW, Rosen RK, Ferris JL, Julson E, Marsland AL, Creswell JD (2017) Mindfulness meditation training and executive control network resting state functional connectivity: a randomized controlled trial. Psychosom Med 79(6):674

Thekkumkara SN, Jagannathan A, Muliyala KP, Murthy P (2022) Psychosocial interventions for prisoners with mental and substance use disorders: a systematic review. Indian J Psychol Med 44(3):211–217

Thimmapuram J, Pargament R, Sibliss K, Grim R, Risques R, Toorens E (2017) Effect of heartfulness meditation on burnout, emotional wellness, and telomere length in health care professionals. J Community Hosp Intern Med Perspect 7(1):21–27

Thimmapuram J, Patel K, Madhusudhan DK, Deshpande S, Bouderlique E, Nicolai V, Rao R (2022) Health-related quality of life outcomes with regular *Yoga* and heartfulness meditation practice: results from a multinational, cross-sectional study. JMIR Form Res 6(5):e37876

Tibbitts DC, Aicher SA, Sugg J, Handloser K, Eisman L, Booth LD, Bradley RD (2021) Program evaluation of trauma-informed *Yoga* for vulnerable populations. Eval Program Plann 88:101946

Tye CS, Mullen PE (2006) Mental disorders in female prisoners. Aust N Z J Psychiatry 40(3):266–271

Unver Y, Yuce M, Bayram N, Bilgel N (2013) Prevalence of depression, anxiety, stress, and anger in Turkish prisoners. J Forensic Sci 58(5):1210–1218

Wallace D, Wang X (2020) Does in-prison physical and mental health impact recidivism? SSM Popul Health 11:100569

Wilper AP, Woolhandler S, Boyd JW, Lasser KE, McCormick D, Bor DH, Himmelstein DU (2009) The health and health care of US prisoners: results of a nationwide survey. Am J Public Health 99(4):666–672

Wimberly AS, Engstrom M (2018) Stress, substance use, and *Yoga* in the context of community reentry following incarceration. J Correct Health Care 24(1):96–103

Wong WP, Coles J, Chambers R, Wu DB, Hassed C (2017) The effects of mindfulness on older adults with mild cognitive impairment. J Alzheimers Dis Rep 1(1):181–193

Yadav GS, Cidral-Filho FJ, Iyer RB (2021) Using heartfulness meditation and brainwave entrainment to improve teenage mental wellbeing. Front Psychol 12:742892

Youngs MA, Lee SE, Mireku MO, Sharma D, Kramer RS (2021) Mindfulness meditation improves visual short-term memory. Psychol Rep 124(4):1673–1686

Chapter 13
Tele-*Yoga* and Its Implications for Digital Health-Emerging Trends in Health and Wellness

Prashant Verma, Manjari Rain, and Gurmeet Singh

Contents

Abstract *Yoga* has been shown to have an essential role in maintenance of non-communicable and communicable diseases. During the COVID-19 pandemic, doctors and medical institutions acknowledged *Yoga* as a natural immune-boosting intervention because of the preventative and therapeutic capabilities it has. To maintain the physical distance, information and services associated with *Yoga* were provided via various forms of telecommunication, which later got defined as tele-*Yoga*. This chapter is based on the importance, future, precautions, and applications of tele-*Yoga* in emerging digital health trends. It aims to synergize the interests of both the service provider and the service recipient, and raise awareness among the general public regarding tele-*Yoga* and digital health. We searched articles related to

---

P. Verma
Interdisciplinary Centre for Swami Vivekananda Studies, Panjab University, Chandigarh, India

M. Rain
Neuroscience Research Lab, Department of Neurology, Postgraduate Institute of Medical Education and Research, Chandigarh, India

G. Singh (✉)
Department of Physical Education, Panjab University, Chandigarh, India
e-mail: gurmeet@pu.ac.in

A. Anand (ed.), *Neuroscience of Yoga*,
https://doi.org/10.1007/978-981-97-2851-0_13

tele-*Yoga* by using keywords like tele-*Yoga*, digital health *Yoga*, mobile *Yoga*, and e-health *Yoga* with search output of 18, 44, 75, and 4 articles, respectively. Most of the articles were reviews and as perspective. However, various research articles including randomized trials show promise for tele-*Yoga* use. The chapter compiles such information and concludes that during the pandemic tele-*Yoga* played a spectacular role in reducing stress and providing overall wellness which continues even after the pandemic. Post-pandemic studies suggest that tele-*Yoga* is very helpful and feasible in structural disorders, management of metabolic disorders, neurodegeneration diseases, rehabilitation of heart attack patients, social issues like domestic violence, and post-COVID symptoms. The market of tele-*Yoga* in the field of digital health is expanding on a daily basis and it is reasonable to anticipate that this growth will continue into the foreseeable future.

Keywords Tele-*Yoga* · Health and wellness · Mobile health · Digital health · Stress

13.1 Introduction

Telecommunication or telecom services have been making significant contribution in providing essential health through telemedicine for decades. Monitoring of patients remotely has become possible by advances in telecom. It reduces repetitive hospital visits, saves time, and makes consultations easier between physician and patient. During COVID pandemic, it was necessary to maintain distance. Greater distance is a barrier to various services and learning. However, there are several activities which cannot be halted under any circumstances. Telecom services like online counseling sessions, camps, workshops, seminars through video conferencing have made learning easier. Therefore, most of the educational, government, and various organizational institutions have started working online, in which telecom technologies have played an important role without disrupting public health guidelines (De' et al. 2020). The pandemic drew the attention of clinicians, researchers, and general population toward health and fitness (Monaghesh and Hajizadeh 2020). Apart from the allopathic medicine, inclusion of traditional medicine, exercise, meditation, and *Yoga* are proving to be very beneficial in both communicable and non-communicable diseases (Snyder et al. 2021). Where the practice of other physical activities or exercises requires a specific ground or area, the practice of *Yoga* can be performed at home or at open spaces such as halls and parks. Various AYUSH (*Ayurveda, Yoga* and Naturopathy, *Unani, Siddha,* and Homeopathy) programs and wellness centers are providing their health care services online. An alliance between *Yoga* and telecom services, coined as Tele-*Yoga*, may play a vital role in the field of health and wellness. Tele-*Yoga* is used nowadays as an inclusive term embracing the delivery of many forms of *Yoga* services and the sharing of information across distances with the assistance of a broad range of technologies. Traditional knowledge of *Yoga* and its therapeutic aspect can be delivered through tele-*Yoga* (Nagendra 2020). The word tele-*Yoga* supplements terms such as online health, telehealth, mobile health, and e-health.

Tele-*Yoga* has played an important role in reducing the stress of not only health care recipients but also health care workers (HCWs) (Nagarathna et al. 2021). Apart from health care workers, other essential service providers such as police and other administrative service workers also report that they practiced *Yoga* using the tele-*Yoga* module in order to boost their immune system and overcome fear (Khadse et al. 2020). In the beginning of pandemic, it was very difficult for the common man to adopt health and wellness service through online mode. Gradually, the use of online health services including tele-*Yoga* has increased because it is fast, easily accessible, overcome geographical barrier, and cost-effective. In this era of competitive examinations, students have started using tele-*Yoga* services at home to reduce their stress. Thus, it would not be overvalued that in present era there is a huge need for tele-*Yoga* in the field of health and wellness (Joshi et al. 2021).

Mental disorders such as anxiety disorders, major depressive disorder, post-traumatic stress disorder (PTSD), schizophrenia, bipolar disorder, disruptive behavior, and other social disorders are very common among adults globally. In 2019, roughly 970 million people worldwide attributed their mental illness to various psychological, biological, economic, environmental, and social causes. In 2020, the pandemic saw an exponential increase of 26% in anxiety and 28% in major depressive disorder (WHO 2022). Although some causes of mental disorders are challenging to address, others can be reduced by providing mind-body interventions such as *Yoga* and meditation. *Yoga* may be helpful in reducing the global burden of mental disorders by increasing the calming effects, attention span, acceptance, and adaptability (Brunner et al. 2017). *Yoga* is an effective intervention that has both preventive and curative properties, which can be helpful in relieving mental disorders and creating a sense of security among people (Tolahunase et al. 2018). Owing to its accessibility, practicality, safety, cost-effectiveness, time-saving, therapeutic, and preventative advantages, tele-*Yoga* is the future of critical health services.

Various researches have proved that most of the non-communicable diseases are not just physical but are affected by psychological health as increasing stress is one of the main reasons. The same has been mentioned in ancient Indian medical texts that consider restlessness of mind as the cause of most diseases. According to the following ayurvedic text:

Janmaantara kritam paapam vyaadhi roopena bhaadate|
Tat shantihi aushadhaihi, daanaihi, japa homa suaarchanaihi||

The root cause of all diseases is uncontrolled mind and most of the diseases are produced by the mind and then they multiply manifold.

13.2 Need for Tele-*Yoga*

Yoga is an ancient Indian knowledge-cum-tradition whose main objective is to take an individual to the highest state of consciousness. According to the archaeological and literary evidence found so far, it has existed from more than 5000 years. This was the time when *Yoga* was made compulsory for the all-round development of the students. *Acharya* (teachers) of almost all Indian traditions used to give special knowledge of earning a livelihood to their students as well as knowledge of health and protection, which is described in the *Indus* Valley Civilization, *Vedic*, *Upanishadic*, *Bodhik*, and *Jain* literature. *Yoga* texts written in medieval history show that even in medieval India, everyone from saints to kings practiced *Yoga* (Pandurangi et al. 2017).

Just as *Yoga* was an integral part of the lifestyle of Indians in ancient times, it is currently being used as a therapy across the world as it has both preventive and therapeutic properties. There are various molecular and neuropsychological studies that confirm the effectiveness of *Yoga* in metabolic disorders, structural disorders, and psychological issues (Yadav et al. 2020; Prathikanti et al. 2017). In 2019, COVID-19 started spreading and soon became a pandemic. Various studies and observations made by clinicians and general public found that an individual with good immunity, healthy lifestyle, and no co-morbidities are less susceptible to the coronavirus (Barone Gibbs et al. 2021; Pérez et al. 2021). People, worldwide, started considering traditional medicine and practicing exercise and *Yoga* at home to protect them from infection following the rules of social isolation. Tele-*Yoga* became the need of the hour for guiding participants online without any physical contact. *Yoga* forms a psychological basis and plays an important role in the rehabilitation of post-COVID-19 (Lai et al. 2020) patients by reducing anxiety and fear in times of isolation (Sharma et al. 2022a). It also improves biological profile and neurological functions, and provides immunity against infections (Dalpati et al. 2022).

As we have learned to live with new strains of corona virus, there is an increase in online activity around the world. While telecom services have made the whole world a global village, the collaboration between service providers and receivers at the international level has further strengthened. Due to the flexibility of workplace, the intensity of this collaboration has also been seen in the field of tele-*Yoga*, which is likely to grow further (Sharma et al. 2020a).

India, the most populated country in the world, is a developing country which lacks health and wellness infrastructure in many small cities and towns (More et al. 2012). Although *Yoga* is becoming popular day by day after the UNESCO declared *Yoga* as a World Heritage and International Day of *Yoga* by the Government of India, there is a dearth of experienced *Yoga* instructors and *Yoga* therapists in small cities, towns, and villages (Mishra et al. 2020).

The threat of a pandemic has exposed the health care system of all countries. At the same time, it has ruined the sports and fitness industries, which require a specialized space and equipment to practice. *Yoga* had maintained its acceptability and adequacy even in adverse conditions like lockdown due to its convenience and lesser requirement of equipment. *Yoga* can be practiced anywhere in ventilated clean room, home, classroom, park, hospital, and any other workplace. Studies have shown that people are more attracted toward *Yoga* and other alternative medicines like *Ayurveda* and *Siddha* during the time of lockdown (Natarajan et al. 2021). People were practicing *Yoga* with their *Yoga* instructors through video calling during the lockdown. Many apps were launched and existing apps were updated during the pandemic to promote group conferencing with large number of participants. Through these telecom services, *Yoga* trainers are providing therapeutic *Yoga* and counseling to their patients and healthy participants (Sharma et al. 2022b).

13.2.1 Stress Management

There are several studies that report a significant reduction in stress and depression after few weeks of *Yoga* intervention (Kwok et al. 2019). Jasti et al. in 2020 reported that tele-*Yoga* is the most effective and safest method of managing stress and improving one's well-being. Ninety-five volunteers (69 women, 26 men) were recruited after online consent in the study. Significant reductions in perceived stress levels were reported by Perceived Stress Scale (PSS), and improvement in the flexibility, movements, and attention basis through the *Yoga* Performance Assessment Scale (YPA) with 92% of people reported that tele-*Yoga* had changed their lives positively after 4 weeks of intervention (Jasti et al. 2020). Another initiative was taken by Anand et al. for the health care workers (HCWs) and other common people at the time of lockdown in March 2020. The *Yoga* session started with a Facebook page "*Yoga* Scholar PGIMER." Three sessions a day for 225 days were provided by experienced *Yoga* experts and eminent national-international masters of *Yoga*, and the masters of other alternative medicine including both practice and theory sessions for awareness of their health and wellness. This *Yoga* knowledge by Facebook page has reached nearly three million people worldwide, including around 670 live recorded videos. They concluded that *Yoga* played an important role in combating lockdown stress for the general public and the extreme occupational stress of HCWs during the onset of the pandemic (Anand et al. 2021). Tele-*Yoga* practices at workplace are efficient in reducing post-occupational stress and anxiety by reducing sympathetic activity (Dalpati et al. 2022). *Yoga* classes through video conferencing are the best way to identify with the ancient *Yoga* in the modern world where distance is a constraint.

13.2.2 Rehabilitation

According to the WHO, rehabilitation is "a set of interventions designed to optimize functioning and reduce disability in individuals with health conditions in interaction with their environment" (WHO 2023). Inpatient and outpatient rehabilitation are two types of therapy. Inpatient rehabilitation refers to the therapy or treatment that is given during the course of treatment in a hospital or clinic, whereas outpatient rehabilitation therapy refers to various interventions given outside the hospital. Outpatient rehabilitation is necessary in various diseases, disorders, and medical events, including injuries, surgeries, strokes, and cardiovascular events, to help people recover from the effects of the disease and restore lost functional abilities. Outpatient rehabilitation therapy includes many types of rehabilitation therapy such as physical, speech, respiratory, and occupational therapies (Yaşar et al. 2021).

Rehabilitation is a core health care service not only for people with chronic or physical disabilities but for anyone with a serious or chronic health condition. Today rehabilitation is not just an option but an essential therapy, but in many places it is a costly intervention (Stożek et al. 2016). While these interventions are far beyond the reach of the general public due to their luxury, mind-body based methods such as *Yoga* and meditation have also been recognized as one of the best rehabilitation therapies in the last few decades. Taking into account all these socioeconomic factors, tele-*Yoga* has been involved in various researches in order to deliver high-quality and cost-effective rehabilitation services on time (James-Palmer et al. 2022).

With all these factors in mind, a controlled pilot study was conducted on 15 moderate-severe patients diagnosed with chronic obstructive pulmonary disease (COPD). COPD is a chronic inflammatory lung disease causing obstruction in airflow, which makes it difficult to breathe. In this study, 8 weeks of tele-*Yoga* rehabilitation practices were provided, which show improved sleep quality, flexibility, strength, and breathing capacity among the practitioners. These practices were effective in increasing stamina as well as dealing with stress and anxiety. Participants appreciated this home-based tele-*Yoga* intervention for symptomatic reliefs (Selman et al. 2015).

A few other rehabilitation studies provide evidence that tele-*Yoga* interventions may be useful in providing symptomatic relief in heart diseases, multiple sclerosis (MS), acute calculous cholecystitis (ACC), and heart failure (HF). A single-blind, parallel, two-arm randomized-controlled study suggests that tele-*Yoga* with breathing exercises and meditation is effective in various chronic diseases (Strömberg et al. 2021). MS is a neurological disorder that interferes with physical activity, which limits their social and outdoor activities. Studies have shown that long-term exercise and consistent exercise are the best ways to manage its complications. One study opines that a set of exercises and *Yoga* practices through telecom services such as voice notes, recorded videos, video conferencing may be the best approach for MS patients. Rimmer et al. aim to compare telerehabilitation and physical rehabilitation in MS in 820 patients (Rimmer et al. 2018; Tracy et al. 2022). Tele-yoga is also helpful and feasible in providing rehabilitation to substance use disorder (Bhargav et al. 2022).

A case study on a 34-year-old woman with ACC evaluated whether various *Ayurvedic* treatments such as *Panchakarma*, Tele-*Yoga* with *Virechana*, and *Ayurvedic* lifestyle modifications (*Ahaar, Vihara, Achar*) can improve the condition of ACC. It is a common disease associated with the gallbladder. According to *Ayurveda*, it is a *Pitta* predominant condition identified as *rudhapathakamala* (*kamala*—a type of jaundice). The study has concluded the feasibility of alternative therapies like *Ayurveda* (telemedicine) and tele-*Yoga* (Metri et al. 2021).

13.2.3 Pain Management

Yoga is a practice based on the mind-body complex that not only flexes and strengthens the muscles but also allows participants to meditate by controlling their breath. Past studies have shown that *Yoga* can be helpful in reducing many other types of chronic pain such as cervical, chronic low back pain, arthritis, fibromyalgia, migraine (Wells et al. 2021; Joyce et al. 2021). The pain is classified under different names according to the intensity and area. Several body structure-related disorders are associated with sedentary lifestyles, excessive use of electronic gadgets, poor posture, low amounts of physical activity, resulting in musculoskeletal pain (Khodadad et al. 2020). Individual tele-*Yoga* practice may be a reliable non-pharmacological treatment to reduce pain, once the type and intensity of pain is diagnosed (Mathersul et al. 2018). Tele-*Yoga* has shown remarkable improvement in chronic pain patients and may act as an effective tool in management of chronic pain (Sharma et al. 2020b).

13.2.4 Domestic Violence

Various studies and local newspapers from different countries and cities reported an increase in domestic violence cases during lockdown (Mazza et al. 2020). Domestic violence was fueled by restricted movement, economical stress, and fear of disease, which can be easily managed by practicing *Yoga* (Vora et al. 2020). In April 2020, a cross-sectional study was conducted between married men (n = 39) and married women (n = 55) according to their socioeconomic status, education level, and their age during lockdown in India. About 7.4% (n = 7) reported that they had faced domestic violence; of these 85.7% (n = 6) reported an increase in the violence during lockdown (Sharma and Khokhar 2022). To overcome these types of incidences counseling with tele-*Yoga* methods can be beneficial.

As we know counseling is a collaborative effort between the consultant and the client. Professional *Yoga* consultants can help clients identify the problem and possible solutions. *Yoga* therapy is used to deal with many psycho-physical issues

(Hegarty and Tarzia 2019). There are a number of technique-oriented practices in *Yoga* that mainly place great emphasis on counseling. These practices include *Asana*, *Pranayama*, meditation, as well as counseling by *Yoga-sutras* (aphorisms on the theory and practice of *Yoga*) provided by sage *Patanjali* (Iyengar 1993). While *Asana* and *Pranayama* help in bringing changes at the physical and mental level, these *Yoga*-sutras can also bring about changes at the philosophical and practical level.

Positive personal attitude can be created by inculcating qualities like empathy and compassion in the clients (Stevens and McLeod 2019). Women abused by their partners report *Yoga* as meaningful in addition to psychotherapy, suggesting that tele-*Yoga* can be a useful tool to provide help to such women, even men with traumatic experience (Clark et al. 2014). Tele-*Yoga* with counseling sessions can be conducted to increase understanding in family relationships.

Table 13.1 summarizes some of the recent studies in tele-*Yoga* for health and wellness including various diseases and psychological disorder or trauma (Selman et al. 2015; Strömberg et al. 2021; Donesky et al. 2017; Addington et al. 2018; Sarah et al. 2019; Malhotra et al. 2022; Rain et al. 2022; Malik et al. 2022; James-Palmer and Daneault 2022; Najafi et al. 2023a; Singh et al. 2023; Günebakan and Acar 2023; Jain Shrimal et al. 2023; Kashyap et al. 2023; Phansikar et al. 2023; Majumdar et al. 2023; Acar et al. 2023; Eyuboglu et al. 2023; Najafi et al. 2023b). Several clinical trials are also going on to investigate the effect of tele-*Yoga* (Singh et al. 2022; Zaccari et al. 2022; Balasubramanian et al. 2022).

13.3 Modes and Methods for Providing Tele-*Yoga*

Informed consent should be obtained from the participants before starting tele-*Yoga* sessions. In case of non-therapeutic *Yoga* session, consent will be valid for ten sessions in a time-span of 1 month while consent from patients for *Yoga* therapy is valid for 2 months. Parents or guardians can provide consents for children and teenagers below 18 years.

Ensuring privacy of *Yoga* practitioners and *Yoga* teachers during online sessions is a challenging task. Although there is a lot of education available about the operation of various video conferencing applications, still there is a lack of awareness regarding online data security and maintaining privacy of individuals. To remove this ignorance and bring awareness in the field of tele-*Yoga*, the Ministry of AYUSH, Government of India, has issued an advisory on tele-*Yoga* practices based on clinical experiments on July 21, 2020 (Department of Integrative Medicine, National Institute of Mental Health and Neurosciences (NIMHANS), Swami Vivekananda Yoga Anusandhana Samasthana (S-VYASA) 2020). The advisory is a collaborative effort from experts of Central Council for Research in *Yoga* and Naturopathy (CCRYN), Department of Integrative Medicine, National Institute of Mental Health and Neurosciences (NIMHANS), and *Swami Vivekananda Yoga Anusandhana Samasthana* (S-VYASA). These consultations are helpful in making tele-*Yoga*

Table 13.1 Recent studies on tele-*Yoga*

S. no.	Author/year	Journal	Methods/participants	Tele-*Yoga* intervention	Variable measures	Outcome
1.	Selman et al. (2015)	*BMC Complementary and Alternative Medicine*	A controlled, nonrandomized trial in patients with heart failure and COPD $N = 7$, tele-*Yoga* intervention $N = 8$, educational control	8 weeks (biweekly, 1 h)	Physiological tests, feasibility, and acceptability	Tele-*Yoga* is acceptable. Participants enjoyed home-based tele-*Yoga*
2.	Donesky et al. (2017)	*Journal of Alternative and Complementary Medicine*	A controlled, nonrandomized trial in patients with heart failure and COPD $N = 7$, tele-*Yoga* intervention $N = 8$, educational control	8 weeks (biweekly, 1 h)	Feasibility and acceptability	Tele-*Yoga* is acceptable and adherent. Dyspnea after exercise improved in *Yoga* group
3.	Addington et al. (2018)	*Complementary Therapies in Medicine*	Single arm-interventional study in women undergoing radiation/chemotherapy for breast cancer $N = 4$	12 weeks (biweekly, 75 min)	Feasibility and acceptability	Tele-*Yoga* has easy access Cancer and technology-related barriers are present

(continued)

Table 13.1 (continued)

S. no.	Author/year	Journal	Methods/participants	Tele-*Yoga* intervention	Variable measures	Outcome
4.	Sarah et al. (2019)	*Complementary Therapies in Clinical Practice*	Randomized, prospective, two-arm controlled study in cardiac rehabilitation patients $N = 115$, telerehabilitation group $N = 113$, control group	3 weeks (daily)	Blood pressure, endothelial function, heart rate, and quality of life	Telerehabilitation by *Yoga* was adherent. Blood pressure and quality of life improved in telerehabilitation group
5.	Sharma et al. (2020b)	*medRxiv*	A prospective, nonrandomized, single-arm clinical trial in chronic pain patients $N = 18$	6 weeks (biweekly)	Pain intensity, pain disability, anxiety, and depression	Improvement in pain intensity and disability, reduction in anxiety and depression score after tele-*Yoga* intervention
6.	Strömberg et al. (2021)	*International Journal of Environmental Research and Public Health*	Single-blind, parallel, two-arm randomized-controlled study in heart failure patients $N = 5$, tele-*Yoga* group $N = 5$, control group	12 weeks (biweekly, 1 h)	Physical function, dynamic balance/mobility, gait speed, Montreal Cognitive Assessment, quality of life, hospital anxiety and depression scale, minimal insomnia symptom scale, and Exercise Motivation Index	Tele-*Yoga* was feasible, safe, and acceptable. The tele-*Yoga* intervention showed some improvements
7.	Bhargav et al. (2022)	*Journal of Family Medicine and Primary Care*	Single-arm study in substance use disorder patients $N = 30$	4 weeks (3 times/week, 60 min)	Any side effects, ease of learning, yoga performance, clinical global impressions scale (CGI–severity), and perceived stress	Tele-yoga leads to improvement in Yoga performance, CGI-severity, and stress

8.	Malhotra et al. (2022)	*International Journal of Yoga*	A qualitative study in geriatric patients having symptoms of depression, anxiety, and stress $N = 30$	Time period was different of each participant	Interview	Tele-*Yoga* was feasible
9.	Rain et al. (2022)	*Frontiers in Public Health*	Two-arm, randomized-controlled study in COVID-positive patients, post-COVID, and health care workers $N = 12$, intervention (tele-*Yoga*) group $N = 12$, control group for each type of participants	15 days (daily, 5 min and 15 min)	Exercise capacity, psychological and biochemical parameters	D-dimer was decreased in post-COVID and HCWs. Exercise capacity was improved by tele-yogic breathing in COVID positive and HCWs
10.	Malik et al. (2022)	*Cureus*	Randomized, two-arm, pilot validation study in COPD patients $N = 30$, tele-*Yoga* group $N = 30$, tele-pulmonary rehabilitation session	3 months (5 days/week, 45 min)	Exercise capacity, COPD assessment test, quality of life, depression, and anxiety	Tele-*Yoga* is better than tele-pulmonary rehabilitation session on the basis of symptom scores, exercise capacity, and depression and anxiety scores
11.	James-Palmer and Daneault (2022)	*Digital Health*	Single-arm study in Parkinson's disease patients $N = 16$	6 weeks (biweekly, 30 min)	Safety and feasibility	Tele-*Yoga* was safe, feasible, usable, and enjoyable

(continued)

Table 13.1 (continued)

S. no.	Author/year	Journal	Methods/participants	Tele-*Yoga* intervention	Variable measures	Outcome
12.	Najafi et al. (2023a)	*Journal of Clinical Medicine*	Single-blinded, randomized-controlled trial in women with relapsing-remitting multiple sclerosis $N = 15$, tele-*Yoga* group $N = 15$, tele-Pilates group $N = 15$, control group	8 weeks (60–70 min)	Cortisol and prolactin serum levels, walking speed, depression, mental health, quality of life, and physical activity	Both tele-*Yoga* and tele-Pilates were beneficial for patients. Cortisol decreased and prolactin increased and improvement in depression, physical activity levels, quality of life, and speed of walking was noted in these groups
13.	Singh et al. (2023)	*Complementary Therapies in Clinical Practice*	A prospective randomized-controlled trial in ankylosing spondylitis patients $N = 57$, tele-*Yoga* group $N = 52$, control group	3 months (60 min)	Bath Ankylosing Spondylitis Functional Index (BASFI), Bath Ankylosing Spondylitis Disease Activity Index (BASDAI), Ankylosing Spondylitis-Quality of Life (ASQOL), and psychological and inflammatory biomarkers	Tele-*Yoga* shows improvement in pain, functional index, and mental health
14.	Günebakan and Acar (2023)	*Irish Journal of Medical Science*	A randomized study in healthy premenopausal women $N = 16$, tele-*Yoga* group $N = 16$, control group	6 weeks (biweekly 45-min)	Menstrual symptoms, quality of life, anxiety, depression, body awareness, and self-esteem	Tele-*Yoga* helps in reduction of menstrual symptoms and depression, improves quality of life and body awareness
15.	JainShrimal et al. (2023)	*Complementary Medicine Research*	Parallel, two-arm, randomized-controlled study in asymptomatic COVID-19 patients $N = 30$, tele-*Yoga* group $N = 30$, control group	7 days (daily, over 30 min)	Physiological variables (heart rate, respiratory rate, oxygen saturation, blood pressure) and psychological variables (stress, mindful attention)	Integrated tele-*Yoga* may have therapeutic benefits as it improves physiological, psychological, and mindfulness parameters

16.	Kashyap et al. (2023)	*Annals of Indian Academy of Neurology*	Open-label, randomized, control study in poststroke cognitive impairment *N* = 30, intervention group *N* = 30, control group	5 days physical *Yoga* followed by 11 weeks of tele-*Yoga* (1–2 times a week, 60 min)	Montreal Cognitive Assessment, Modified Rankin Scale, Clinician-Rated Dimensions of Psychosis Symptom Severity, Frontal Assessment Battery, Caregiver Burden Scale, and event-related potential	Significant improvement in cognitive ability and reduction in caregiver burden were observed in *Yoga* group
17.	Phansikar et al. (2023)	*Journal of Behavioral Medicine*	Two-arm, randomized-controlled study in stressed adults *N* = 43, tele-*Yoga* group *N* = 43, waitlisted control group	8 weeks (3 days/week, 50 min)	Adherence, enjoyment, safety, stress, anxiety, and neuropsychological tests	Tele-*Yoga* was safe and enjoyable, reduced stress, and improved cognitive function
18.	Majumdar et al. (2023)	*Frontiers in Public Health*	Randomized-controlled study in mild-to-moderate COVID-19 patients *N* = 113, adjunct tele-*Yoga* group *N* = 112, standard care group	14 days (2 times daily of 15- and 10-min session)	COVID outcome scale, clinical status, mortality, stress, inflammatory markers, HbA1c, blood hemogram, and kidney function markers	Tele-yoga improved clinical status of COVID-19 patients
19.	Acar et al. (2023)	*Journal of Complementary and Integrative Medicine*	Randomized-controlled study in ankylosing spondylitis *N* = 28, tele-*Yoga* group *N* = 27, waitlist control group	8 weeks (3 times/week)	Physical function, disease activity, spinal mobility, flexibility, muscular endurance, exercise capacity, balance, sleep quality, stress, depression, anxiety, quality of life, and mindfulness	Tele-yoga improved disease activity, spinal mobility, flexibility, muscular endurance, balance, sleep quality, stress, depression, and quality of life

(continued)

Table 13.1 (continued)

S. no.	Author/year	Journal	Methods/participants	Tele-*Yoga* intervention	Variable measures	Outcome
20.	Eyuboglu et al. (2023)	Heart and Lung	Randomized-controlled study in obstructive sleep apnea syndrome $N = 28$, tele-*Yoga* group $N = 27$, waitlist control group	12 weeks (60 min/day, 3 days/week)	Inspiratory and expiratory respiratory muscle strength (MIP, MEP), cardiopulmonary exercise test, Corsi Block Tapping Test (CBTT), and Stroop color word interference test-TBAG form (SCWT), 6-minute walk test (6MWT), Pittsburgh Sleep Quality Index (PSQI), and Epworth Sleepiness Scale (ESS)	Tele-yoga improved MIP, exercise test parameters, CBTT and Stroop TBAG test scores, sleep duration, sleep efficiency, sleep quality of PSQI, and ESS scores
21.	Najafi et al. (2023b)	*Multiple Sclerosis and Related Disorders*	Randomized-controlled trial in multiple sclerosis patients with relapsing-remitting MS (RRMS) and secondary progressive MS (SPMS) $N = 26$, tele-Yoga group $N = 29$, tele-Pilates group $N = 27$, control group	8 weeks (3 times/week)	Physical factors, mental factors, and quality of life	Tele-yoga and tele-Pilates, both improved quality of life, mental and physical health in RRMS and SPMS

interesting, enjoyable, and effective in implementing *Yoga* therapy. At the same time, the confidential interests of both the tele-*Yoga* provider and receiver can be protected by the rules mentioned herein. As per the advisory, the instructor should be YCB (*Yoga* certification board) certified with additional tele-*Yoga* certificate. *Yoga* instructor with basic *Yoga* knowledge (with 200 h of training) and tele-*Yoga* certificate should teach healthy participants only. *Yoga* therapy to patients can only be provided by individual with graduation or post-graduation in *Yoga* therapy with an additional tele-*Yoga* certification. *Yoga* therapist should work under the supervision of a tele-*Yoga* therapy consultant, who are registered medical practitioners and have certification of *Yoga* therapy. It is advised to keep the ratio of *Yoga* therapist and patients at 1:5 or less. Thus, in a session of ten patients, one therapist can conduct session and other can monitor the correctness of *Yoga* practiced by the patients and be in contact with them.

According to the execution of tele-*Yoga* and its purpose, tele-*Yoga* practices can be divided into three different categories, which are as follows:

1. **Tele-*Yoga* practice:** Tele-*Yoga* practices can be designed for healthy individuals for their health and well-being, in which beginner–intermediate level *Asana*, various breathing exercises (*Pranayama*), meditation, and *Yoga nidra* can be added to their practice.
2. **Tele-*Yoga* therapy:** The tele-*Yoga* therapy module should be designed for various patients that require a *Yoga* therapist. It can be practiced both collectively and individually. This may include specific *Asana*, *Pranayama*, meditation, and *Yoga nidra*, which may be included in *Yoga* therapy at an early stage.
3. **Tele-*Yoga* training:** Tele-*Yoga* training is required for *Yoga* students who are pursuing degree, diploma, or any certificate course in *Yoga*. In this situation, it is necessary for them to practice *Yoga* and acquire knowledge of the subject, both for academic purpose and to become a *Yoga* professional.

The quality of *Yoga* can be better after adopting tele-*Yoga* in these three different ways, but some policies should be practiced by both tele-*Yoga* practitioners and providers to maintain their privacy (Department of Integrative Medicine, National Institute of Mental Health and Neurosciences (NIMHANS), Swami Vivekananda Yoga Anusandhana Samasthana (S-VYASA) 2020). Several policies are mentioned in the advisory, and a few of them are depicted in Table 13.2. Tele-*Yoga* advisory recommends 30–45 min for each session. Advisory suggests that tele-*Yoga* should be on an appointment basis only, and details such as time, date, frequency, and duration of sessions should be discussed in advance.

It is important to document date, time, duration, and number of administered tele-*Yoga* sessions, including the reason for opting tele-*Yoga*. The details related to participants such as any difficulty felt by the participant at any time point, need of participant, any recommendations, and any referral to another professional should be documented and maintained. If any tele-*Yoga* session is temporarily stopped or discontinued, then the reason for it should be recorded. It is recommended to take time-to-time feedback from participants for improving the tele-*Yoga* sessions according to participant's need.

Table 13.2 Policies among tele-*Yoga* seekers and providers

For practitioners/patients	For instructors/therapists
1. No video recording will be done without the prior permission of the *Yoga* instructors	1. The instructor shall ensure during the online consent itself that neither party shall record any video during the exercise
2. Practitioners cannot share any data or recordings sent by *Yoga* instructors to anyone	2. No image or contact of practitioners may be shared on any other social media platform without the permission of practitioners
3. Practitioners cannot share any information or video meeting links with anyone else	3. Instructors should not allow anyone who has not submitted their online consent
4. Practitioners must follow the instructions of the instructor carefully	4. Instructors have to ensure safety, data privacy, security, and confidentiality. He/she should not share personal contact and information with the practitioners

The detailed policies are available at *https://cdn.ymaws.com/www.iayt.org/resource/collection/B3DF2972-93AF-4A8E-9466-585C9A9F8F40/Final_Advisory_on_Tele-Yoga_Services_Aarti.pdf*

13.4 Precautions for Tele-*Yoga*

Although *Yoga* exercises/*Asana* can be done through tele-*Yoga*, but some *Yoga* practices require the physical presence of a *Yoga* teacher so that the practitioner can practice the *Yoga* exercises properly and avoid any possible injuries. Certain exercises such as arm, shoulder, and head standing; excessive forward- or backward-bend postures (such as *Poorna Chakrasana*), *Asana* involving balance on only one leg (such as *vrikshasana*, *garudasana*), and other advance postures should not be practiced in absence of a *Yoga* trainer. *Asana*, other than meditation or relaxation postures, should be restricted to less than 3 min. *Shatkarma* practice should not be performed, except *Trataka* and *Kapalbhati*. The strokes for *Kapalbhati* and *Bhastrika* in online practice should be less than 40 strokes/min and 20 strokes/min, respectively.

It is essential to completely understand an *Asana* before attempting it. It is suggested that if a participant is not able to understand a *Yoga* protocol or a specific *Asana*, then he/she should not practice it before confirming every step from the *Yoga* instructor. Every individual has his/her own speed and understanding of a posture. No one should increase the speed or try to stretch more than his/her capability to match with other participants; this might cause injury. It should be understood that perfection in performing *Yoga* will be achieved gradually.

13.5 Disadvantages and Limitations of Tele-*Yoga*

Tele-*Yoga* is empowering the health and wellness sector, but it also involves a plethora of downsides. Some disadvantages are listed below:

1. Strong telecom and internet network: Providing *Yoga* services via telecom requires both parties (provider and receiver) to have a strong network connection; otherwise, the classroom cannot operate properly. Although most of the people's devices have the next-generation network (4G), but sometimes it also breaks down. Mode of communication plays an important role in *Yoga* practices, but poor networks can impede audio-visual communication. Limited access to internet and electricity is also challenging in remote areas.
2. Environment: *Hatha Yogic* texts show the importance of environment in *Yoga* practice, in which special attention is required including ventilated room, optimal temperature, etc. Simultaneously, it has also been seen that in group *Yoga* classes, practitioners are inspired by each other to practice *Yoga* more effectively and avoid absence from classes.
3. Interaction: The face-to-face process of interaction builds trust and familiarity between the therapist/teacher and patients/students, which makes them comfortable and honest. Showing compassion, listening actively, spending proper time with the patients all these also help in the speedy recovery of the patient. These parameters are absent or minimal in the online session.
4. Discipline: *Yoga* is more of a discipline that is not incorporated in patients/students merely by listening; it is a set of processes that cannot be learned through lectures alone. More important than memorizing and remembering *Yoga* is that it reflects on one's behavior and character. *Maharishi Patanjali* described *Yoga* discipline as the absence of fluctuations in the mind for which he performed the eight limbs of *Yoga*. *Yoga* is an ancient Indian traditional practice of providing spiritual benefits and taught under the *Guru-Shishya Parampara* (relationship and tradition between teacher and student). Under this tradition, the teacher carefully observes the mistakes of his student and help the disciple to rectify them, for this physical presence of a teacher is important. The method of tele-*Yoga* services limits learning to a teacher-centered method only.
5. Limited aspects of therapy: For quick improvement in the patient's condition, along with *Yoga* therapy, some components of naturopathy such as soil therapy, water therapy, *abhyanga, shirodhara*, etc., along with acupressure/*marma* therapy are constructive. At the same time, it is also important to monitor the patient's condition regularly, but all this is not possible through tele-*Yoga*.
6. Misunderstanding: The problem of coordination between instructions and practiced *Yoga* exercises has also been noticed due to lack of proper face and body visualization in online *Yoga* classes. Misunderstandings are also common among *Yoga* trainers and practitioners sometimes due to bad internet and language problems.
7. Expensive internet and shortage of electricity: In African and Central Asian countries such as Malawi, Benin, Chad, Yemen, Botswana, it can be very expensive to use a tele-*Yoga* service due to high internet rates. Rates for 1 GB of data in these countries are more than $20, while it is common for 1-h *Yoga* class to consume 1 GB of data. There are also some countries where there is a severe shortage of power supply; in such a situation, tele-*Yoga* services cannot be provided due to lack of device charging.

13.6 Clinical Implications

Tele-*Yoga* can protect us from unusual and unexpected events in the future by reducing the psycho-emotional side effects of COVID-19 or any other diseases which can emerge in future with ever-changing global environment and lifestyle. This technique may also be used as an effective non-pharmacological intervention for patients with chronic pain, muscular dystrophy, stroke survivors, and more, where patients are constraint to their home and are unable to attend physical sessions. Through tele-*Yoga* one can get sessions from renowned *Yoga* trainers all around the globe without travelling and following their routine lifestyle.

13.7 Future Directions for Research

The use of tele-*Yoga* has increased after COVID pandemic. However, the efficacy of tele-*Yoga* needs to be established, especially in comparison to physical *Yoga*. Although tele-*Yoga* is acceptable and has positive effects almost similar to physical *Yoga*, participants prefer physical *Yoga* (Brinsley et al. 2021). Randomized control trials in large sample size and in those diseases or disorders where role of physical *Yoga* is established may be used in determining benefits of tele-*Yoga*. Dua et al. have concluded from their narrative review that tele-yoga is feasible as well as beneficial and have shown to improve symptoms of dyspnea, stress, anxiety, and depression (Ruchi et al. 2023). It is essential to investigate it to understand whether tele-*Yoga* can replace physical *Yoga* or not. Many trials are designed to study the same (Rimmer et al. 2018; Singh et al. 2022; Zaccari et al. 2022; Balasubramanian et al. 2022; Nilkantham et al. 2023). The efficiency of tele-*Yoga* should also be evaluated in healthy participants to understand whether it is comparable to physical *Yoga*.

13.8 Conclusion

In summary, various studies including randomized controlled trials over the past decade suggest that tele-*Yoga* services can reduce symptoms of anxiety and depression, thereby improving sleep quality. With this, aspirants sitting anywhere in the world can easily get *Yoga* classes as well as personal training advice, thereby saving time as well as resources and giving health and wellness to the general public very easily. Apart from having unlimited benefits of tele-*Yoga*, it also has many limitations which makes it slightly less efficient than physical *Yoga* classes. However, following the AYUSH advisory should reduce the limitations and improve efficiency. Right now tele-*Yoga* research studies are short and have limited sample sizes, so there is a need for well-designed and adequately powered studies that include diverse form of *Yoga* practices via tele-*Yoga*.

References

Acar Y, Ilçin N, Sarı İ (2023) The effects of tele-Yoga in ankylosing spondylitis patients: a randomized controlled trial. J Complement Integr Med 29(11):727–737. https://doi.org/10.1089/jicm.2023.0022

Addington EL, Sohl SJ, Tooze JA, Danhauer SC (2018) Convenient and live movement (CALM) for women undergoing breast cancer treatment: challenges and recommendations for internet-based *Yoga* research. Complement Ther Med 37:77–79. https://doi.org/10.1016/j.ctim.2018.02.001

Anand A, Tyagi R, Khosla R, Bali P, Rain M, Maity K, Verma P, Gupta SJ, Nadholta P, Kaur N, Sharma K, Tripathi S, Avti P, Singh A (2021) Integrative approach to COVID-19: an Indian Facebook recipe for mental health. Ann Neurosci 28(3–4):183–190. https://doi.org/10.1177/09727531211052971

Balasubramanian S, Harper J, Sterba KR, Viswanathan R, Eldredge-Hindy H (2022) Protocol for the pilot study of group video yogic breathing app in breast cancer survivors. Int J Aayush Tradit Med 2(2):38–57

Barone Gibbs B, Kline CE, Huber KA, Paley JL, Perera S (2021) Covid-19 shelter-at-home and work, lifestyle and well-being in desk workers. Occup Med 71(2):86–94. https://doi.org/10.1093/occmed/kqab011

Bhargav H, Sharma S, Mahadevan J, Bokde R (2022) Feasibility of adjunct tele-yoga therapy in patients with substance use disorders. J Family Med Prim Care 11(6):3386–3387. https://doi.org/10.4103/jfmpc.jfmpc_2194_21

Brinsley J, Smout M, Davison K (2021) Satisfaction with online versus in-person *Yoga* during COVID-19. J Altern Complement Med 27(10):893–896. https://doi.org/10.1089/acm.2021.0062

Brunner D, Abramovitch A, Etherton J (2017) A *Yoga* program for cognitive enhancement. PLoS One 12(8):e0182366. https://doi.org/10.1371/journal.pone.0182366

Clark CJ, Lewis-Dmello A, Anders D, Parsons A, Nguyen-Feng V, Henn L, Emerson D (2014) Trauma-sensitive *Yoga* as an adjunct mental health treatment in group therapy for survivors of domestic violence: a feasibility study. Complement Ther Clin Pract 20(3):152–158. https://doi.org/10.1016/j.ctcp.2014.04.003

Dalpati N, Jena S, Jain S, Sarangi PP (2022) *Yoga* and meditation, an essential tool to alleviate stress and enhance immunity to emerging infections: a perspective on the effect of COVID-19 pandemic on students. Brain Behav Immun Health 20:100420. https://doi.org/10.1016/j.bbih.2022.100420

De' R, Pandey N, Pal A (2020) Impact of digital surge during Covid-19 pandemic: a viewpoint on research and practice. Int J Inf Manag 55:102171. https://doi.org/10.1016/j.ijinfomgt.2020.102171

Department of Integrative Medicine, National Institute of Mental Health and Neurosciences (NIMHANS), Swami Vivekananda *Yoga* Anusandhana Samasthana (S-VYASA) (2020) Advisory on Tele-*Yoga* Services. Central Council for Research in *Yoga* & Naturopathy (CCRYN), Ministry of AYUSH, Government of India, New Delhi. https://cdn.ymaws.com/www.iayt.org/resource/collection/B3DF2972-93AF-4A8E-9466-585C9A9F8F40/Final_Advisory_on_Tele-*Yoga*_Services_Aarti.pdf

Donesky D, Selman L, McDermott K, Citron T, Howie-Esquivel J (2017) Evaluation of the feasibility of a home-based Tele*Yoga* intervention in participants with both chronic obstructive pulmonary disease and heart failure. J Altern Complement Med 23(9):713–721. https://doi.org/10.1089/acm.2015.0279

Eyuboglu F, Inal-Ince D, Karamancioglu B, Vardar-Yagli N, Kaya-Utlu D, Dalkilic O, Somay G (2023) Effect of tele-yoga on aerobic capacity, respiratory muscle strength, and cognitive performance in patients with obstructive sleep apnea syndrome. Heart Lung 62:157–167. https://doi.org/10.1016/j.hrtlng.2023.07.005

Günebakan Ö, Acar M (2023) The effect of tele-*Yoga* training in healthy women on menstrual symptoms, quality of life, anxiety-depression level, body awareness, and self-esteem during COVID-19 pandemic. Ir J Med Sci 192(1):467–479. https://doi.org/10.1007/s11845-022-02985-0

Hegarty K, Tarzia L (2019) Identification and management of domestic and sexual violence in primary care in the #MeToo era: an update. Curr Psychiatry Rep 21(2):12. https://doi.org/10.1007/s11920-019-0991-6

Iyengar BK (1993) Light on the *Yoga* sutras of *Patanjali*. Aquarian/Thorsons

Jain Shrimal P, Maharana S, Dave A, Raghuram N, Thulasi A (2023) Efficacy of integrated tele *Yoga* intervention on physiological and psychological variables in asymptomatic COVID-19 positive patients: a randomized control trial. Complement Med Res 30(2):151–160. https://doi.org/10.1159/000528832

James-Palmer AM, Daneault JF (2022) Tele-*Yoga* for the management of Parkinson disease: a safety and feasibility trial. Digit Health 8:20552076221119327. https://doi.org/10.1177/20552076221119327

James-Palmer A, Anderson EZ, Daneault JF (2022) Remote delivery of *Yoga* interventions through technology: scoping review. J Med Internet Res 24(6):e29092. https://doi.org/10.2196/29092

Jasti N, Bhargav H, George S, Varambally S, Gangadhar BN (2020) Tele-*Yoga* for stress management: need of the hour during the COVID-19 pandemic and beyond? Asian J Psychiatry 54:102334. https://doi.org/10.1016/j.ajp.2020.102334

Joshi A, Kaur M, Arora S, Bhatt A, Sharma P, Kaur H, Kumar K, Arora M, Malhotra B, Anshuman A (2021) A pilot evaluation of Swasthya Pahal program using SMAART informatics framework to support NCD self-management. Mhealth 7:55. https://doi.org/10.21037/mhealth-20-110

Joyce C, Roseen EJ, Keysor JJ, Gross KD, Culpepper L, Saper RB (2021) Can *Yoga* or physical therapy for chronic low back pain improve depression and anxiety among adults from a racially diverse, low-income community? A secondary analysis of a randomized controlled trial. Arch Phys Med Rehabil 102(6):1049–1058. https://doi.org/10.1016/j.apmr.2021.01.072

Kashyap M, Rai NK, Singh R, Joshi A, Rozatkar AR, Kashyap PV, Mishra S, Mudda S (2023) Effect of early *Yoga* practice on post stroke cognitive impairment. Ann Indian Acad Neurol 26(1):59–66

Khadse PA, Gowda GS, Ganjekar S, Desai G, Murthy P (2020) Mental health impact of COVID-19 on police personnel in India. Indian J Psychol Med 42(6):580–582. https://doi.org/10.1177/0253717620963345

Khodadad B, Letafatkar A, Hadadnezhad M, Shojaedin S (2020) Comparing the effectiveness of cognitive functional treatment and lumbar stabilization treatment on pain and movement control in patients with low back pain. Sports Health 12(3):289–295. https://doi.org/10.1177/1941738119886854

Kwok JYY, Kwan JCY, Auyeung M, Mok VCT, Lau CKY, Choi KC, Chan HYL (2019) Effects of mindfulness *Yoga* vs stretching and resistance training exercises on anxiety and depression for people with Parkinson disease: a randomized clinical trial. JAMA Neurol 76(7):755–763. https://doi.org/10.1001/jamaneurol.2019.0534

Lai KSP, Watt C, Ionson E, Baruss I, Forchuk C, Sukhera J, Burhan AM, Vasudev A (2020) Breath regulation and yogic exercise an online therapy for calm and happiness(BREATH) for frontline hospital and long-term care home staff managing the COVID-19 pandemic: a structured summary of a study protocol for a feasibility study for a randomised controlled trial. Trials 21(1):648. https://doi.org/10.1186/s13063-020-04583-w

Majumdar V, Manjunath NK, Nagarathna R, Panigrahi S, Kanchi M, Sahoo S, Nagendra HR, Giridharan A, Reddy M, Nayak R (2023) Adjunct tele-yoga on clinical status at 14 days in hospitalized patients with mild and moderate COVID-19: a randomized control trial. Front Public Health 11:1054207. https://doi.org/10.3389/fpubh.2023.1054207

Malhotra V, Sampath A, Javed D, Bharshankar R, Mishra S, Singh V, Singh D, Kulkarni A, Gautam N, Rastogi R (2022) *Yoga* as an escape from depreciating mental health due to COVID 19: a qualitative study analyzing the factors associated with mental status based on the experiences of geriatric population's participation in an online program during COVID 19 lockdown in India. Int J Yoga 15(3):230–239

Malik S, Dua R, Krishnan AS, Kumar S, Kumar S, Neyaz O, Bhadoria AS (2022) Exercise capacity in patients with chronic obstructive pulmonary disease treated with tele-*Yoga* versus tele-pulmonary rehabilitation: a pilot validation study. Cureus 14(11):e30994. https://doi.org/10.7759/cureus.30994

Mathersul DC, Mahoney LA, Bayley PJ (2018) Tele-*Yoga* for chronic pain: current status and future directions. Global Adv Health Med 7:2164956118766011. https://doi.org/10.1177/2164956118766011

Mazza M, Marano G, Lai C, Janiri L, Sani G (2020) Danger in danger: interpersonal violence during COVID-19 quarantine. Psychiatry Res 289:113046. https://doi.org/10.1016/j.psychres.2020.113046

Metri K, Patra S, Ramakrishna KK, Salvi K, Naik J, Nagaratna R (2021) Management of acute calculus cholecystitis with integrated *Ayurveda* and *Yoga* intervention: a case report. J Ayurveda Integr Med 12(1):187–190. https://doi.org/10.1016/j.jaim.2020.12.014

Mishra AS, Sk R, Hs V, Nagarathna R, Anand A, Bhutani H, Sivapuram MS, Singh A, Nagendra HR (2020) Knowledge, attitude, and practice of *Yoga* in rural and urban India, KAPY 2017: a nationwide cluster sample survey. Medicines (Basel) 7(2):8. https://doi.org/10.3390/medicines7020008

Monaghesh E, Hajizadeh A (2020) The role of telehealth during COVID-19 outbreak: a systematic review based on current evidence. BMC Public Health 20(1):1193. https://doi.org/10.1186/s12889-020-09301-4

More NS, Bapat U, Das S, Alcock G, Patil S, Porel M, Vaidya L, Fernandez A, Joshi W, Osrin D (2012) Community mobilization in Mumbai slums to improve perinatal care and outcomes: a cluster randomized controlled trial. PLoS Med 9(7):e1001257. https://doi.org/10.1371/journal.pmed.1001257

Nagarathna R, Anand A, Rain M, Srivastava V, Sivapuram MS, Kulkarni R, Ilavarasu J, Sharma MNK, Singh A, Nagendra HR (2021) *Yoga* practice is beneficial for maintaining healthy lifestyle and endurance under restrictions and stress imposed by lockdown during COVID-19 pandemic. Front Psychiatry 12:613762. https://doi.org/10.3389/fpsyt.2021.613762

Nagendra HR (2020) *Yoga* for COVID-19. Int J Yoga 13(2):87–88. https://doi.org/10.4103/ijoy.IJOY_27_20

Najafi P, Hadizadeh M, Cheong JPG, Mohafez H, Abdullah S, Poursadeghfard M (2023a) Effects of tele-Pilates and tele-*Yoga* on biochemicals, physical, and psychological parameters of females with multiple sclerosis. J Clin Med 12(4):1585. https://doi.org/10.3390/jcm12041585

Najafi P, Hadizadeh M, Cheong JPG, Motl RW, Abdullah S, Mohafez H, Poursadeghfard M (2023b) Effects of tele-exercise training on physical and mental health and quality of life in multiple sclerosis: do the effects differ by modality and clinical disease course? Mult Scler Relat Disord 80:105129. https://doi.org/10.1016/j.msard.2023.105129

Natarajan S, Anbarasi C, Sathiyarajeswaran P, Manickam P, Geetha S, Kathiravan R, Prathiba P, Pitchiahkumar M, Parthiban P, Kanakavalli K, Balaji P (2021) Kabasura Kudineer (KSK), a poly-herbal Siddha medicine, reduced SARS-CoV-2 viral load in asymptomatic COVID-19 individuals as compared to vitamin C and zinc supplementation: findings from a prospective, exploratory, open-labeled, comparative, randomized controlled trial, Tamil Nadu, India. Trials 22(1):623. https://doi.org/10.1186/s13063-021-05583-0

Nilkantham S, Majumdar V, Singh A (2023) Scientific yoga module for hypothyroidism: a study protocol for tele-yoga RCT. Contemp Clin Trials Commun 33:101157. https://doi.org/10.1016/j.conctc.2023.101157

Pandurangi AK, Keshavan MS, Ganapathy V, Gangadhar BN (2017) *Yoga*: past and present. Am J Psychiatry 174(1):16–17. https://doi.org/10.1176/appi.ajp.2016.16080853

Pérez G, de Sevilla G, Barceló Guido O, De la Cruz MP, Blanco Fernández A, Alejo LB, Montero Martínez M, Pérez-Ruiz M (2021) Adherence to a lifestyle exercise and nutrition intervention in university employees during the COVID-19 pandemic: a randomized controlled trial. Int J Environ Res Public Health 18(14):7510. https://doi.org/10.3390/ijerph18147510

Phansikar M, Gothe N, Hernandez R, Lara-Cinisomo S, Mullen SP (2023) Feasibility and impact of a remote moderate-intensity *Yoga* intervention on stress and executive functioning in working adults: a randomized controlled trial. J Behav Med 46:720. https://doi.org/10.1007/s10865-022-00385-4

Prathikanti S, Rivera R, Cochran A, Tungol JG, Fayazmanesh N, Weinmann E (2017) Treating major depression with *Yoga*: a prospective, randomized, controlled pilot trial. PLoS One 12(3):e0173869. https://doi.org/10.1371/journal.pone.0173869

Rain M, Puri GD, Bhalla A, Avti P, Subramaniam B, Kaushal V, Srivastava V, Mahajan P, Singh M, Pandey N, Malhotra P, Goel S, Kumar K, Sachdeva N, Maity K, Verma P, Dixit N, Gupta SJ, Mehra P, Nadholta P, Khosla R, Ahuja S, Anand A (2022) Effect of breathing intervention in patients with COVID and healthcare workers. Front Public Health 10:945988. https://doi.org/10.3389/fpubh.2022.945988

Rimmer JH, Thirumalai M, Young HJ, Pekmezi D, Tracy T, Riser E, Mehta T (2018) Rationale and design of the tele-exercise and multiple sclerosis(TEAMS) study: a comparative effectiveness trial between a clinic- and home-based telerehabilitation intervention for adults with multiple sclerosis(MS) living in the deep south. Contemp Clin Trials 71:186–193. https://doi.org/10.1016/j.cct.2018.05.016

Ruchi D, Saloni M, Ramesh K, Indrajeet S, Sunil K, Anissa MA, Girish S (2023) Effect of Teleyoga before COVID-19 and during pandemic: a narrative review. Recent Adv Antiinfective Drug Discov 18(3):178–187. https://doi.org/10.2174/2772434418666230223104608

Sarah S, Wolfgang MB, Claudia P (2019) Effect of telerehabilitation on long-term adherence to *Yoga* as an antihypertensive lifestyle intervention: results of a randomized controlled trial. Complement Ther Clin Pract 35:148–153. https://doi.org/10.1016/j.ctcp.2019.02.001

Selman L, McDermott K, Donesky D, Citron T, Howie-Esquivel J (2015) Appropriateness and acceptability of a tele-*Yoga* intervention for people with heart failure and chronic obstructive pulmonary disease: qualitative findings from a controlled pilot study. BMC Complement Altern Med 15:21. https://doi.org/10.1186/s12906-015-0540-8

Sharma P, Khokhar A (2022) Domestic violence and coping strategies among married adults during lockdown due to coronavirus disease (COVID-19) pandemic in India: a cross-sectional study. Disaster Med Public Health Prep 16(5):1873–1880. https://doi.org/10.1017/dmp.2021.59

Sharma K, Anand A, Kumar R (2020a) The role of *Yoga* in working from home during the COVID-19 global lockdown. Work 66(4):731–737. https://doi.org/10.3233/WOR-203219

Sharma N, Modi D, Nathwani A, Pandya B, Joshi J (2020b) Tele-*Yoga* therapy for patients with chronic pain during COVID-19 lockdown: a prospective nonrandomized single arm clinical trial. medRxiv 2020–07. https://doi.org/10.1101/2020.07.16.20154229

Sharma N, Sahni PS, Sharma US, Kumar J, Garg R (2022a) Effect of *Yoga* on the stress, anxiety, and depression of COVID-19-positive patients: a quasi-randomized controlled study. Int J Yoga Therap 32(2022):Article 8. https://doi.org/10.17761/2022-D-22-00013

Sharma SK, Telles S, Gandharva K, Balkrishna A (2022b) *Yoga* instructors' reported benefits and disadvantages associated with functioning online: a convenience sampling survey. Complement Ther Clin Pract 46:101509. https://doi.org/10.1016/j.ctcp.2021.101509

Singh J, Jha M, Metri K, Mohanty S, Singh A, Tekur P (2022) A study protocol for a randomised controlled trial on the efficacy of *Yoga* as an adjuvant therapy for patients with ankylosing spondylitis amidst COVID-19 pandemic. Adv Integr Med 9(1):75–79. https://doi.org/10.1016/j.aimed.2021.11.001

Singh J, Metri K, Tekur P, Mohanty S, Singh A, Raghuram N (2023) Tele-*Yoga* in the management of ankylosing spondylitis amidst COVID pandemic: a prospective randomized controlled trial. Complement Ther Clin Pract 50:101672. https://doi.org/10.1016/j.ctcp.2022.101672

Snyder S, Silva RF, Whisenant MS, Milbury K (2021) Videoconferenced *Yoga* interventions for cancer patients and their caregivers during the COVID-19 pandemic: a report from a clinician's perspective. Integr Cancer Ther 20:15347354211019111. https://doi.org/10.1177/15347354211019111

Stevens K, McLeod J (2019) *Yoga* as an adjunct to trauma-focused counselling for survivors of sexual violence: a qualitative study. Br J Guid Counsel 47(6):682–697. https://doi.org/10.1080/03069885.2018.1472368

Stożek J, Rudzińska M, Pustułka-Piwnik U, Szczudlik A (2016) The effect of the rehabilitation program on balance, gait, physical performance and trunk rotation in Parkinson's disease. Aging Clin Exp Res 28(6):1169–1177. https://doi.org/10.1007/s40520-015-0506-1

Strömberg A, Thylén I, Orwelius L, Klompstra L, Jaarsma T (2021) Tele-*Yoga* in long term illness-protocol for a randomised controlled trial including a process evaluation and results from a pilot study. Int J Environ Res Public Health 18(21):11343. https://doi.org/10.3390/ijerph182111343

Tolahunase MR, Sagar R, Faiq M, Dada R (2018) *Yoga*- and meditation-based lifestyle intervention increases neuroplasticity and reduces severity of major depressive disorder: a randomized controlled trial. Restor Neurol Neurosci 36(3):423–442. https://doi.org/10.3233/RNN-170810

Tracy TF, Young HJ, Lai B, Layton B, Stokes D, Fry M, Mehta T, Riser ES, Rimmer J (2022) Supporting successful recruitment in a randomized control trial comparing clinic and home-based exercise among adults with multiple sclerosis. Res Involv Engag 8(1):35. https://doi.org/10.1186/s40900-022-00366-6

Vora M, Malathesh BC, Das S, Chatterjee SS (2020) COVID-19 and domestic violence against women. Asian J Psychiatry 53:102227. https://doi.org/10.1016/j.ajp.2020.102227

Wells RE, O'Connell N, Pierce CR, Estave P, Penzien DB, Loder E, Zeidan F, Houle TT (2021) Effectiveness of mindfulness meditation vs headache education for adults with migraine: a randomized clinical trial. JAMA Intern Med 181(3):317–328. https://doi.org/10.1001/jamainternmed.2020.7090

WHO (2022) Mental disorders. https://www.who.int/news-room/fact-sheets/detail/mental-disorders. Accessed 26 Aug 2022

WHO (2023) Rehabilitation. https://www.who.int/news-room/fact-sheets/detail/rehabilitation. Accessed 3 Mar 2023

Yadav R, Yadav RK, Khadgawat R, Pandey RM, Upadhyay AD, Mehta N (2020) Randomized controlled trial of a 12-week *Yoga*-based (including diet) lifestyle vs. dietary intervention on cardio-metabolic risk factors and continuous risk score in Indian adults with metabolic syndrome. Behav Med 46(1):9–20. https://doi.org/10.1080/08964289.2018.1538098

Yaşar MF, Yakşi E, Kurul R, Balcı M (2021) Comparison of outpatient and inpatient spa therapy in knee osteoarthritis. Int J Biometeorol 65(10):1675–1682. https://doi.org/10.1007/s00484-021-02122-z

Zaccari B, Loftis JM, Haywood T, Hubbard K, Clark J, Kelly UA (2022) Synchronous telehealth *Yoga* and cognitive processing group therapies for women veterans with posttraumatic stress disorder: a multisite randomized controlled trial adapted for COVID-19. Telemed J E Health. https://doi.org/10.1089/tmj.2021.0612

Chapter 14
Patanjali's RajYoga Meditation: An Indian Classical Intervention Technique

Pavitra Singh, Pradeep Kumar, and Krishan Kumar

Contents

Abstract Meditation helps in exhausting routine stress and spreading inner peace. Numerous meditation techniques are practiced worldwide, one of them is *RajYoga*, the oldest but latest form of meditation, where Raj means "supreme or highest" and *Yoga* means "connection" (Raj S, Tiwari R, Int Res J Eng Technol. 6:4206–4210, 2019). In *RajYoga* meditation, there is a connection with supreme power which enlightens the person to come across his latent potentialities. *Yoga* is considered a therapy, which leads to a healthy lifestyle harmony and ensures mental peace. *Yoga* is one of the Indian classical philosophies and Sage *Patanjali* was the first one to write a text on it, he systematically codified the *Yoga Sutra* (i.e., aphorism in *Sanskrit*) emphasizing the eight steps of *Yoga*, also called *Asthanga Yoga* or *RajYoga* (Prabhavananda S, Isherwood C, *Patanjali Yoga* sutras: translated with a new commentary. Sri Ramakrishna Math, Chennai, 1948). Eight steps are explained in terms of ***Yama, Niyama, Asana, Pranayama*, Pratyahara, *Dharana, Dhyana*, and**

P. Singh · P. Kumar (✉)
Department of Psychology, Central University of Haryana, Pali, Haryana, India
e-mail: pradeeppsy@cuh.ac.in

K. Kumar
Department of Psychiatry, Post Graduate Institute of Medical Education and Research, Chandigarh, India

A. Anand (ed.), *Neuroscience of Yoga*,
https://doi.org/10.1007/978-981-97-2851-0_14

***Samadhi*. It is realized that the eight steps of meditation may be practiced in a tailored form.** The ancient system of *Yoga* may be used to make it applicable to the general population for finding the solution to mental health problems (Balodhi JP, NIMHANS J. 4:133–138, 1986). **Therefore,** this chapter includes a description of eight steps along with detailed commentaries. Every step is explained with the help of simple exercises. These commentaries are helpful for laypersons to understand and practice the steps of *RajYoga* Meditation. Hence, it may be helpful as an intervention technique.

Keywords Meditation · Yoga · RajYoga meditation · Asthanga Yoga · Patanjali Yoga Sutra

14.1 Introduction

If we look at our lives, we are living a life of emotional dependence, where we blame people and situations for how we are feeling. From the past two to three decades people are running the marathon of their life even without training themselves, which leads to emotional drainage and ultimately collapse. To improve the quality of life in times of adversity, there is a need to blend happiness, spiritual awareness, and emotional intelligence (Mayer 2000). An emotional immunity system has to be built up to protect from invisible crises going inside, meditation can play a major role in it. *David Fink* described the term "Average Middle-Class Neurotic." It was said that the normal population falls into this category, as most of the time there is a feeling of restlessness, tension, fear, forgetting of important things, difficulty in sleeping at times, and experience of palpitation, etc., where *RajYoga* meditation alone can tremendously help in overcoming these symptoms (Naragatti and Hiregoudar 2019). There are few Organizations or Meditation Centers where *RajYoga* Meditation is practiced in tailored form; Brahma Kumaris World Spiritual University, The Divine Life Society by *Swami Sivananda Maharaj, Sahaj Marg-Raja Yoga* by *Shri Ram Chandra Mission,* etc.

Among the oldest known texts, *Rigveda* where the concept of *Yoga* was mentioned and refined in the Ancient Scripture of India *Shrimad Bhagavad Gita,* different forms of *Yoga* were described, namely, *Jnana Yoga, Bhakti Yoga, Karma Yoga, and RajYoga* (Raj 2018). Even Sage *Patanjali* discussed *RajYoga/Asthanga Yoga* in his sacred text containing 196 Aphorisms in *Sanskrit* known as *Yoga Sutras. Yoga sutras* are bifurcated into four chapters, namely, *Samadhi Pada* (51 *Sutras*) teaches about the definition, purpose, and obstacles in *Yoga*, the importance of practice, and detachment. *Sadhana Pada* (55 *Sutras*) explains techniques to reach the ultimate goal of *Yoga* through the practice of eight steps/limbs, this chapter talks about the first five steps and next chapter *Vibhuti Pada* (56 *Sutras*) discusses the remaining three steps, *dharana*, *dhyana*, and *samadhi*. The last *Kaivalya Pada* (34 *Sutras*) explains ultimate liberation (Saraswati and Saraswati 2002).

The eight limbs of *Yoga* are designed to discipline lifestyle, mind, and body (Rao 2012). These eight steps/limbs are as follows: (a) ***Yama*: includes** five social disciplines, considered as a code of restraint and self-regulation which are related to the outer world, namely, *Ahimsa* (not to harm mentally, verbally, and physically), *Satya* (to remove the layers of untruth, mental and audible speech to be cohesive), *Asteya* (not to obtain anything unfairly or to gain the benefit at anyone else's expense), *Brahmacharya* (to live a disciplined life and to moderate sexual impulses), and *Aparigraha* (being non-possessive and stepping toward detachment). (b) ***Niyama:* also includes** five personal disciplines or the binding rules for self which focus on training the body and mind. These are, namely, *Saucha* (mind and body hygiene, purer the thoughts, so it is the mind, similarly purer the food, so is the body), *Santosha* (contentment and acceptance of reality in every form, and not to react when the situation does not fit our expectations), *Tapas* (austerity and commitment to the self for crossing all physical and mental hardship to accomplish the goal), *Svadhyaya* (study of self, pathway for personal growth), and *Ishvara Pranidhana* (surrendering toward divinity and perceiving divinity in everything). *Yoga*'s first two limbs, *Yamas* and *Niyamas*, are like interwoven threads of righteous behavior and mindfulness. (c) ***Asana*: in** *Sanskrit* means a posture, here it is related to meditation posture, which makes the body fit for long sitting. (d) ***Pranayama*:** suggests breathing exercises with awareness. There are four different ways to practice *Pranayama*, pause after breathing inside and outside, focus during the inhalation, and exhalation. (e) ***Pratyahara:*** focuses on the detachment of senses from the outer world, their images, and impressions in the mind. (f) ***Dharana*** is the concentration or focus of the mind on one fixed object. (g) ***Dhyana*:** is much more than meditation; it is sustained and continuous attention. (h) ***Samadhi*: means freedom from all kinds of suffering, it is the highest state of meditation** (Prabhavananda and Isherwood 1948; Rao 2012). These are the eight steps to be followed systematically, as they serve as the guidelines for living a meaningful and purposeful life.

14.2 Psychophysiological Rationale to Treat the Disorders

RajYoga Meditation is an esoteric and useful form of meditation that enables one to live a healthy and happy life. It can be used for a variety of physical and mental well-being issues and is an effective tool for conventional therapy (Shanker et al. 2018). Additional advantages of *RajYoga* Meditation are it harmonizes breathing and normalizes pulse rate, it lowers medication dosage, psychosomatic disease symptoms, and hypertension (Vahia et al. 1973), anxiety (Datey et al. 1969), and psychoneurotic conditions (Kiran et al. 2017a). Blood lactate levels drop, Systolic blood pressure returns to normal, immunity rises, and muscle tension decreases (Olson 2013). It was even discovered that *RajYoga* meditation causes neural plasticity in the brain (Goyal et al. 2019). It modifies the cortisol level in surgical patients (Kiran et al. 2017b). Another study discovered that *RajYoga* is a successful treatment to help patients with their obsessions and compulsions (Varambally and

Gangadhar 2012; Mehta et al. 2020). A structured yogi counseling therapy based on *Patanjali*'s Eight-Limbed Model is required for various psychiatric illnesses (Singh and Pandey 2015), it even improves sleep quality and subjective well-being (Singh and Kumar 2022). It is the need of the hour to make meditation therapy based on *Patanjali*'s *Yoga Sutras*, including all the eight components of *Yoga* in meditative form. It is important to simplify the ancient system of *Yoga* so that it can be applied to find solutions to mental health problems (Balodhi 1986).

14.3 Techniques

Meditation is a skill that can be acquired, no one is born with an aptitude for meditation. To learn any skill, two essential components must be present, i.e., Instructions and Practice. It is essential to gain mastery over instructions, through practice. *Patanjali*'s *Yoga Sutras* provide the best theoretical framework for meditation practice. *Patanjali* lays out a systematic approach to control the mind's fluctuations and for appropriate connections. To build a strong foundation and to move toward enlightenment, this eight-limbed meditative model can be practiced in the form of short exercises in daily routines.

There are ten steps in the framework of the meditation program, which are as follows:

S. no.	Themes	Techniques
1.	Initiation[a]	*Om* chanting
2.	*Yamas*	Meditation
3.	*Niyamas*	Meditation
4.	*Asana*	Physical exercise with meditation
5.	*Pranayama*	Breathing exercise with meditation
6.	*Pratyahara*	Meditation
7.	*Dharana*	Meditation
8.	*Dhyana*	Meditation
9.	*Samadhi*	Meditation
10.	Realization[a]	Silence

[a] Themes are not part of the eight limbs; only steps of starting (initiation) and ending (realization) the intervention technique

14.3.1 Initiation

To initiate meditation practice, begin with the sound of OM. OM is more than just the invitation to begin a meditation practice, it is a primordial sound that is mentioned in *Hindu* Philosophy (Alpert 2015). It is a great practice that energies the body and mind at the same time.

Start the journey with *OM*. Take a deep breath and relax. Sit comfortably, and keep your hands over your lap with one hand resting on the other hand. Close your eyes gently and relax. Break down the *AUM* syllable to A-A-U-U-U-M-M followed by silence and back again. Take a deep breath inhale deeply and then while releasing the breath Say "A" and "U" with an open mouth and next "M" pursing the lips together. Continue with OM chanting keeping it in sync with your breath relax and try to note all the thoughts and sensations that arise while you chanted OM (Alpert 2015).

14.3.2 Building a Strong Foundation Before the Journey Inside

***Yamas* and *Niyamas*:** The first two limbs of *Yoga* are *Yamas* and *Niyamas*, whereas *Yamas* are the avoidance of immoral acts and *Niyamas* are the performance of moral disciplines. For example, if a player gets involved in conflicts or disagreements right before a match, the argument is likely to be forgotten as he concentrates on the game or it may affect the performance slightly. However, if a meditator has some disagreement right before sitting to practice, the effects of that argument will be noticeable certainly. You can't meditate if your mind is distracted or preoccupied somewhere else. Meditation is an introspective and subtle activity, unlike any game.

Here are some practices to inculcate the yamas and niyamas in daily routine which are compiled together:

14.3.2.1 YAMAS-Ahimsa

Ahimsa means nonviolence. Ahimsa is when the person is free from the thought of him even at the thought level, as the thoughts become words; words become action. So, we need to work on our thoughts and the best way to imbibe the virtue of peace is through affirmations, i.e., by transforming the negative thoughts into positive ones. The more we embrace nonviolence more we emit the vibrations of peace, love, and compassion (Vigneswar 2021).

Exercise I. Practice *Ahimsa* (SCC: Stop Check and Change):

Stop: Reflect upon any incident that disturbed you today or within a week, anything undesirable happened where you felt your response was not appropriate. Any mild disagreement with someone. Relive that scene again for a few minutes and then immediately stop. Stop and move to the next step for checking and realization.

Check: Enter that scene as a third person and mentally observe the scene from a distance. Reflect upon it, do you want to change something? Was everything ok…? Can that incidence be crossed with a different response?

Change: Mentally transform your negative response into a constructive one with positive affirmations, of what you want to be. There are no fixed affirmations for practicing, it may be part of our faith or may come from any spiritual tradition. Allow yourself to sink deeply into positive affirmations and program your mind.

SCC does not change the situation, nor does it influence the response of another person, the only change it can bring is a change in the attitude or feelings of the practitioner. They feel much lighter and relaxed, and it makes them realize that they have the power to transform their thoughts (The Janki Foundation for Global Health Care 2007).

Satya: *Satya* means truthfulness. Audible and mental speech must be cohesive in truth. There are various layers on people, the layer of untruth, the layer of vices, and the layer of ignorance. So, we need to introspect and go deeper to connect with our eternal being. As mentioned in Sankhya Philosophy there are two principles of nature *Prakriti* (Body/Matter) and Purusha (Pure Consciousness/Soul), we consider the body as permanent and live life that way but the truth is the body will die one day, whereas we give less importance to the important factor, i.e., soul. We need to raise our consciousness and be in the awareness of Purusha which is the ultimate truth.

Exercise II. **Practice *Satya***: Take the journey inside by concentrating on consciousness. Remove the layer of untruth and the layer of ignorance.

Imagine that nothing exists outside this room. You are insulated from the outside world and ready to explore your inner world. Turn all your attention inward and concentrate your consciousness. Feel the detachment from the physical environment and physical body. Feel the natural stillness and peacefulness surrounding you. Visualize the waves of peace gently washing over you and removing the layers of ignorance, untruth, and vices. Feel that peace, just peace all around.

Asteya: *Asteya* means not to steal and *steya* means mental theft when someone seeks to gain any benefit at anyone else's expense. Stealing is much more subtle, it starts with cognition and ends with action. At the cognitive level, if someone unnecessarily spends too much time seeking approval or comfort from others or creating drama over minor concerns, that person is stealing the energy or attention of family or friends and engaging in emotional theft. The one who misuses or overuses the resources of mother earth is stealing the sacred resources from other mankind is also considered theft. One should practice *Asteya* to become a yogi, as it increases concentration.

Exercise III. **Practice**: Contemplate the things you have possessed and which are not yours. Release the thought of holding or possessing the things of others. Trust the law of the universe whatever comes itself, let it come, and now reflect upon the following thoughts: free yourself from all the unnecessary desires and detach yourself from everything. Feel contented with what you have, all your relationships, your career, and your life. Everything is exactly as it's meant to be. Radiate peace and power, as you are on the journey to empower yourself.

Brahmacharya: The person who wants to reach the highest consciousness should move beyond the consciousness of the body. One should not seek fleeting pleasure by overindulging the sense and should know how to moderate and regulate sexual impulses.

Exercise IV. **Practice *Brahmacharya* (Inside Out)**: Practice the exercise of focusing and defocusing to move beyond the consciousness of the body. Bring all your attention inward, visualize a point of light in the brain, and feel the detachment from the outside world. Concentrate on one point and feel your existence inside. After a few seconds try to defocus by putting all the attention outward. Visualize the light on tip of the nose and try to see the body outwardly and feel the external environment and activate your external senses. Repeat the exercise for a few minutes. This practice will help in bringing the realization of soul and body separately.

Aparigraha: *Parigraha* means possession. When the person tries to possess everything like money, clothes, people, status, etc., gradually the feeling of attachment arises and which hinders the journey of *Yoga*. Every tangible thing is destructible, even this body too we need to realize this and should practice detachment from everything we have.

Exercise V. **Practice *Aparigraha*** **(Rumination with Personification)**: Choose a bird as an object for personification. See your soul as a bird, in the garden. Birds are free, they don't have a fixed place to live. They fly away wherever they want to.

See yourself as a bird sitting on a branch. Tag all the unnecessary thoughts to the branch…the fear of loss, attachment, disappointment, and anxiety. List down these thoughts and introspect upon them.

If you want to fly, you can't till the time you don't leave that branch. Ask yourself, do you want to detach yourself? If yes, then make a firm thought, spread the wings of detachment, feel the happiness and lightness inside, flap your wings and fly.

See yourself flying high, leaving the branch and whole tree beneath, giving up everything that is not required, flying symbolizes giving upon everything that weighs you down, don't hold yourself with mental limitations, you are a free bird, see yourself flying high above in the sky, over the rivers, detached and free, feel it and live in it.

Now slowly begin to deepen your breath. As you become aware of your surrounding come back…and reenter the world with a new perspective, refreshed and renewed.

14.3.2.2 NIYAMAS

Let's begin with a meditation on *Niyamas*. Disconnect yourself from everything around you and practice inculcating new virtues.

Saucha: *Saucha* means to clean, cleaning of mind and body. Over consumption of negative impressions pollutes the mind, the same way that overeating and contaminated food pollutes the body, and both are to be taken care of. It is said that you are what you think, thoughts are the food for the mind, the purer the information purer the thoughts, and purer the mind, similarly purer the food purer the body.

Exercise VI. Practice *Saucha*: Begin the meditation by sitting in a comfortable position, back straight and shoulder rolled back. Take a deep breath while expanding the stomach and ribcage fully, hold your breath to the top and feel the vibrations in the body and exhale deeply. Repeat and try to expand more widely releasing all the tension and worry with the breath. Turn your focus inward and give yourself a mini-vacation, traveling to a place where all is your breath.

Now, imagine a beautiful scene inside your head, a scenery of a sky with clouds, where clouds symbolize thoughts and a clear blue sky symbolizes a clean mind. As the thoughts pass by, push them aside or allow them to pass it through and try to keep the sky clean as possible. Do not judge or label your thoughts as they pass through, just focus all your awareness on your breath and the clean sky. Feel grateful when you show up with a clear mind. The more you appreciate more it is easy in the future to practice. Take the clarity of your mind with you and open your eyes back to the world around you, when you feel ready.

Santosha: *Santosha* means contentment. Contentment is learning to understand and accept reality for what it is and then responding wisely rather than reacting when reality does not fit our expectations. A complaining and dissatisfied person cannot build a harmonious connection with anyone. Only the self-contented person builds deep connections with the right people at right time. Much energy is saved when I accept everything on this earth including myself. Gratitude and acceptance are to be practiced day in and day out. Breath in peace and acceptance and breath out gratitude.

Exercise VII. Practice *Santosha*: Sit comfortably and take a deep breath and disconnect yourself from everything around you and practice inculcating the new virtue of *Santosha*, i.e., satisfaction with whatever you have. Check up on your thoughts and see your thoughts coming and crossing over your mind. Take a deep breath and breath in peace and acceptance and exhale out all the complaints and discontentment that you have held till now.

Visualize the red light of love and acceptance, spreading throughout your body and up to the whole universe. Visualize the list of people and things you want to thank and keep passing this ball of red light to them. Feel the level of peace and lightness at the moment when you accept everyone and everything just as it is. Realize everything is exactly right how it is meant to be.

Tapas: *Tapas* derives from the word Tap, which means "to apply one's heart," to achieve the goal by crossing all the physical and mental hardship. Tapas is like an inner flame that burns away all the impurities and transforms knowledge into wisdom. Practice believing in your great power and ability to take control of your life and acknowledge your fullest potential.

Exercise VIII. Practice *Tapas*: Take a deep breath and relax. It's time for commitment, commit to bearing all the hardships on this journey, either at a physical or cognitive level.

Visualize the flame burning on your hand. Feel the warmth of it and also feel the tingling sensation over the hand. Commit to yourself for tolerating every circumstance, whether it may be cold, hot, hunger, thirst, praise or insult, or any kind of difficulty it may be, and to continue with the journey inside till the time you have committed. Believe you have the power to take control of your life and your thoughts, and you can do it for sure.

Svadhaya: *Svadhyaya* means self-study, a pathway to self-awareness or personal growth. The only way to turn knowledge into wisdom is by using knowledge in our life and living with it. Practice, choosing your source of wisdom, and commit to reflecting on it and exploring it deeply. When the old pattern of awareness arises transform them with reflection and insight.

Exercise IX. Practice *Svadhyaya*: Take a deep breath and relax. It's time to introspect and know yourself more. Try to identify and list your strengths and weaknesses. Your strengths will provide you with the right direction to find your true passion and purpose. Ask yourself can you use these strengths and make a difference in the world around you? What motivates you? List down the things which take you to the flow state or the things you are naturally good at. Reflect and introspect.

Exercise IX. a-Scan your day: Reflect upon your day till now and while doing this accept all the challenges of the day that came across or that are going to come, introspect on the things that went wrong and prepare yourself for the next day. Take a deep breath inhale peace and exhale the experiences of the day that no longer serve you.

Ishvara Pranidhana: *Ishvara* means supreme *Purusha* (being) or supreme cosmic energy and *pranidhana* means to surrender. *Ishvara pranidhana* one has to practice surrendering to the divine and seeing the divinity in everything, moreover feeling it within yourself too. When there is awareness of divine existence, it becomes easy to embrace non-harming, non-stealing, purity of mind, contentment, non-possessiveness, truthfulness, etc.

Exercise X. Practice *Ishvara Pranidhana*: Visualize yourself as a divine being and see a star right in front of you, full of divinity and bright light. Surrender all your fear, anxiety, and tensions to that divine star in the form of light, and thank him for this human body, mind, family, friends, and profession, and thank him for this life and existence. Surrender all the burden of this life to him and visualize the bright white light coming toward you from that star that showers you with love and power. Feel the lightness as if you are fully charged. Be in it and live with it.

14.3.2.3 Asana

Asana means posture or seat. *Asana* is a steady and comfortable posture, taking a seat crossed-legged on the ground for introspection and meditation. It is an act of holding still, being silent, and looking within. To make the body relax let's remove blockages that allow the free flow of energy in the body and mind.

Exercise XI. Loosening up Drill: There is a list of a few exercises clubbed together to be performed with imagery meditation, i.e., visualizing the body part with white/red light and feel as if it is being charged and immune with these loosening up exercises. All these exercises are based on *Sukshma Vyayama,* an ancient yogic posture technique introduced by *Swami Dhirendra Brahmachari*, a student of *Maharishi Kartika.* They are considered a powerful exercise that can activate the *pranic* body, the circuit through which the *prana* flows (Krishnan and Singh 2016).

Vision 360°: Eyes Movements based on *Nethra Sanchalana*: Sit in a comfortable posture and relax. Blink your eyes a few times. Move the eyes up and down, left and right, from the left top corner to the right down corner and from the right top corner to the left down corner without moving the head. Now move the eyes in a circular motion both clockwise and anti-clockwise (Krishnan and Singh 2016). Repeat the practice 4–5 times. Rub both the palm and visualize the energy being created by each rub and put them on the eyes. Feel as if your eyes have been charged.

Head-2-Shoulder: Neck Exercise based on *Greeva Sanchalana*: Sit in a comfortable posture and relax. Slowly move the head down as much as possible, tense the muscles, and move the head back for a few seconds and feel the stretch. Inhale while moving the head backward and exhale while moving forward. In the same manner, move the head to the right and touch the ear to the right shoulder and then to the left shoulder. Now move the head clockwise and anti-clockwise (Krishnan and Singh 2016). Repeat the practice 4–5 times. This exercise releases tension, heaviness, and stiffness in the head, neck, and shoulder. Visualize the white light over your neck while doing the exercise and feel as if your neck has been charged.

Warmup Arms: Shoulder Rotation based on *Skandha Chakra*: Sit in a comfortable position either a base position with outstretched legs or *sukhasana.* Place the right-hand fingers on the right shoulder and the left fingers on the left shoulder. Stretch the elbows outwardly and move upward, downward, forward, and backward. Then move the arms clockwise and anti-clockwise (Krishnan and Singh 2016), 4–5 rounds, and visualize the white light over your shoulders and your shoulder has been charged.

Elbow Bending based on *Kehuni Naman*: Sit in a base position or *sukhasana.* Stretch the arms in front of the body at the level of the shoulder, while the palm facing upward, and then bend the elbow by touching the fingers to the shoulder (Krishnan and Singh 2016). Repeat 4–5 times with both arms and visualize the white light in your elbow and feel as if it has been charged.

Hand Clenching based on *Mushtika Bandhana*: Sit in a base position or *sukhasana.* Bring the arms at the level of the shoulder and open the hands, palm down and stretch the fingers as wide as possible, then make a tight fist by closing the fingers with the thumb inside. Now repeat it by again opening the hand, stretching the fingers wide apart, and closing it tightly (Krishnan and Singh 2016). Do this at least 5 times and visualize the red light of power on your hands and feel the charging on your palm.

Wrist Bending based on *Manibandha Naman*: Hold the arms straight in front at shoulder level. Open up your palm facing it downward, now move the hand up and down from the wrist. Don’t bend your elbow throughout the practice (Krishnan and Singh 2016). Do it 5 times and visualize the red light flowing through your palm and spreading all over.

Wrist Rotation based on *Manibandha Chakra*: Hold the arms straight in front at shoulder level. Make the fist with the right hand by keeping the thumb inside. Slowly rotate the fist by moving the wrist and make a large circle clockwise and anti-clockwise without moving the elbow (Krishnan and Singh 2016). Practice this 5 times on each side and finally rotate the fist in the opposite direction and practice in each direction.

Power Walk Toe Bending based on *Padanguli Naman*: Sit in the base position with legs outstretched by keeping the feet apart. Place the hands on the sides and use the arms to support the trunk. Keep the spine erect. Move the toes of both feet slowly backward and forward by keeping the feet in the upright position (Krishnan and Singh 2016). Hold each position for a few seconds and visualize the red light over your feet. Repeat this 4–5 times.

Ankle Rotation based on *Goolf Chakra*: Remain in the base position. Slowly rotate the foot from the ankle, clockwise and anti-clockwise one by one. Then repeat the same with both feet together keeping them in close contact with each other (Krishnan and Singh 2016). Now again rotate the feet but in opposite direction and keep visualizing the red light over your feet.

Ankle Bending based on *Goolf Naman*: Sit in the base position. Keep the feet apart. Move the feet forward and backward bending the ankle joint. Stretch the feet and try to touch the floor down and then bring them back toward the body and visualize the red light passing from the feet, stretch it, and spread it outward (Krishnan and Singh 2016). Do it at least 4–5 times.

Outstretch (The Complete Stretch): Sit in the base position. Take a deep breath and exhale it, twice. Now stretch the hands upward above your head while breathing in deeply and visualize the inhalation of white light and hold for 5 s, stretch upward and then exhale all the toxins and negativity out while breathing out, bend downward from the waist as much as possible. Try to bring your chest toward your thighs and hold for a second and again breathe in. Repeat this twice or thrice.

Now sit in a comfortable position with crossed or folded legs.

***Vajrasana* (Thunderbolt or Diamond Pose)**: It is a simple *Yoga Asana* that makes the body strong as a diamond. People of any age can perform this *Asana*. It is the only *Asana* that can be performed after having the meal, it is beneficial for overall health. Initially, a beginner can start with 3–4 min and can gradually increase the timings with practice up to 20 min or more.

Sit on the floor by folding the knees and ankles backward, by keeping the feet's bottom upward and buttocks resting on the thighs and heals. Keep your back and head straight and now put your right palm on the right thigh and left palm on the left thigh, breathe in deeply, and breathe out slowly (Pathare 2021).

***Padmasana* (Lotus Position)**: *Padmasana* is considered the best pose for meditation which calms the mind and body; therefore, it is considered the basis for many *Yoga* poses. It is vital in improving digestion, easing menstrual pain, and reducing muscular tension (Chilukoti 2015). Sit comfortably and stretch out your legs and heels touching the ground. Now slowly move your legs, bend your knees and place the right foot on the opposite thighs, i.e., left thighs and left feet on the right thigh till both your legs are in the crisscrossed locked position. Place your hand in the *mudra* position and keep your head and spine straight.

Sukhasana **(Easy Pose)**: As the name so is the pose, comfortable, and the most basic one. Simply sit with crossed legs with your back straight and in the most comfortable position. it is the most basic pose for meditation and *Yoga* practices (Varambally and Gangadhar 2012; Pizer 2021).

14.3.2.4 Pranayama

Pranayama has been derived from two words "Prana" which means life force and "Ayama" which means extension. *Pranayama* is the technique where the mechanism of life force (breath) can be regulated, activated, and charged to attain a higher state of energy, even the effects of the mental and emotional state of breathing. There are four aspects of *Pranayama*: Inhalation, Exhalation, Internal Breath Retention, and External Breath Retention.

Exercise XII. Rhythmic Breathing 1:1:1 based on *Nadi Sodhana/Anuloma-Viloma Pranayama*. Sit in *Padmasana* or *Sukhasana* and breathe in and exhale once. Block your right nostril with your right thumb and breath in deeply filling up your lungs with air from your left nostril. Block your left nostril with the middle finger of your right hand and hold your breath for a while. Lift the thumb from the right nostril and exhale through the right nostril. This practice should be done slowly and with the rhythm of normal breathing. Breathing exercise helps in the regulation of the prana in the body (Arya 2017). Expiration time should be in proportion 1:1:1. Inhale all the positivity, i.e., white light, and hold for 1 s and see it spreading throughout the body and then exhale out with all the toxins of the body and mind. Repeat it 10–15 times.

14.3.2.5 PRATYAHARA

"Prati" means away and *"Ahara"* means food. *Pratyahara* means withdrawing from the senses and drawing the attention inside and internalizing by being mindful. Whatever is there in the mind has a deep impact on our entire life, even at the cellular level, i.e., the smallest element of the body. Whatever the person sees, tastes, touches, smells, or hears if one surrender to an outside stimulus or toward senses, that person will live a life being pulled here and there, seeking some new pleasurable stimulus. When the mind is withdrawn from the outside world, the senses are automatically withdrawn from external stimuli. Senses do not follow the objects or external things themselves; they are followed by the mind.

Exercise XIII. Step In-Step Out: Step In, into the internal world by zooming in. Close your eyes and visualize the area or room wherever you are sitting right now, see it through the mind's eye. See the big picture for a few seconds, and visualize the door and the walls of the room. As you zoom in, see the things kept around you and as you zoom in again, now see yourself sitting in the room after a few seconds again zoom in and see your forehead. Now step in more, visualize the star inside the head and focus on your consciousness, stay calm and still for a few seconds and feel the dissociation from the outside environment. Keep breathing normally.

Step Out, to the external world by zooming out. Move your consciousness out from the body to the outside environment, step-wise gradually in the same way as you entered inside. Be in the picture and the scenario being associated with the scene.

14.3.2.6 *Dharana*, *Dhyana*, AND *samadhi* (SANYAM)

The first four limbs—*Yamas, Niyamas, Asana,* and *Pranayama*—are considered outer limbs, and the remaining four are considered inner limbs. These inner limbs *dharana*, *dhyana*, and *samadhi* are called *Sanyam* (control) or the pathway to the true light of knowledge. Through this one can fully step into the journey of self-awareness and discovery of who we truly are.

There is no dividing line between these three stages. So, we are going to practice them together in one go.

Dharna is binding the mind to one place, object, or idea. *Dharna* is an effort to bring the mind to one's meditation object repeatedly and this at times develops into *Dhyana*. It is an effortless flow of awareness toward a particular object and *Dhyana* gradually develops into *Samadhi*. We can simply divide this practice into three parts:

First: Attention and stabilizing the mind at a point. Visualizing the light inside the head.

Exercise XIV. **Centration Point Concentration**: Being seated with cross-legs, have a lighted candle or light at eye level just 2–3 feet from your gaze. Stare it without blinking and without creating any thoughts. Then close your eyes and follow the image of the flame or light for 2–3 min. Just follow whatever you see with your inner gaze and feel the consciousness inside (based on *Tratak*).

Second: Concentration on the consciousness and shifting to the object of meditation, i.e., highest consciousness (universal powerhouse). The highest form of meditation is not to stabilize the mind on an object or just to relax the body and reduce stress and anxiety. The highest form is recognizing true consciousness and being in that state (soul-conscious).

Third: Be in the consciousness of light and forget everything around even your body. Emerging your consciousness with the light. Connecting to your highest self and attaining the full potential, i.e., Super Consciousness. It is *Satchitananda, Sat* (Being), *Chit* (Consciousness), and *Ananda* (Bliss).

Exercise 14.2. **Infinite Bliss**: Take a deep breath and relax. Breathe in and let it out, disconnect yourself from everything around you. Now take your attention to the center of your forehead and visualize a sparkling star. Visualize yourself as a being of light, you are the energy which takes charge of your body. Now visualize another star right in front of you radiating peace and love bigger than you and brighter than you. Feel the connection with the powerhouse of the universe and visualize that you are in the canopy of the supreme power. You are being charged by the peace and love, feel that peace inside and spread it around you, enjoy the feeling of blissfulness, feel it and be in it. I am a peaceful, lovely, and blissful soul. *Om Shanti.*

**** SILENCE****

Silence to Silence the Mind: Be silent for 30 s to 1 min and after that open your eyes. Feel the change and live your day with a positive mindset. Silence after meditation helps to internalize the new experience and new thought process, it allows the practitioner to live in a blissful flow state. Silence is the way to introspection, it helps in awakening and bringing awareness within. It is a powerful way of self-transformation; it helps to think more vividly (Ludwig 2021).

14.4 Clinical Implication

Literature reveals that there are numerous studies conducted to see the impact of *RajYoga* meditation on physiological and psychological variables. *RajYoga* meditation is primarily studied in terms of psychological factors, including stress, anxiety, depression, well-being, quality of life, positive thinking, etc. The literature also suggests that *RajYoga* can remove the symptoms of neurobiological and psychiatric disorders, which suggests that mental health professionals should consider the advantages of spiritual practices for patients and should use these as complementary intervention techniques for treatment (Singh and Pandey 2015).

14.5 Future Direction for Research

Ancient Indian Texts such as *Vedas, Bhagavad Gita, Patanjali Yoga Sutra, Arthashastra*, etc., contain ample wisdom in them which serve to make an effective meditation paradigm or therapy. It is even suggested to tailor the ancient system to make it suitable and easily applicable for modern society, for finding solutions to mental health problems (Balodhi 1986). *RajYoga* meditation is the oldest form of meditation and the newest one ought to be explored more and used worldwide.

14.6 Conclusion

Meditation is for stabilizing the fluctuations of the mind and to help in moving beyond the everyday consciousness. *RajYoga* meditation develops willpower and displays physiological benefits too. It is a complete package of physical and mental exercises; it is not only a spiritual path but the path of lifelong learning and applying that learning in day-to-day life (Olson 2013). Research regarding the effects of meditation usually is divided into three major areas that deal with the effect of meditation on personality variables, psychophysiological correlates of meditation, and the use of meditation as a therapy (West 1979). However, to our knowledge, there is much less work done by interweaving all eight limbs together as a meditative therapy. Therefore, it is an effort to bring *Patanjali*'s eight-limbed model into practice. *RajYoga* meditation should be practiced worldwide, irrespective of age, profession, and mental and physical health (Singh et al. 2021). Thus, the researcher suggests mental health professionals to use these techniques as complementary to other treatments. The suggested therapy is for complete change, i.e., it changes a person's physical, psychological, and socio-ethical outlook. Sage *Patanjali* has given techniques not only for altering maladaptive behaviors but a path for overall personality growth.

References

Alpert YM (2015) Mastering the Om: a guide for beginners (Internet). Yoga J (cited 2023 Jan 10). https://www.*Yoga*journal.com/*Yoga*-101/sanskrit/mastering-om/

Arya K (2017) *Yoga* education. Friends Publication, Daryaganj

Balodhi JP (1986) Perspective of rāja*Yoga* in its application to mental health. NIMHANS J 4(2):133–138

Chilukoti B (2015) Step-by-step guide to perform padm*Asana* or lotus pose. The Health Site. https://www.thehealthsite.com/fitness/step-by-step-guide-for-padm*Asana*-or-lotus-pose-b1115-349247/

Datey KK, Deshmukh SN, Dalvi CP, Vinekar SL, Datey KK (1969) Shavasan: a yogic exercise in the management of hypertension. Angiology 20(6):325–333

Goyal AK, Bakshi J, Rani S, Didi BA, Anand A (2019) Is *RajYoga* helpful in maintaining patient's biochemical and hematological profile during breast cancer treatment?. J Complement Integr Med 16(2)

Kiran U, Ladha S, Makhija N, Kapoor PM, Choudhury M, Das S, Gharde P, Malik V, Airan B (2017a) The role of *RajYoga* meditation for modulation of anxiety and serum cortisol in patients undergoing coronary artery bypass surgery: a prospective randomized control study. Ann Card Anaesth 20(2):158

Kiran U, Ladha S, Makhija N, Kapoor PM, Choudhury M, Das S, et al, editors (2017b) The role of *RajYoga* meditation for modulation of anxiety and serum cortisol in patients undergoing coronary artery bypass surgery: a prospective randomized control study. Ann Card Anesth 20(2):158

Krishnan GA, Singh S (2016) An illustration guide to *Yoga* practice. Jaico Publishing House

Ludwig S (2021) The power of silence (Internet). Canyon Ranch (cited 2023 Jan 10). https://www.canyonranch.com/well-stated/post/the-power-of-silence/

Mayer JD (2000) Spiritual intelligence or spiritual consciousness? Int J Psychol Relig 10(1):47–56

Mehta K, Mehta S, Chalana H, Singh H, Thaman RG (2020) Effectiveness of *RajYoga* meditation as an adjunct to first-line treatment in patients with obsessive compulsive disorder. Indian J Psychiatry 62(6):684

Naragatti S, Hiregoudar NK (2019) Brahma kumaris sahaj raj-*Yoga* meditation-as a tool to manage various levels of stress. J Adv Res Ayurveda *Yoga* Unani Siddha Homeopathy 6(1 and 2):1–9. issn:2394–6547

Olson C (2013) Understanding the brahma kumaris by frank. J Hindu Stud Whaling 6(2):237–238

Pathare S (2021) Health benefits of vajr*Asana* pose and how to do it (Internet). Healthily Me (cited 2023 Jan 10). https://www.healthifyme.com/blog/vajr*Asana*-pose/

Pizer A (2021) How to do easy pose (Sukh*Asana*) in *Yoga*. https://www.verywellfit.com/easy-pose-sukh*Asana*-3567124

Prabhavananda S, Isherwood C (1948) Patanjali Yoga sutras: translated with a new commentary. Sri Ramakrishna Math, Chennai, pp 17–19

Raj R (2018) *Yoga* meditation and spiritual education. Int J Innov Res Adv Stud 5(7):63–66

Raj S, Tiwari R (2019) Effect of *Yoga Asana* on self-esteem and achievement motivation of adolescents. Int Res J Eng Technol 6(3):4206–4210

Rao NG (2012) *Patanjali*'s ashtanga *Yoga* & human life—an overview (cited 2023 Jan 10), pp 25–27. https://esamskriti.com/essays/pdf/8limbs.pdf

Saraswati S, Saraswati SN (2002) Four chapters on freedom: commentary on the *Yoga* sutras of *Patanjali.* Nesma Books India, pp 400

Shanker S, Sadhana S, Lal B, Agarwal G, Bute J (2018) Study to evaluate the impact of *RajYoga* meditation on psychosomatic aspects of hypertensive patients. J Med Sci Clin Res 6(11):462–468

Singh P, Kumar P (2022) Meditation: a correlational study between subjective well-being and quality of sleep among *RajYoga* meditators. Int J Indian Psychol 10(4):206–212

Singh B, Pandey S (2015) Spiritual healing and *RajYoga* meditation in schizo-affective disorder, depressive type-a case report. Int J Social Sci Res 4(3):261–263

Singh A, Sharma SB, Rana A, Amandeep K, Kaur A, Rani B (2021) Effectiveness of *RajYoga* meditation: a quasi-experimental study for holistic healing in cancer patients. Environ Conserv J 22(1 and 2):35–40

Values in Healthcare: a spiritual approach, 2nd edn. The Janki Foundation for Global Health Care, London (2007)

Vahia NS, Doongaji DR, Jeste DV, Kapoor SN, Ardhapurkar I, Nath RS (1973) Further experience with the therapy based upon *Patanjali*'s concepts in treating psychiatric disorders. Indian J Psychiatry 15(1):32–37

Varambally S, Gangadhar BN (2012) *Yoga*: a spiritual practice with therapeutic value in psychiatry. Asian J Psychiatry 5(2):186–189

Vigneswar J (2021) The art of removing and creating habits—part 2 (Internet). Heartfulness Magazine. Heartfulness (cited 2023 Jan 10). https://www.heartfulnessmagazine.com/the-art-of-removing-and-creating-habits-part-2/

West M (1979) Meditation. Br J Psychiatry 135(5):457–467

Chapter 15
Yoga as a Tool to Achieve Cardiovascular Resonance and Improve Heart Rate Variability

Harinder K. Bali, Navdeep S. Sidhu, and Sumandeep Kaur

Contents

Abstract Cardiovascular diseases (CVDs) are a major global health problem. Perturbations in the tone of the autonomic nervous system (ANS), with decreased parasympathetic activity and enhanced sympathetic activity, are significant contributors towards the emergence of many CVD risk factors like dyslipidemia, hypertension, insulin resistance and diabetes mellitus, and are associated with worse outcomes in those with established CVDs. Heart rate variability (HRV) is the most frequently used clinical measure of cardiac autonomic balance, with optimum HRV indicating good health and well-being. The traditional Indian practice of *Yoga* has been explored as a means to activate cardiovascular resonance properties, thus enhancing HRV and improving cardiac autonomic balance. In this chapter we discuss the importance of optimum HRV in health and disease, the concept of cardio-

H. K. Bali (✉)
Paras Hospitals, Panchkula, Haryana, India
e-mail: drhk.bali@parashospitals.com

N. S. Sidhu
Department of Cardiology, All India Institute of Medical Sciences, Bilaspur, Himachal Pradesh, India

S. Kaur
College of Nursing, All India Institute of Medical Sciences, Bilaspur, Himachal Pradesh, India

A. Anand (ed.), *Neuroscience of Yoga*,
https://doi.org/10.1007/978-981-97-2851-0_15

vascular resonance, and *Yoga* as a tool to activate cardiovascular resonance and achieve optimum HRV.

Keywords *Yoga* · Heart rate variability · Cardiovascular resonance · Cardiac autonomic balance · Cardiovascular diseases

15.1 Introduction

Cardiovascular diseases (CVDs) are the foremost cause of worldwide morbidity and mortality, being responsible for nearly one-third of adult deaths worldwide (Roth et al. 2020). A large proportion of these CVDs can be prevented by addressing modifiable risk factors like hypertension, diabetes mellitus, dyslipidemia, tobacco use, physical inactivity, insulin resistance and obesity. The development and progression of these risk factors is intricately related to unhealthy lifestyle habits, thus underscoring the need for a healthy lifestyle for the prevention of CVDs. Recent research has also stressed the contribution of chronic psychosocial stress and negative affective states towards causing autonomic imbalance with enhanced sympathetic tone and decreased parasympathetic tone, which potentiate many cardiovascular risk factors including insulin resistance, diabetes mellitus, hypertension and dyslipidemia (Thayer et al. 2010).

Yoga is a centuries-old mind-body practice which originated in India nearly 4000–4500 years ago and has gained worldwide popularity. It is a unique blend of light physical exercises with specific postures (*Asana*), breathing regulation (*Pranayama*) and relaxation/meditation (*Dhyana*) to promote physical and mental well-being. A growing body of evidence suggests beneficial effects of *Yoga* when used as an adjunctive therapy in a variety of diseased states including musculoskeletal pain, arthritis, insulin resistance, diabetes mellitus, dyslipidemia, hypertension, heart failure, atrial fibrillation, asthma, anxiety disorders and post-traumatic stress disorders (Büssing et al. 2012; Bali 2013; Cramer et al. 2015; Akella et al. 2020). Many of these beneficial effects of *Yoga* are possibly related to strengthening of the parasympathetic system and amelioration of the autonomic imbalance seen commonly in many diseased states (Thayer et al. 2010; Büssing et al. 2012).

15.2 Heart Rate Variability, Cardiovascular Resonance and *Yoga*

15.2.1 Heart Rate Variability as a Marker of Cardiac Autonomic Balance

The autonomic nervous system (ANS), consisting of sympathetic and parasympathetic arms, plays a major role in maintaining the internal homeostasis of the body by regulating several key processes including body temperature, heart rate, blood pressure, digestion and electrolyte balance. It responds to various internal and external environmental challenges and is involved in production of beneficial adaptive changes so as to meet environmental challenges and maintain overall health, optimal functioning and resilience (Speer et al. 2020).

In normal health, the activity of the two branches of the ANS are in a state of dynamic balance so as to maintain body homeostasis. Disturbed autonomic balance with increased sympathetic activity and/or decreased parasympathetic activity is associated with various pathological conditions including many CVDs (Thayer et al. 2010). The autonomic imbalance can be viewed as a final common pathway in many diseased conditions, leading to excess morbidity and mortality.

Although there is no gold-standard test for the measurement of cardiac autonomic balance, heart rate variability (HRV) is the most valid, reliable and most commonly used tool to measure cardiac autonomic balance. HRV is a measurement of dynamic oscillations in the interval between consecutive heartbeats and between successive instantaneous heart rates (Fig. 15.1). These oscillations are the result of a complex interplay of heart–brain interactions and the ANS dynamics. HRV is assessed by using electrocardiographic (ECG) or photoplethysmography (PPG) sensors to measure the inter-beat interval (IBI). Variations in the heart rate usually reflect the varying activity of the two arms of the ANS. During normal physiological states, the chronotropic tone of the heart is primarily under the control of the sinoatrial (SA) node, which has direct innervation from the fibres of the sympathetic and

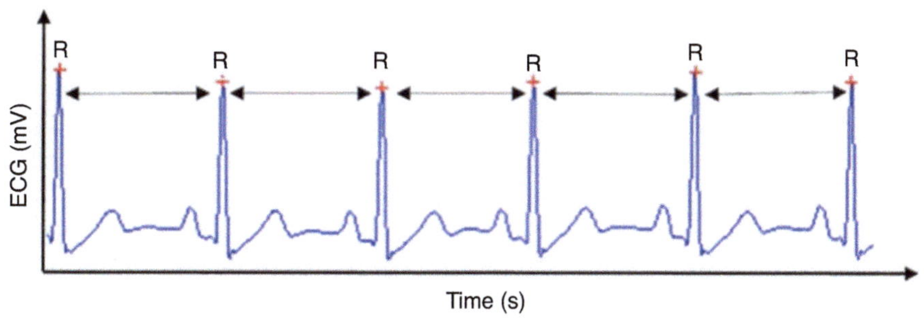

Fig. 15.1 The calculation of heart rate variability (HRV) based on RR intervals of the QRS complexes from an electrocardiographic signal (Laborde et al. 2017), open access, used under CC-BY

parasympathetic (via the vagus nerve) arms of the ANS. Although these arms of the ANS have opposing effects on the sinus rate, with sympathetic stimulation enhancing the heart rate and parasympathetic stimulation decreasing the heart rate, there is a peculiar asymmetry in the action of these systems. The parasympathetic effects have shorter latency periods with a rapid onset of action as compared to sympathetic effects, due to faster kinetics of acetylcholine (ACh), the principal neurotransmitter of this system, as compared to slower kinetics of noradrenaline, which mediates the sympathetic effects (Thayer et al. 2010; Speer et al. 2020).

15.2.2 Heart Rate Variability Analysis

HRV analysis can be performed by using linear methods (like frequency domain analysis, time domain analysis) and non-linear methods like the Poincaré plot. Linear methods are the most commonly used tools in clinical and research practice (Heart rate variability: standards of measurement, physiological interpretation and clinical use 1996; Sassi et al. 2015).

During the time domain analysis, the indices of HRV are presented as units of time (milliseconds, ms). This involves calculation of heart rate at any point of time or measurement of time interval between consecutive normal R waves on an ECG (NN intervals). The most commonly measured parameters include standard deviation of all NN intervals (SDNN) and the root mean square of differences in the successive NN intervals, known as the root mean square of successive differences (RMSSD). SDNN is regarded as a good marker of overall HRV, whereas the RMSSD is used to estimate parasympathetic or vagal activity. Other time domain parameters include NN50, the number of adjacent NN intervals which have a difference of greater than 50 ms, and pNN50, the percentage of adjacent NN intervals having a difference of greater than 50 ms (Thomas et al. 2019; Shaffer et al. 2014; Draghici and Taylor 2016; Laborde et al. 2017).

In the frequency-domain analyses, fast Fourier transformation or power density spectral analysis is used to quantify oscillations of RR intervals and to obtain a break-up of various component rhythms which operate at varied frequencies. The three major components of the HRV spectrum distinguished from short-term ECG recordings include: high frequency (HF) band (frequency range of 0.15–0.4 Hz), low frequency (LF) band (frequency range of 0.04–0.15 Hz) and very low frequency (VLF) band (frequency <0.04 Hz) (Heart rate variability: standards of measurement, physiological interpretation and clinical use 1996). The HF band, also called the respiratory band, is a reflection of the parasympathetic or vagal activity and is primarily related to the heart rate variations corresponding to the respiratory cycle. These variations, also called respiratory sinus arrhythmia (RSA), are characterized by an enhanced sinus rate in inspiration and decreased sinus rate in expiration. During inspiration, the vasomotor centre (VMC) in the brainstem inhibits the parasympathetic or vagal outflow with a resultant increase in the heart rate. In expiration, there is restoration of the vagal outflow with a resultant slowing of the heart rate. The degree of these oscillations is variable, being influenced by a host of intrinsic and extrinsic factors, and are accentuated by slow and deep breathing (Shaffer et al. 2014). The origin and utility of LF band oscillations has been a subject of considerable debate and controversy, with different studies citing varied contributions of the sympathetic and parasympathetic systems towards its genesis (Triggiani et al. 2016; Choudhary et al. 2016; Habek et al. 2016; Samuel Sundar Doss et al. 2016; Sá et al. 2016; Reyes del Paso et al. 2013). Whatever the origin of the LF band, it has been generally recognized that one important component of the LF band is the oscillations close to 0.1 Hz (which correspond to a 10-s rhythm). These oscillations are related to the occurrence of arterial pressure Meyer waves, which are dependent on an intact arterial baroreflex (BR) function for their occurrence. An increase in BP causes an increased activation of these baroreceptors (primarily those located in the aortic arch and carotid sinus), which is relayed to the brainstem VMC. This results in an inhibition of the sympathetic outflow along with augmentation of the parasympathetic outflow. By causing a coupling of the sympathetic inhibition with parasympathetic stimulation, these reflexes potently reduce the BP when it is too elevated and elevate it when it is too low. Baroreflex gain is usually measured as the instantaneous change in heart rate as a function of change in BP. Decreased BR gain is associated with poor regulatory capability and ageing (Shaffer et al. 2014). The ratio of LF:HF band has been suggested to be representative of sympatho–vagal balance, with a high value indicating sympathetic preponderance and a lower value indicating parasympathetic dominance. But this view has received significant criticism in the literature, due to controversies related to the analysis and interpretation of the LF band (Thomas et al. 2019). Analysis and interpretation of VLF band oscillations is even more controversial than for the LF band. Hence, this band is not generally reported in most HRV studies (Heart rate variability: standards of measurement, physiological interpretation and clinical use 1996; Thomas et al. 2019; Shaffer et al. 2014).

15.2.3 Heart Rate Variability During Health and Disease

It is generally accepted that a low resting heart rate and a high HRV is associated with a healthy autonomic balance, robust cardiovascular health and a good overall physiological state. An optimum HRV is a marker of good integrative function of various body systems and is associated with the body's inherent flexibility, adaptability and resilience. While high levels of HRV are counter-productive to efficient energy utilization and good physiological functioning, too little variation is an indicator of energy depletion and pathology. Low HRV is indicative of a less adaptive autonomic system, less robust regulatory mechanisms of the body and poorer overall health (Shaffer et al. 2014).

Low HRV has been linked to a higher rate of future adverse health events and lower overall survival. Many studies have shown an association of low HRV with increased mortality and morbidity and mortality in patients with myocardial infarction (MI) (La Rovere et al. 1998; Bigger Jr et al. 1992). Also, reduced HRV is associated with higher incidence of heart failure and arrhythmias (Sandercock and Brodie 2006; Andresen and Bruggemann 1998). In patients with a known CVD including MI and heart failure, patients with low HRV have higher adverse events on follow-up and poorer prognosis as compared to those with high HRV. Even in persons without CVD, low HRV is associated with a nearly 40% increased risk of an adverse cardiovascular event as compared to a population with high HRV (Hillebrand et al. 2013). A low HRV has also been linked with the emergence of many cardiac risk factors including hypertension, insulin resistance, diabetes mellitus and dyslipidemia (Thayer et al. 2010). Besides cardiovascular diseases, a low HRV has been associated with many other diseases including anxiety, depression, post-traumatic stress disorder, sudden infant death syndrome, asthma, gastrointestinal disorders, inflammatory diseases, chronic kidney disease and cancer (Gorman and Sloan 2000; Hamilton and Alloy 2016; Kazuma et al. 1997; Lehrer and Gevirtz 2014; Tracey 2007; Brotman et al. 2010; Zhou et al. 2016).

15.3 Concept of Cardiovascular Resonance

The cardiovascular system has been shown to exhibit resonance characteristics at frequencies close to 0.1 and 0.03 Hz. The resonance property at about 0.1 Hz frequency is of particular interest, as it has shown significant utility in the management of disorders involving autonomic imbalance by training patients to voluntarily breathe at resonant frequency so as to achieve high-amplitude oscillations in the cardiovascular physiological parameters (Lehrer et al. 2000, 2003; Vaschillo et al. 2002; Song and Lehrer 2003).

Resonance is a physical characteristic exhibited by all oscillating feedback systems having a constant delay. Such an oscillating feedback system, when stimulated at a particular frequency (known as the resonant frequency of the system), results in the production of very high-amplitude oscillations of a single frequency with overshadowing or recruitment of other frequencies. The resonant frequency of a system is the property of the length of delay in the system. This is similar to the Larsen effect in acoustics, where high pitched sound of a particular frequency is produced by keeping a loudspeaker at a close distance from a microphone and stimulating the system with a sound.

Similar resonant properties are exhibited by the cardiovascular system, as elicited by the pioneering works of Lehrer, Vaschillo and co-workers (Lehrer et al. 2000, 2003; Vaschillo et al. 2002; Song and Lehrer 2003). These researchers demonstrated that breathing rates of frequency of 0.1 Hz (six breaths per minute) using HRV biofeedback results in the production of high-amplitude oscillations in the heart rate (HR) and blood pressure (BP), along with increased amplitude of the LF band of HRV and baroreflex gain.

The baroreflex system (Fig. 15.2) is primarily responsible for the property of resonance in the cardiovascular system. An increase in BP causes baroreflex

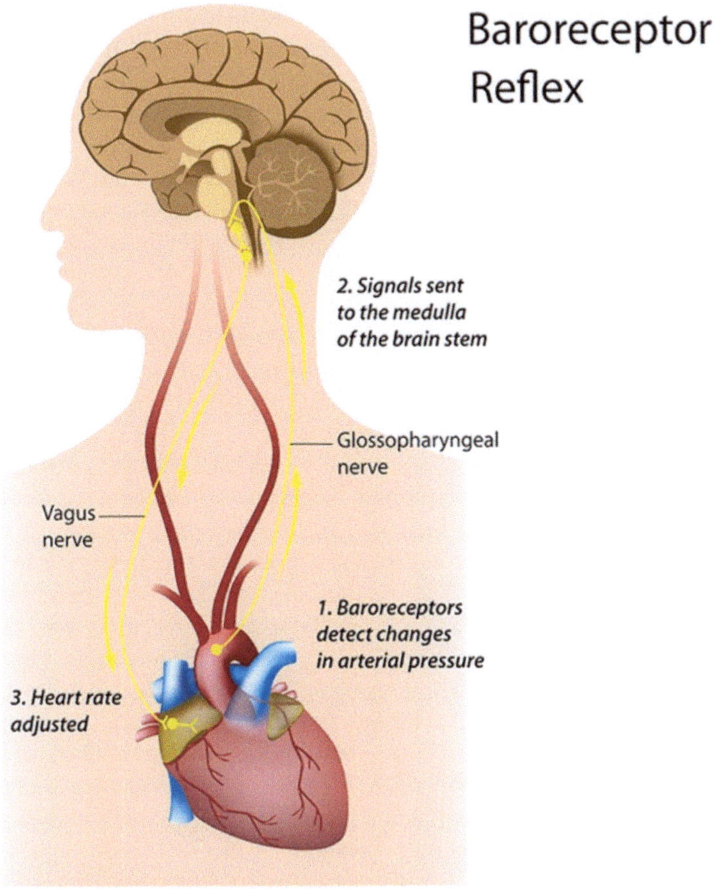

Fig. 15.2 The pathway of baroreceptor reflex (11, Credit: Alila Sao Mai/shutterstock.com), open access, used under CC-BY

activation which leads to a decrease in the HR, thus eventually resulting in decrease in BP. Similarly, a decrease in BP is associated with an increase in HR, thus eventually leading to an increase in BP. Thus, as a result of the baroreflex mechanism, a series of coordinated oscillations are produced in these two parameters of the cardiovascular system. However, the oscillations in these parameters are not perfectly synchronous. There is a delay of about 4–7 s between full swing changes in the HR and full swing changes in the BP. This delay is attributed to the amount of blood contained in the vasculature with a smaller contribution from diameter and stiffness of the blood vessels. This delay varies with the height and gender of an individual, being higher in men and in taller individuals, and is nearly 5 s in an average-sized adult. This resonance frequency of the circulatory system is a function of this delay and varies from 4.5 to 6.5 breaths per minute in adults to 6.5 to 9.5 in children (due to smaller body size in children). The average resonance frequency in an adult is typically around 6 breaths per minute (Shaffer and Meehan 2020).

According to the two-closed loop system model of Vaschillo, the baroreflex system consists of two interconnected closed loops, the first one involving the HR and BP control system and the second one involving the vascular tone and BP control system, with the BP control system serving as the interconnecting link (Vaschillo et al. 2002). Oscillations in any of the BP, HR or vascular tone variables produce oscillations in the other two components. Thus, when an individual breathes at the resonant frequency (i.e. six breaths per minute), HRV amplitudes are maximized. During breathing at this frequency, the HR oscillations are in phase (0-degree relationship) with respiration and 180° out of phase with BP oscillations resulting in strong interaction between the respiratory and baroreflex influences on the HRV and production of very high-amplitude oscillations of a single frequency. In this case, the individual inhales at the resonance frequency just as the HR starts to increase in the baroreflex arch, resulting in augmentation of the baroreflex effect by the effects of the RSA. This is analogous to pushing the swing of a child just as it returns to you, thus maximizing the arc of the swing. These phasic relationships between HR, BP and respiration occurring during breathing at resonant frequency result in most efficient gas exchange in the alveoli, thus maximizing the oxygen saturation (Bernardi et al. 2001a; Yasuma and Hayano 2004; Vaschillo et al. 2004).

The resonant properties of the cardiovascular system can be activated by simulation of an individual's resonant frequency using paced breathing at 0.1 Hz, rhythmic muscle tension stimulation at 0.1 Hz and emotional stimulation at resonant frequency (Vaschillo et al. 2008, 2011). The most commonly used methods to activate cardiovascular resonance involve breathing practices using HRV biofeedback and *Yoga* practices, so as to breathe at the resonance frequency of an individual. With practice, an individual can be trained to breathe at his/her resonance frequency so as to achieve cardiovascular resonance. Continued practice over several months resets the baroreflex gain, keeping it sustained even during periods of normal daily activity. This enhanced baroreflex gain potentiates the ability of the body to regulate the cardiovascular functions and respiratory gas exchange more efficiently (Shaffer et al. 2014). In HRV biofeedback, an individual receives feedback about HR and HRV during periods of slow breathing, with an aim to maximize the HRV by breathing at the resonant frequency. HRV biofeedback has been shown to have a positive impact on general body resilience, cardiovascular function, and cognitive and physical performance. It is also a promising modality in the management of various diseases including hypertension, anxiety, stress, fibromyalgia, chronic muscle pain, asthma and irritable bowel syndrome, and has been associated with positive outcomes during cardiac rehabilitation (Gevirtz 2013).

15.4 Cardiovascular Resonance, Heart Rate Variability and *Yoga*

Similar to HRV biofeedback training, *Yoga* practices have been thought to be potentially useful in achieving resonance in the cardiovascular system and thus promote health and well-being. Several studies have explored the effect of *Yoga* practices on the functioning of the autonomic nervous system. Many studies have reported the occurrence of parasympathetic dominance during and after these practices including breathing exercises, relaxation, mediation and integrated *Yoga* practices. Experienced *Yoga* practitioners have been shown to have enhanced parasympathetic tone as compared to sedentary persons, non-*Yoga* practitioners and even long-term regular runners (Muralikrishnan et al. 2012; Friis and Sollers III 2013; Satin et al. 2014). In patients with heart failure regular *Yoga* practices have been demonstrated to reduce the sympathetic tone and augment the parasympathetic tone (Krishna et al. 2014).

Yoga breathing exercises have been the focus of major research as a means to modulate cardiovascular autonomic functioning. Breathing functions as a dynamic link between mind and body and *Yoga* breathing techniques (*Pranayama*) are among the most important *Yoga* practices. Various types of *Yoga* breathing techniques include unilateral nostril breathing, alternate nostril breathing, *Bhramari Pranayama* (humming bee breathing), *Kapalbhati* (frontal brain cleansing breathing), *Bhastrika* (rapid shallow breathing), *Savitri* (slow rhythmic deep breathing), *Kukuriya* (dog panting) and *Pranav* (meditation while chanting sacred mantra "Om"). The various techniques of *Pranayama* involve voluntarily controlled breathing exercises which affect the respiratory rhythm with shortening or prolongation of breaths, and sometimes periods of breath-holding. This involves application of voluntary effort to control/modify the breathing pattern, which is normally under the control of the autonomic nervous system. Evidence suggests that such voluntary efforts to control breathing, when practised sufficiently (usually for 6–10 weeks), have the potential to induce persistent and long-lasting changes in breathing patterns (Joshi et al. 1992; Telles et al. 1993; Villien et al. 2005).

Different types of *Yoga* breathing techniques have been shown to produce different effects on the cardiovascular autonomic functioning (Nivethitha et al. 2016, 2017, 2021; Tyagi and Cohen 2016). The available evidence in the literature suggests that the effects of yogic breathing techniques on cardiac autonomic tone and HRV are primarily dependent on the rate of breathing in the exercise employed. *Kapalbhati* exercises with high breath rates in the range of 60–120/min have been associated with increased sympathetic tone, decreased parasympathetic tone and decreased baroreflex sensitivity.

Of particular interest are the slow breathing yogic exercises like slow-paced *Bhramari*, slow-paced *Bhastrika* and *Savitri*, which have been investigated as a means to achieve cardio-respiratory coupling and enhance HRV. Slow breathing in these yogic breathing exercises at 4–7 breaths/min (especially at 6 breaths/min) has been shown to activate cardiovascular resonance, increase baroreflex sensitivity and improve cardiac autonomic balance. Similar resonance effects have been reported with the chanting of mantras and some meditative processes, with slow breathing at around 6 breaths/min being the fundamental component driving such effects (Peng et al. 1999, 2004; Bernardi et al. 2001b).

Major studies exploring the role of yogic slow breathing exercises on HRV are summarized in Table 15.1.

15.5 Clinical Implications and Future Directions for Research

Slow breathing yogic exercises may be used as a physiological measure to draw upon the cardiac vagal reserve with important implications in both health and disease. Improved cardiac autonomic balance and optimum HRV achieved through these exercises has many potential therapeutic applications. It can be used in healthy individuals as a means to achieve good mental and physical health and prevent the development (or attenuate the effects) of various cardiac risk factors including hypertension, dyslipidemia, insulin resistance and diabetes mellitus. It can be used in improving the quality of life in patients with heart failure and ischemic heart disease and possibly improve the long-term prognosis in these patients.

However, the studies relating to *Yoga* and HRV have been limited by poor study designs, variable *Yoga* practices used, small number of participants and lack of standardized methods of measurement and reporting. Moreover, the majority of the studies have been reported from India, thus limiting their generalizability.

Well-designed large-scale studies are needed to further explore these possible therapeutic implications in the future.

15.6 Conclusion

Impaired cardiac autonomic balance has been suggested to play a causative role in a number of cardiac diseases. Heart rate variability (HRV) is a valid marker of cardiac autonomic balance. Optimum HRV is an indicator of robust cardiovascular health and correlates with general well-being, adaptability and resilience. *Yoga* practices, especially slow yogic breathing exercises (at around 6 breaths/min), can be used to activate resonance properties in the cardiovascular system, resulting in improved cardiac autonomic balance and optimum HRV.

Table 15.1 Major studies related to yogic slow breathing and heart rate variability (HRV) (Tyagi and Cohen 2016)

Author, year	Study design	Population	Comparison arm	Intervention	Duration of intervention	HRV outcome (only statistically significant results are reported as '↑' increase and '↓' decrease)	Other outcomes
Peng et al. (1999)	Single practice session	*Kundalini Yoga* meditators ($n = 4$), Chi meditators ($n = 8$)	Baseline vs during intervention	*Kundalini* group: slow breathing and chanting; Chi group: slow breathing and meditation	Single session	High-amplitude HR oscillations in LF band during intervention	–
Bernardi et al. (2001b)	Single practice in one session	Healthy non-*Yoga* practising adults, $n = 23$	During intervention vs after intervention	Mantra chanting, Ave Maria rosary	Single session	High-amplitude HR oscillations in LF band during intervention	↓ BR and ↑ BRS during both interventions
Peng et al. (2004)	Multiple practices in single session	*Kundalini Yoga* practitioners, $n = 10$	Baseline vs during intervention	Three meditation protocols: relaxation response, breath of fire, segmented breathing	10 min each protocol	↑ LF power, ↑ HR-respiratory coupling during slow breathing (relaxation response and segmented breathing)	↑ HR during segmented breathing, breath of fire
Jovanov, (2005)	Single practice in one session	One *Yoga* practitioner	Baseline vs during intervention	ANB at 1 breath/min	10 min	↓ LF power, ↓ HF power, ↑ VLF power, ↑ LF/HF ratio	↑ HR during intervention
Raghuraj and Telles (2008)	Multiple practices during multiple sessions	Regular *Yoga* practitioners, $n = 21$	Baseline vs during intervention	RNYB, LNYB, ANB	40 min for each session	↑ LF power, ↓ HF power, ↑ LF/HF ratio during ANB	↑ HR, ↓ SBP, ↓ DBP during ANB; ↓ BR during all breathing practices

(continued)

Table 15.1 (continued)

Author, year	Study design	Population	Comparison arm	Intervention	Duration of intervention	HRV outcome (only statistically significant results are reported as '↑' increase and '↓' decrease)	Other outcomes
Selvaraj et al. (2008)	Single practice in one session	Regular *Yoga* practitioners, $n = 8$	Baseline vs during intervention	*Shambhavi Mahamudra* (4 stages: *Sukha Pranayama*, *AUM* chanting, rapid breathing and relaxed sitting)	19 min	↑ LF power, ↑ LF:HF ratio, ↓ RMSSD, ↓ pNN50	↑ HR during *Sukha Pranayama*, *AUM* chanting and rapid breathing, ↓ HR in relaxed sitting stage
Jaju et al. (2011)	NRCT	Patients with COPD ($n = 11$), Healthy controls ($n = 6$)	Pre- vs post intervention	ANB with 5 breaths/min	3 months (30 min/day, at least 5 days/week)	No significant difference in either group in frequency domain	Improved respiratory function in both groups
Bhimani et al. (2011)	Cohort study	Healthy MBBS students, $n = 59$	Pre- vs post intervention	*Pranayama* practices (*Kapalbhati, Bahya, Sukha, Surya Bhedan, Ujjayi, Sitkari, Sitali*)	2 months	↓ LF power, ↑ HF power, ↓ LF:HF ratio, no significant change in time domain parameters	No change in HR
Ghiya and Lee (2012)	Multiple practices during multiple sessions	Healthy non-*Yoga* practitioners, $n = 20$	Pre- vs post intervention	ANB and paced breathing at 5 breaths/min	30 min	↑ TF power, ↑ LF power, ↑ HF power	–
Muralikrishnan et al. (2012)	Multiple practices during one session	Regular *Yoga* practitioners ($n = 14$), non-*Yoga* practitioners ($n = 14$)	Comparison between groups	*Isha Yoga* with deep breaths at 6/min	1 min	↓ LF power, ↑ HF power, ↑ SDNN, ↑ NN50, ↑ pNN50 in *Yoga* practitioners as compared to controls	–

Raghvendra et al. (2013)	Multiple practices during multiple sessions	*Yoga* trainees, $n = 50$	Baseline vs during intervention	Relaxation with breath regulation; breath regulation	12 min/ session	↑ LF power, ↓ HF power, ↑ LF:HF ratio, ↓ VLF power during both interventions	↓ HR during relaxation with breath regulation
Telles et al. (2014)	Multiple practices during one session	Experimental group: regular *Yoga* practitioners ($n = 26$), control group: non- *Yoga* practitioners ($n = 15$)	Baseline vs during intervention	Experimental group: ANB and BAW; control group: quiet sitting	25 min	No change in frequency domain in either intervention, ↑ RMSSD, ↑ NN50 during ANB, no change in time domain parameters during BAW and quiet sitting	↓ BR, SBP during ANB
Kharya et al. (2014)	RCT	Non-*Yoga* practitioners, $n = 50$	Pre- vs post intervention	*Sudarshan kriya* (slow breathing, fast breathing and Om chanting; $n = 19$), *Prana-Yoga* ($n = 16$), Controls: leisure walking ($n = 20$)	5 months	No significant difference in frequency domain in either group	Significant improvement in depression, anger and stress management in *Sudarshan kriya* group
Bhaskar et al. (2017)	Single practice in one session	Regular *Yoga* practitioners, $n = 30$	Pre- vs post intervention	Long *Sudarshan kriya Yoga* (cyclic slow, medium and fast breaths)	1 h	↓ LF power, ↑ HF power, ↓ LF:HF ratio, ↑ SDNN, ↑ RMSSD, ↑ pNN50	↓ HR following interventions

Abbreviations: *ANB* alternate nostril breathing, *BAW* breath awareness, *BR* breath rate, *BRS* Baroreflex sensitivity, *COPD* chronic obstructive pulmonary disease, *DBP* diastolic blood pressure, *HF* high frequency, *HR* heart rate, *HRV* heart rate variability, *LF* low frequency, *LNYB* left nostril yogic breathing, *NN50* number of adjacent NN intervals which differ from each other by more than 50 ms, *NRCT* non-randomized controlled trial, *pNN50* proportion of NN50, *RCT* randomized controlled trial, *RMSSD* root mean square of successive differences of intervals, *RNYB* right nostril yogic breathing, *SBP* systolic blood pressure, *SDNN* standard deviation of all NN intervals, *TF* total frequency, *VLF* very low frequency

References

Akella K, Kanuri SH, Murtaza G, Della Rocca DG, Kodwani N, Turagam MK et al (2020) Impact of *Yoga* on cardiac autonomic function and arrhythmias. J Atr Fibrillation 13(1):2408. https://doi.org/10.4022/jafib.2408

Andresen D, Bruggemann T (1998) Heart rate variability preceding onset of atrial fibrillation. J Cardiovasc Electrophysiol 9:S26–S29

Bali HK (2013) *Yoga*—an ancient solution to a modern epidemic. Ready for prime time? Indian Heart J 65(2):132–136. https://doi.org/10.1016/j.ihj.2013.03.002

Bernardi L, Gabutti A, Porta C, Spicuzza L (2001a) Slow breathing reduces chemoreflex response to hypoxia and hypercapnia, and increases baroreflex sensitivity. J Hypertens 19(12):2221–2229. https://doi.org/10.1097/00004872-200112000-00016

Bernardi L, Sleight P, Bandinelli G, Cencetti S, Fattorini L, Wdowczyc-Szulc J et al (2001b) Effect of rosary prayer and *Yoga* mantras on autonomic cardiovascular rhythms: comparative study. BMJ 323(7327):1446–1449. https://doi.org/10.1136/bmj.323.7327.1446

Bhaskar L, Kharya C, Deepak KK, Kochupillai V (2017) Assessment of cardiac autonomic tone following long sudarshan kriya *Yoga* in the art of living practitioners. J Altern Complement Med 23(9):2016–2391. https://doi.org/10.1089/acm.2016.0391

Bhimani NT, Kulkarni NB, Kowale A, Salvi S (2011) Effect of *Pranayama* on stress and cardiovascular autonomic function. Indian J Physiol Pharmacol 55:370–377

Bigger JT Jr, Fleiss JL, Steinman RC, Rolnitzky LM, Kleiger RE, Rottman JN (1992) Frequency domain measures of heart period variability and mortality after myocardial infarction. Circulation 85:164–171. https://doi.org/10.1161/01.cir.85.1.164

Brotman DJ, Bash LD, Qayyum R, Crews D, Whitsel EA, Astor BC et al (2010) Heart rate variability predicts ESRD and CKD-related hospitalization. J Am Soc Nephrol 21(9):1560–1570. https://doi.org/10.1681/ASN.2009111112

Büssing A, Michalsen A, Khalsa SB, Telles S, Sherman KJ (2012) Effects of *Yoga* on mental and physical health: a short summary of reviews. Evid Based Complement Alternat Med 2012:165410. https://doi.org/10.1155/2012/165410

Choudhary AK, Alam T, Jiwane R, Kishanrao SS (2016) A comparative analysis of dietary habits on sensory motor association and heart rate variability during menstrual cycle. J Clin Diagn Res 10(1):CC04–CC08. https://doi.org/10.7860/JCDR/2016/16421.7068

Cramer H, Lauche R, Haller H, Dobos G, Michalsen A (2015) A systematic review of *Yoga* for heart disease. Eur J Prev Cardiol 22(3):284–295. https://doi.org/10.1177/2047487314523132

Draghici AE, Taylor JA (2016) The physiological basis and measurement of heart rate variability in humans. J Physiol Anthropol 35(1):22. https://doi.org/10.1186/s40101-016-0113-7

Friis AM, Sollers JJ III (2013) *Yoga* improves autonomic control in males: a preliminary study into the heart of an ancient practice. J Evid Based Complement Alternat Med 18:176–182. https://doi.org/10.1177/2156587212470454

Gevirtz R (2013) The promise of heart rate variability biofeedback: evidence-based applications. Biofeedback 41(3):110–120. https://doi.org/10.5298/1081-5937-41.3.01

Ghiya S, Lee CM (2012) Influence of alternate nostril breathing on heart rate variability in non-practitioners of yogic breathing. Int J Yoga. 5(1):66–69. https://doi.org/10.4103/0973-6131.91717

Gorman JM, Sloan RP (2000) Heart rate variability in depressive and anxiety disorders. Am Heart J 140(4 Suppl):77–83. https://doi.org/10.1067/mhj.2000.109981

Habek M, Crnošija L, Lovrić M, Junaković A, Krbot Skorić M, Adamec I (2016) Sympathetic cardiovascular and sudomotor functions are frequently affected in early multiple sclerosis. Clin Auton Res 26(6):385–393. https://doi.org/10.1007/s10286-016-0370-x

Hamilton JL, Alloy LB (2016) Atypical reactivity of heart rate variability to stress and depression across development: systematic review of the literature and directions for future research. Clin Psychol Rev 50:67–79. https://doi.org/10.1016/j.cpr.2016.09.003

Heart rate variability: standards of measurement, physiological interpretation and clinical use. Task Force of the European Society of Cardiology and the North American Society of Pacing and Electrophysiology. Circulation 1996;93(5):1043–1065

Hillebrand S, Gast KB, de Mutsert R, Swenne CA, Jukema JW, Middeldorp S et al (2013) Heart rate variability and first cardiovascular event in populations without known cardiovascular disease: meta-analysis and dose-response meta-regression. Europace 15(5):742–749. https://doi.org/10.1093/europace/eus341

Jaju DS, Dikshit MB, Balaji J, George J, Rizvi S, Al-Rawas O (2011) Effects of *Pranayama* breathing on respiratory pressures and sympathovagal balance of patients with chronic airflow limitation and in control subjects. Sultan Qaboos Univ Med J 11:221–229

Joshi LN, Joshi VD, Gokhale LV (1992) Effect of short term '*Pranayama*' practice on breathing rate and ventilatory functions of lung. Indian J Physiol Pharmacol 36(2):105–108

Jovanov E (2005) On spectral analysis of heart rate variability during very slow yogic breathing. Conf Proc IEEE Eng Med Biol Soc 2005:2467–2470. https://doi.org/10.1109/IEMBS.2005

Kazuma N, Otsuka K, Matsuoka I, Murata M (1997) Heart rate variability during 24 hours in asthmatic children. Chronobiol Int 14(6):597–606. https://doi.org/10.3109/07420529709001450

Kharya C, Gupta V, Deepak KK, Sagar R, Upadhyav A, Kochupillai V et al (2014) Effect of controlled breathing exercises on the psychological status and the cardiac autonomic tone: Sudarshan Kriya and Prana-*Yoga*. Indian J Physiol Pharmacol 58:211–221

Krishna BH, Pal P, Pal GK, Balachander J, Jayasettiaseelon E, Sreekanth Y et al (2014) Effect of *Yoga* therapy on heart rate, blood pressure and cardiac autonomic function in heart failure. J Clin Diagn Res 8:14–16. https://doi.org/10.7860/JCDR/2014/7844.3983

La Rovere MT, Bigger JT Jr, Marcus FI, Mortara A, Schwartz PJ (1998) Baroreflex sensitivity and heart-rate variability in prediction of total cardiac mortality after myocardial infarction. ATRAMI (Autonomic Tone and Reflexes After Myocardial Infarction) Investigators. Lancet 351:478–484. https://doi.org/10.1016/s0140-6736(97)11144-8

Laborde S, Mosley E, Thayer JF (2017) Heart rate variability and cardiac vagal tone in psychophysiological research—recommendations for experiment planning, data analysis, and data reporting. Front Psychol 8:213. https://doi.org/10.3389/fpsyg.2017.00213

Lehrer PM, Gevirtz R (2014) Heart rate variability biofeedback: how and why does it work? Front Psychol 5:756. https://doi.org/10.3389/fpsyg.2014.00756

Lehrer PM, Vaschillo E, Vaschillo B (2000) Resonant frequency biofeedback training to increase cardiac variability: rationale and manual for training. Appl Psychophysiol Biofeedback 25(3):177–191. https://doi.org/10.1023/a:1009554825745

Lehrer PM, Vaschillo E, Vaschillo B, Lu SE, Eckberg DL, Edelberg R et al (2003) Heart rate variability biofeedback increases baroreflex gain and peak expiratory flow. Psychosom Med 65(5):796–805. https://doi.org/10.1097/01.psy.0000089200.81962.19

Muralikrishnan K, Balakrishnan B, Balasubramanian K, Visnegarawla F (2012) Measurement of the effect of Isha *Yoga* on cardiac autonomic nervous system using short-term heart rate variability. J Ayurveda Integr Med 3(2):91–96. https://doi.org/10.4103/0975-9476.96528

Nivethitha L, Mooventhan A, Manjunath NK (2016) Effects of various *Prāṇāyāma* on cardiovascular and autonomic variables. Anc Sci Life 36(2):72–77. https://doi.org/10.4103/asl.ASL_178_16

Nivethitha L, Manjunath NK, Mooventhan A (2017) Heart rate variability changes during and after the practice of Bhramari *Pranayama*. Int J Yoga 10(2):99–102. https://doi.org/10.4103/0973-6131.205518

Nivethitha L, Mooventhan A, Manjunath NK (2021) Evaluation of cardiovascular functions during the practice of different types of yogic breathing techniques. Int J Yoga 14(2):158–162. https://doi.org/10.4103/ijoy.IJOY_61_20

Peng CK, Mietus JE, Liu Y, Khalsa G, Douglas PS, Benson H et al (1999) Exaggerated heart rate oscillations during two meditation techniques. Int J Cardiol 70(2):101–107. https://doi.org/10.1016/s0167-5273(99)00066-2

Peng CK, Henry IC, Mietus JE, Hausdorff JM, Khalsa G, Benson H et al (2004) Heart rate dynamics during three forms of meditation. Int J Cardiol 95(1):19–27. https://doi.org/10.1016/j.ijcard.2003.02.006

Raghavendra B, Telles S, Manjunath N, Deepak K, Naveen K, Subramanya P (2013) Voluntary heart rate reduction following *Yoga* using different strategies. Int J Yoga. 6(1):26–30. https://doi.org/10.4103/0973-6131.105940

Raghuraj P, Telles S (2008) Immediate effect of specific nostril manipulating *Yoga* breathing practices on autonomic and respiratory variables. Appl Psychophysiol Biofeedback 33(2):65–75. https://doi.org/10.1007/s10484-008-9055-0

Reyes del Paso GA, Langewitz W, Mulder LJ, van Roon A, Duschek S (2013) The utility of low frequency heart rate variability as an index of sympathetic cardiac tone: a review with emphasis on a reanalysis of previous studies. Psychophysiology 50(5):477–487. https://doi.org/10.1111/psyp.12027

Roth GA, Mensah GA, Johnson CO, Addolorato G, Ammirati E, Baddour LM et al (2020) Global burden of cardiovascular diseases and risk factors, 1990–2019: update from the GBD 2019 study. J Am Coll Cardiol 76(25):2982–3021. https://doi.org/10.1016/j.jacc.2020.11.010

Sá JC, Costa EC, da Silva E, Tamburús NY, Porta A, Medeiros LF et al (2016) Aerobic exercise improves cardiac autonomic modulation in women with polycystic ovary syndrome. Int J Cardiol 202:356–361. https://doi.org/10.1016/j.ijcard.2015.09.031

Samuel Sundar Doss D, Anandhalakshmi S, Rekha K, Akhil AK (2016) Effect of smoking on heart rate variability in normal healthy volunteers. Asian J Pharm Clin Res 9:230–234

Sandercock GR, Brodie DA (2006) The role of heart rate variability in prognosis for different modes of death in chronic heart failure. Pacing Clin Electrophysiol 29:892–904. https://doi.org/10.1111/j.1540-8159.2006.00457.x

Sassi R, Cerutti S, Lombardi F, Malik M, Huikuri HV, Peng CK et al (2015) Advances in heart rate variability signal analysis: joint position statement by the e-Cardiology ESC Working Group and the European Heart Rhythm Association co-endorsed by the Asia Pacific Heart Rhythm Society. Europace 17(9):1341–1353. https://doi.org/10.1093/europace/euv015

Satin JR, Linden W, Millman RD (2014) *Yoga* and psychophysiological determinants of cardiovascular health: comparing *Yoga* practitioners, runners, and sedentary individuals. Ann Behav Med 47(2):231–241. https://doi.org/10.1007/s12160-013-9542-2

Selvaraj N, Shivplara NB, Bhatia M, Santhosh J, Deepak KK, Anand S (2008) Heart rate dynamics during Shambhavi Mahamudra—a practice of Isha *Yoga*. J Complement Integr Med 5:1–9. https://doi.org/10.2202/1553-3840.1137

Shaffer F, Meehan ZM (2020) A practical guide to resonance frequency assessment for heart rate variability biofeedback. Front Neurosci 14:570400. https://doi.org/10.3389/fnins.2020.570400

Shaffer F, McCraty R, Zerr CL (2014) A healthy heart is not a metronome: an integrative review of the heart's anatomy and heart rate variability. Front Psychol 5:1040. https://doi.org/10.3389/fpsyg.2014.01040

Song HS, Lehrer PM (2003) The effects of specific respiratory rates on heart rate and heart rate variability. Appl Psychophysiol Biofeedback 28(1):13–23. https://doi.org/10.1023/a:1022312815649

Speer KE, Semple S, Naumovski N, McKune AJ (2020) Heart rate variability for determining autonomic nervous system effects of lifestyle behaviors in early life: a systematic review. Physiol Behav 217:112806. https://doi.org/10.1016/j.physbeh.2020.112806

Telles S, Nagarathna R, Nagendra HR, Desiraju T (1993) Physiological changes in sports teachers following 3 months of training in *Yoga*. Indian J Med Sci 47(10):235–238

Telles S, Sharma SK, Balkrishna A (2014) Blood pressure and heart rate variability during *Yoga*-based alternate nostril breathing practice and breath awareness. Med Sci Monit Basic Res 20:184–193. https://doi.org/10.12659/MSMBR.892063

Thayer JF, Yamamoto SS, Brosschot JF (2010) The relationship of autonomic imbalance, heart rate variability and cardiovascular disease risk factors. Int J Cardiol 141(2):122–131. https://doi.org/10.1016/j.ijcard.2009.09.543

Thomas BL, Claassen N, Becker P, Viljoen M (2019) Validity of commonly used heart rate variability markers of autonomic nervous system function. Neuropsychobiology 78(1):14–26. https://doi.org/10.1159/000495519

Tracey KJ (2007) Physiology and immunology of the cholinergic anti-inflammatory pathway. J Clin Invest 117(2):289–296. https://doi.org/10.1172/JCI30555

Triggiani AI, Valenzano A, Del Percio C, Marzano N, Soricelli A, Petito A et al (2016) Resting state Rolandic mu rhythms are related to activity of sympathetic components of autonomic nervous system in healthy humans. Int J Psychophysiol 103:79–87. https://doi.org/10.1016/j.ijpsycho.2015.02.009

Tyagi A, Cohen M (2016) *Yoga* and heart rate variability: a comprehensive review of the literature. Int J Yoga 9(2):97–113. https://doi.org/10.4103/0973-6131.183712

Vaschillo E, Lehrer P, Rishe N, Konstantinov M (2002) Heart rate variability biofeedback as a method for assessing baroreflex function: a preliminary study of resonance in the cardiovascular system. Appl Psychophysiol Biofeedback 27(1):1–27. https://doi.org/10.1023/a:1014587304314

Vaschillo E, Vaschillo B, Lehrer P (2004) Heartbeat synchronizes with respiratory rhythm only under specific circumstances. Chest 126(4):1385–1386; author reply 1386–7. https://doi.org/10.1378/chest.126.4.1385-a

Vaschillo EG, Bates ME, Vaschillo B, Lehrer P, Udo T, Mun EY et al (2008) Heart rate variability response to alcohol, placebo, and emotional picture cue challenges: effects of 0.1-Hz stimulation. Psychophysiology 45(5):847–858. https://doi.org/10.1111/j.1469-8986.2008.00673.x

Vaschillo EG, Vaschillo B, Pandina RJ, Bates ME (2011) Resonances in the cardiovascular system caused by rhythmical muscle tension. Psychophysiology 48(7):927–936. https://doi.org/10.1111/j.1469-8986.2010.01156.x

Villien F, Yu M, Barthélémy P, Jammes Y (2005) Training to *Yoga* respiration selectively increases respiratory sensation in healthy man. Respir Physiol Neurobiol 146(1):85–96. https://doi.org/10.1016/j.resp.2004.11.010

Yasuma F, Hayano J (2004) Respiratory sinus arrhythmia: why does the heartbeat synchronize with respiratory rhythm? Chest 125(2):683–690. https://doi.org/10.1378/chest.125.2.683

Zhou X, Ma Z, Zhang L, Zhou S, Wang J, Wang B et al (2016) Heart rate variability in the prediction of survival in patients with cancer: a systematic review and meta-analysis. J Psychosom Res 89:20–25. https://doi.org/10.1016/j.jpsychores.2016.08.004

Correction to: Neuroscience of Yoga

Akshay Anand

Correction to:
A. Anand (ed.), *Neuroscience of Yoga*, https://doi.org/10.1007/978-981-97-2851-0

The front matter of this book has been revised to acknowledge editorial contributions. Specifically, Dr. Pooja Nadholta is now listed as Editorial Assistant on the copyright page, and the reviewer list has been updated with the correct names. This book has been updated to reflect these changes.

The updated version of this book can be found at
https://doi.org/10.1007/978-981-97-2851-0

A. Anand (ed.), *Neuroscience of Yoga*,
https://doi.org/10.1007/978-981-97-2851-0_16

MIX
Papier aus verantwortungsvollen Quellen
Paper from responsible sources
FSC® C105338

If you have any concerns about our products,
you can contact us on
ProductSafety@springernature.com

In case Publisher is established outside the EU,
the EU authorized representative is:
Springer Nature Customer Service Center GmbH
Europaplatz 3, 69115 Heidelberg, Germany

Printed by Libri Plureos GmbH
in Hamburg, Germany